Freshwater Ecosystems—Biodiversity and Protection

Freshwater Ecosystems—Biodiversity and Protection

Guest Editors

Dubravka Čerba
Filip Stević
Djuradj Milošević
Maja Raković

MDPI

Basel • Beijing • Wuhan • Barcelona • Belgrade • Novi Sad • Cluj • Manchester

Guest Editors

Dubravka Čerba
National Water Reference
Laboratory of Slovakia
Water Research Institute
Bratislava
Slovakia

Filip Stević
Department of Biology
Josip Juraj Strossmayer
University of Osijek
Osijek
Croatia

Djuradj Milošević
Faculty of Sciences and
Mathematics
University of Niš
Niš
Serbia

Maja Raković
Institute for Biological
Research "Siniša Stanković"
National Institute of the
Republic of Serbia
University of Belgrade
Belgrade
Serbia

Editorial Office

MDPI AG
Grosspeteranlage 5
4052 Basel, Switzerland

This is a reprint of the Special Issue, published open access by the journal *Water* (ISSN 2073-4441), freely accessible at: https://www.mdpi.com/journal/water/special_issues/HXAB1RZ25G.

For citation purposes, cite each article independently as indicated on the article page online and as indicated below:

Lastname, A.A.; Lastname, B.B. Article Title. *Journal Name* **Year**, *Volume Number*, Page Range.

ISBN 978-3-7258-6262-7 (Hbk)
ISBN 978-3-7258-6263-4 (PDF)
https://doi.org/10.3390/books978-3-7258-6263-4

Cover image courtesy of Dubravka Čerba

Contents

water

MDPI

Editorial

Freshwater Ecosystems—Biodiversity and Protection

Dubravka Čerba [1,*], Maja Raković [2], Filip Stević [3] and Djuradj Milošević [4]

1 National Water Reference Laboratory of Slovakia, Department of Assessment and Aquatic Ecosystems Research, Water Research Institute, 81249 Bratislava, Slovakia
2 Department of Hydroecology and Water Protection, Institute for Biological Research "Siniša Stanković"—National Institute of the Republic of Serbia, University of Belgrade, 11000 Belgrade, Serbia
3 Department of Biology, Josip Juraj Strossmayer University of Osijek, 31000 Osijek, Croatia
4 Department of Biology and Ecology, Faculty of Sciences and Mathematics, University of Niš, Višegradska 33, 18000 Niš, Serbia
* Correspondence: dubravka.cerba@vuvh.sk

1. Introduction to the Special Issue

In recent times, there have been growing concerns about biodiversity and habitat protection in all ecosystems, especially freshwater ecosystems, since some of the strongest negative anthropogenic influences are evident in lakes, rivers, ponds, and floodplains [1–3]. Globally, efforts are being made to protect biodiversity, reduce habitat destruction, and ensure vital ecological services through the assessment and monitoring of aquatic habitats and communities, as well as the development of protection, revitalization, and restoration plans for natural freshwater ecosystems [2–4]. To achieve these goals, it is essential to understand how ecosystems function, the status of present communities, and the main environmental factors influencing them [5,6]. Furthermore, biotic elements or multi-indicator approaches can be instrumental in the protection and management of freshwater systems [5–7].

It is vital to approach these issues from different angles and disciplines because freshwater bioassessment is complex process. Certain taxonomic groups, such as Ephemeroptera, Plecoptera, Trichoptera (EPT), oligochaetes, and chironomids, are particularly important as indicators of water quality, reflecting both good and degraded status [5–7]. With increasing anthropogenic pressures and climate change, there is a pressing need for regular monitoring of aquatic ecosystems and for updating and improving applied bioassessment methods. Large-scale international initiatives, such as the Joint Danube Survey, play a critical role in comprehensive data collection and analysis regarding aquatic biodiversity, biotic community structure, chemical status, pollution, microplastics, hydromorphological alterations, and other ecosystem parameters [8,9].

Microplastic pollution has emerged as a major concern in recent years, affecting organisms across trophic levels and potentially impacting human health [8]. Molecular approaches such as eDNA metabarcoding have advanced our ability to accurately assess biodiversity. For instance, Pleše and Buj assessed endangered lamprey species in Croatia, providing insights into their phylogeny, ecology, and conservation needs. Similarly, Šimunović et al. studied the phytoplankton community in a karstic lake, using both morphological and molecular methods to evaluate the applicability of eDNA metabarcoding as a biomonitoring tool.

The aim of this Special Issue is to highlight contemporary challenges in freshwater ecology, including biodiversity loss, environmental stressors—both natural and anthropogenic—biotic community structure and functional complexity, and the application of molecular analyses in hydrobiological research and bioassessment. The contributions

https://doi.org/10.3390/w17213109

gathered here provide valuable insights from diverse studies and demonstrate the importance of integrating classical ecological approaches with modern molecular, chemical, and biotic assessment tools [1–12].

2. A Summary of the Special Issue

This Special Issue presents 11 research articles focused on the ecology, biodiversity, and monitoring of freshwater ecosystems under increasing anthropogenic pressures. The studies cover a wide range of topics, including phytoplankton assessment using morphological and molecular approaches, ecological differentiation of oligochaete communities, and conservation of *Unio crassus*. Other contributions explore the application of the EPT index for water quality assessment, microplastic pollution in benthic organisms, and the genetic diversity and conservation requirements of lampreys in Croatia. Research on meiofaunal assemblages in Baiyangdian Lake (China) further advances understanding of ecosystem health and restoration priorities. Together, these studies provide valuable insights into freshwater biodiversity and highlight the importance of integrating classical and modern approaches for effective ecological assessment and management.

Šimunović et al. presented a study characterizing the phytoplankton community of a natural karstic lake by combining and comparing morphological and molecular approaches to check the applicability of eDNA metabarcoding as a biomonitoring tool. A total of 51 phytoplankton taxa were found using the morphological approach, whilst the molecular approach discovered 97 ASVs that corresponded to the algal community. The comparability of both approaches in describing phytoplankton communities is evident in the designation of centric diatoms, dinoflagellates and cryptophytes as descriptive taxa. The authors also showed that both approaches proved reliable in detecting functional groups (Lo, C, X2, X3) with similar ecological demands. It was confirmed that eDNA metabarcoding is an applicable tool for biodiversity monitoring of a natural karst lake and should be used as a feasible supplement to traditional microscopy in phytoplankton community assessments.

Atanacković et al. examined the distribution of oligochaete species and their ecological differentiation with respect to environmental factors: altitude, temperature, oxygen concentration, conductivity, total organic carbon, and waterbody type. Although they are widespread, differentiation of oligochaete communities in four waterbody types and altitudinal groups was observed through alpha and beta diversity. The study showed that the total beta diversity decreased with a decrease in waterbody size and with an increase in the size of substrate particles, river flow velocity, and altitude. Communities from small mountain rivers and streams and large and medium rivers with coarser substrates differed from other oligochaete communities. The described research represents an important step for using oligochaetes more reliably and effectively, as they are a necessary BQE (biological quality element) in the biological validation of waterbody typology in routine monitoring practice.

Tomović et al. contributed to the knowledge on *Unio crassus* Philipson, 1788, a species with high priority for conservation. The research covered a variety of waterbody types throughout Serbia, and distribution data were considered over three time periods from 1953 to 2019. The paper summarizes all the available literature data, field research and information obtained during the review of the malacological material collection of the Natural History Museum in Belgrade. The results show a positive population trend, which is reflected in an extension of the distribution area and an increase in population density. The study also revealed better insights into the habitat requirements and limiting factors of the species (e.g., substrate characteristics, waterbody types, altitude, and some nutrients seem to be of great importance for the occurrence of the species).

Tubić et al. evaluated the significance of the EPT index in the water quality assessment of three types of waterbodies in hilly and mountainous regions of Serbia. The authors compared the obtained values of biological indices used for the water quality assessment according to the national legislation with the overall status assessment represented by the ecological quality classes (EQC). The results indicate that the EPT index is an excellent indicator of changes in water quality and an important tool for the ecological categorization of waterbodies in mountainous regions.

Stanković et al. studied microplastics in chosen benthic organisms (*Corbicula* spp., *Limnodrilus hoffmeisteri* (Claparede, 1862), and *Polypedilum nubeculosum* (Meigen, 1804)), obtained during the Joint Danube Survey 4 (JDS4) expedition. Alkaline and enzymatic protocols were performed for tissue degradation, followed by filtering through glass microfiber filters (mesh size 0.5 μm), and MP particles were photographed, measured, and counted. After μ-ATR-FTIR spectroscopy analysis, the particles were characterized as polycarbonate (PC), polyethylene terephthalate (PET), polypropylene–polyethylene copolymer (PP-PE), nylon (polyamide-PA) and cellophane, with PET being dominant. New knowledge on microplastics in aquatic environments and organisms is crucial for the protection of nature.

Pleše and Buj addressed the pressing issue of protecting endangered lamprey species in Croatia, a crucial element in preserving biodiversity, particularly in the face of increasing human-induced impacts on natural ecosystems due to global warming. The study aimed to bridge the knowledge gap by assessing the genetic diversity and structure of identified lamprey species and lineages in Croatia using the gene for cytochrome b. The research revealed four distinct lineages within the species *Eudontomyzon vladykovi* Oliva and Zanandrea, 1959, confirmed the presence of the species *Eudontomyzon danfordi* Regan, 1911, in Croatia, and provided important insights into the intricate relationships and conservation needs of lampreys, providing a basis for future discussions involving additional genetic markers. By gaining a comprehensive understanding of the taxonomy, ecology, and genetic diversity of lampreys, we can ensure their conservation and that of associated ecosystems.

Cao et al. conducted their research on the Baiyangdian Lake, North China Plain, which plays a pivotal role in maintaining the regional ecological balance and biodiversity. The aim was to evaluate the density, spatiotemporal patterns, and habitat response dynamics of meiofauna, primarily comprising freshwater nematodes (91.78%), ostracods, and copepods. The study indicated that the distribution and abundance of meiofauna were significantly affected by environmental factors, with water depth and ammonia nitrogen levels being potential key determinants. The authors evaluated the "health status" of the Baiyangdian ecosystem, which can later aid the protection of biodiversity of this area, and can also provide scientific support for its ecological restoration and governance as well as the assessment of ecological service functions.

Together, these studies demonstrate the value of integrating traditional ecological approaches with molecular, chemical, and biotic assessment tools. They contribute to a better understanding of ecosystem responses and inform effective conservation and management strategies, advancing global efforts to preserve freshwater biodiversity and maintain sustainable aquatic environments.

Author Contributions: Conceptualization: D.Č., M.R., F.S. and D.M.; writing—original draft preparation: D.Č. and M.R.; writing—review and editing: D.Č., M.R., F.S. and D.M. All authors have read and agreed to the published version of the manuscript.

Funding: This research received no external funding.

Acknowledgments: The Guest Editors thank all authors for their contributions and the reviewers for their constructive comments. Special thanks are also extended to the editorial team of *Water* for their professional assistance and support.

Conflicts of Interest: The Guest Editors served as editors for this Special Issue but had no role in the peer-review or decision-making processes for papers in which they were authors. The authors declare no other conflicts of interest.

List of Contributions:

1. Šimunović, M.; Kulaš, A.; Žutinić, P.; Gligora Udovič, M. Phytoplankton Diversity of a Natural Karst Lake Combining Morphological and Molecular Approaches. *Water* **2023**, *15*, 1379. https://doi.org/10.3390/w15071379.
2. Atanacković, A.; Popović, N.; Marinković, N.; Tomović, J.; Đuknić, J.; Stanković, J.; Paunović, M. Effects of Environmental Factors on the Distribution and Diversity of Aquatic Oligochaetes. *Water* **2023**, *15*, 3873. https://doi.org/10.3390/w15223873.
3. Tomović, J.; Simić, V.; Petrović, A.; Atanacković, A.; Zorić, K.; Paunović, M.; Raković, M. Distribution Range of the Endangered Species Unio crassus Philipsson, 1788 in Serbia (Western Balkans Region), Historical and Recent Data. *Water* **2023**, *15*, 4248. https://doi.org/10.3390/w15244248.
4. Tubić, B.; Andjus, S.; Zorić, K.; Vasiljević, B.; Jovičić, K.; Čanak Atlagić, J.; Paunović, M. Aquatic Insects (Ephemeroptera, Plecoptera and Trichoptera) Metric as an Important Tool in Water Quality Assessment in Hilly and Mountain Streams. *Water* **2024**, *16*, 849. https://doi.org/10.3390/w16060849.
5. Stanković, J.; Milošević, D.; Paunović, M.; Jovanović, B.; Popović, N.; Tomović, J.; Atanacković, A.; Radulović, K.; Lončarević, D.; Raković, M. Microplastics in the Danube River and Its Main Tributaries—Ingestion by Freshwater Macroinvertebrates. *Water* **2024**, *16*, 962. https://doi.org/10.3390/w16070962.
6. Pleše, S.; Buj, I. Tackling the Phylogeny of Lampreys—Insight from the Croatia's Danube Basin. *Water* **2024**, *16*, 1153. https://doi.org/10.3390/w16081153.
7. Cao, Y.; Mu, J.; Pan, Z.; Ma, F.; Liu, J.; Dong, H.; Zhang, W.; Wang, L. Study on the Spatiotemporal Distribution Characteristics of Meiofauna in Baiyangdian Lake and Its Influencing Factors. *Water* **2024**, *16*, 1959. https://doi.org/10.3390/w16141959.
8. Sotomayor, G.; Alvarado, A.; Romero, J.; López, C.; Aguilar, M.; Forio, M.; Goethals, P. Limnological Characteristics and Relationships with Primary Productivity in Two High Andean Hydroelectric Reservoirs in Ecuador. *Water* **2024**, *16*, 2012. https://doi.org/10.3390/w16142012.
9. Zhang, Y.; He, W.; Wang, L.; Ou, D.; Qiu, J.; Li, W.; Huang, H. Environmental DNA Metabarcoding as a Promising Conservation Tool for Monitoring Fish Diversity in Dongshan Bay, China. *Water* **2025**, *17*, 452. https://doi.org/10.3390/w17030452.
10. Liu, X.; Ho, L.; De Cock, A.; De Saeyer, N.; Pham, K.; Panique-Casso, D.; Forio, M.; Goethals, P. Water Quality and Its Influence on Waterbird Habitat Distribution: A Study Along the Lieve River, Belgium. *Water* **2025**, *17*, 595. https://doi.org/10.3390/w17040595.
11. Cai, W.; Wen, X.; Zhao, Y.; Wu, X.; Zheng, H.; Chen, J.; Hu, Z.; Zhong, Q.; Wu, J. Untangling the Characteristics and Ecological Processes of Microbial Community Assembly in the Source Area of the East Route of the South-to-North Water Diversion Project in China Under Different Water Periods. *Water* **2025**, *17*, 649. https://doi.org/10.3390/w17050649.

References

1. Dudgeon, D.; Arthington, A.H.; Gessner, M.O.; Kawabata, Z.I.; Knowler, D.J.; Lévêque, C.; Naiman, R.J.; Prieur-Richard, A.H.; Soto, D.; Stiassny, M.L.J.; et al. Freshwater biodiversity: Importance, threats, status and conservation challenges. *Biol. Rev.* **2006**, *81*, 163–182. [CrossRef] [PubMed]
2. Vörösmarty, C.J.; McIntyre, P.B.; Gessner, M.O.; Dudgeon, D.; Prusevich, A.; Green, P.; Glidden, S.; Bunn, S.E.; Sullivan, C.A.; Liermann, C.R.; et al. Global threats to human water security and river biodiversity. *Nature* **2010**, *467*, 555–561. [CrossRef] [PubMed]
3. Allan, J.D.; Castillo, M.M. *Stream Ecology: Structure and Function of Running Waters*, 2nd ed.; Springer: Dordrecht, The Netherlands, 2007.

4. Poff, N.L.; Allan, J.D.; Bain, M.B.; Karr, J.R.; Prestegaard, K.L.; Richter, B.D.; Sparks, R.E.; Stromberg, J.C. The natural flow regime: A paradigm for river conservation and restoration. *Bioscience* **1997**, *47*, 769–784. [CrossRef]
5. Bonada, N.; Prat, N.; Resh, V.H.; Statzner, B. Developments in aquatic insect biomonitoring: A comparative analysis of recent approaches. *Annu. Rev. Entomol.* **2006**, *51*, 495–523. [CrossRef] [PubMed]
6. Heino, J.; Mykrä, H.; Kotanen, J. Environmental and spatial controls of freshwater invertebrate assemblages across Europe. *Freshw. Biol.* **2008**, *53*, 796–813.
7. Roussel, J.M.; Paillisson, J.M.; Tréguier, A.; Petit, E. The downside of eDNA as a biomonitoring tool: Lessons from the literature. *Environ. Sci. Technol.* **2015**, *49*, 344–352.
8. Wagner, M.; Lambert, S. Freshwater microplastics: Emerging ecological threats and methods for detection. *Environ. Pollut.* **2018**, *234*, 331–341.
9. Woodward, G.; Perkins, D.M.; Brown, L.E. Climate change and freshwater ecosystems: Impacts across multiple levels of organization. *Philos. Trans. R. Soc. B* **2010**, *365*, 2093–2106. [CrossRef] [PubMed]
10. Deiner, K.; Bik, H.M.; Mächler, E.; Seymour, M.; Lacoursière-Roussel, A.; Altermatt, F.; Creer, S.; Bista, I.; Lodge, D.M.; de Vere, N.; et al. Environmental DNA metabarcoding: Transforming how we survey animal and plant communities. *Mol. Ecol.* **2017**, *26*, 5872–5895. [CrossRef] [PubMed]
11. Maitland, P.S.; Renaud, C.B. Lampreys: Conservation and Status of 2nd Order Freshwater Vertebrates. *Fish Fish.* **2011**, *12*, 89–116.
12. Bista, I.; Carvalho, G.R.; Walsh, K.; Seymour, M.; Hajibabaei, M.; Lallias, D.; Creer, S. Annual monitoring of freshwater biodiversity using eDNA metabarcoding. *Mol. Ecol.* **2017**, *26*, 5432–5447.

water

MDPI

Article

Phytoplankton Diversity of a Natural Karst Lake Combining Morphological and Molecular Approaches

Maja Šimunović [1], Antonija Kulaš [2], Petar Žutinić [2] and Marija Gligora Udovič [2,]*

[1] Paying Agency for Agriculture, Fisheries and Rural Development, 10000 Zagreb, Croatia; schimica@gmail.com
[2] Department of Biology, Faculty of Science, University of Zagreb, 10000 Zagreb, Croatia; antonija.kulas@biol.pmf.hr (A.K.); petar.zutinic@biol.pmf.hr (P.Ž.)
* Correspondence: marija.gligora.udovic@biol.pmf.hr

Abstract: Phytoplankton are considered to be one of the most sensitive indicators of the ecological status of lakes. Nowadays, it is essential to recognize the prospects of the molecular approach (eDNA metabarcoding) in phytoplankton community assessments and combine them with the existing traditional microscopy-based morphological approach before its standardization. In this study, the aim was to characterize the phytoplankton community of a natural karstic lake by combining and comparing the morphological and molecular approach to check the applicability of eDNA metabarcoding as a biomonitoring tool. A total of 51 phytoplankton taxa were found using the morphological approach, whilst the molecular approach discovered 97 ASVs that corresponded to the algal community. The comparability of both approaches in describing phytoplankton communities is evident in the designation of centric diatoms, dinoflagellates and cryptophytes as descriptive taxa. Furthermore, both approaches proved reliable in detecting functional groups (**Lo**, **C**, **X2**, **X3**) with similar ecological demands. Moreover, the results have shown that euphotic zone samples can be reliably exchanged by composite samples to provide an accurate characterization of phytoplankton communities in the euphotic zone. It was confirmed that eDNA metabarcoding is an applicable tool for biodiversity monitoring of a natural karst lake and should be used as a feasible supplement to traditional microscopy in the phytoplankton community assessments, with regards to the drawbacks of each method.

Keywords: phytoplankton; eDNA; karst lake; diversity; metabarcoding; functional groups

1. Introduction

Natural lakes in Croatia are a phenomenon mainly associated with the karst landforms of the Dinaric ecoregion. The Dinaric ecoregion is a part of the Mediterranean Basin, well recognized as one of the Earth's biodiversity hotspots [1]. Each karst lake is a unique freshwater ecosystem with its own geological, physical and chemical characteristics [2]. Looking from this perspective, biodiversity protection, conservation and sustainable management of freshwater karstic ecosystems require a set of applicable practices based on the best available technologies and establishment of relevant measures founded on up-to-date information and a quality knowledge database.

Phytoplankton play an essential role as the foundation of the food web and as primary producers ubiquitous in all aquatic ecosystems, being especially dominant in the pelagic zone of lakes [3]. As a key Biological Quality Element prescribed by the Water Framework Directive [4], phytoplankton are considered to be one of the most sensitive indicators of a lake's ecological status because they respond rapidly to any environmental changes [5,6]. Phytoplankton biotic metrics used to assess the ecological status of surface water bodies are commonly constructed on traditional morphology-based microscopic identification of taxa, which is time-consuming, costly and has limitations in terms of reproducibility, comparability and applicability in biomonitoring programs [7,8]. Compared to the morphological approach, environmental DNA (eDNA) metabarcoding has been recognized as a tool with the potential to revolutionize biomonitoring and bioassessment, as it offers numerous advantages that include higher accuracy in species identification, increased detection of cryptic diversity and genetic variability, as well as high automation potential including high spatial and temporal resolution [7,9,10]. However, eDNA metabarcoding suffers from several setbacks, one being the incompleteness of the current reference database, which could potentially be compensated for by adding representative sequences of local species from the studied aquatic ecosystems [8]. This method may also be less suitable for estimating abundance, may not provide information on the age or size structure of a population [10] and needs to be standardized before it can be adequately applied in routine monitoring [11]. Nevertheless, the combination of microscopy and molecular methods can provide great improvements in phytoplankton community assessments [7,8,12]; therefore, it is important to recognize the potential of eDNA-based methods and harmonize them with the existing traditional morphological approach.

The aim of this study was to gain a better insight into phytoplankton diversity in the natural karst Lake Visovac (Krka River, southern Croatia) in order to achieve an adequate implementation of a reliable ecological status assessment. Specifically, the aims were to: (a) describe the horizontal and vertical distribution of phytoplankton in the Lake Visovac by applying the traditional morphological approach and the molecular eDNA metabarcoding approach using the amplicon sequencing of hypervariable region V9 of the 18S rRNA gene to investigate total eukaryotic phytoplankton diversity, (b) determine if composite samples could be used as a relevant substitute for discrete sampling in the characterization of phytoplankton community in the euphotic zone, (c) compare the results obtained by morphological and molecular approaches, and (d) establish the applicability of eDNA metabarcoding as a biomonitoring tool in Lake Visovac. The study results will enable establishing a functional system for monitoring changes in the environment and a contribution to the protection of unique karst ecosystems such as the Krka River.

2. Materials and Methods

2.1. Study Area

Krka River rises at the base of Dinara Mountain near the city of Knin in Croatia. It is a 72.5 km long karstic river situated in the central part of the eastern Adriatic coast. Its course is distinguished by alternating lotic and lentic parts as well as tufa deposits forming barrages and cascades. Lake Visovac has a volume of 103×10^6 m^3 and origins from the post-Würm period with the formation of the final and the largest tufa barrier in the Krka River hydrosystem, named Skradinski Buk [13]. Following the provisions of the national typology, Lake Visovac is classified as a medium-sized, medium-depth lowland lake on the carbonate substrate [14].

2.2. Sampling and Methods

Phytoplankton composition and biomass were investigated during August 2018 on 10 sampling stations (V1 to V10) along the limnetic and littoral zone of Lake Visovac (Figure 1). The study area extended from the northern part of the Lake close to Roški slap waterfall (V1) to the southern part immediately before Skradinski Buk waterfall (V8) and near the confluence of the tributary Čikola River (V9 and V10). Stations were divided into three groups according to their geographic position on the Lake: upper (V1, V2), central (V3, V4, V5 and V6) and lower stations (V7, V8, V9 and V10). Station V8 was excluded from the analyses due to shallow maximum depth (3 m).

The total number of samples was defined by the maximum depth of each station, Secchi depth, thermocline depth and mixing depth of the water column with respect to the temperature difference. Water column transparency (Z_{SD}) was determined with a Secchi disc and used for the calculation of euphotic zone depth (Z_{EU}) by multiplying with a standardized factor (2.5 × Secchi depth) for the Mediterranean geographical region [15]. The biological and chemical water samples were collected using the vertical sampler (Hydro-Bios Apparatebau Gmbh, Altenholz, Germany). For the morphological analysis of phytoplankton, discrete samples were taken at 5 m depth intervals (from the surface to the bottom) together with the composite samples taken from the euphotic zone on all stations. With respect to the euphotic zone depth, the discrete samples were divided into those from the euphotic and those from the aphotic zones, and means were calculated for each station, which were used in all further analyses. Grouping of discrete samples into euphotic zone and aphotic zone sets was also applied to spatially compare the phytoplankton community across sampling stations in Lake Visovac. For the DNA analysis of phytoplankton, composite and aphotic zone samples were taken at all stations.

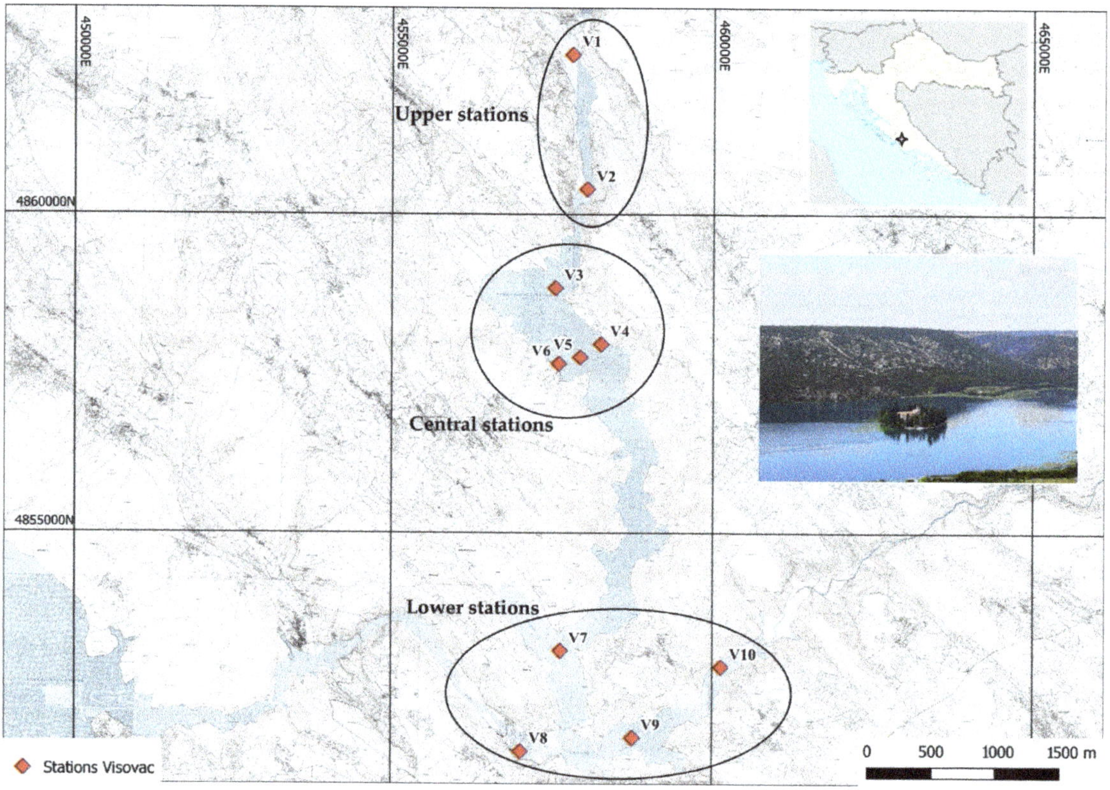

Figure 1. Sampling stations on Lake Visovac during August 2018.

Composite samples for chemical analysis of water were collected together with the phytoplankton samples and stored at −20 °C until laboratory processing. Chemical analysis included quantification of nitrate (NO_3^--N), nitrite (NO_2^--N), ammonium (NH_4^+-N), total nitrogen (TN-N), orto-phosphate (PO_4^{3-}-P), total silica (SiO_2), total inorganic carbon (TIC), dissolved inorganic carbon (DIC), total organic carbon (TOC), dissolved organic carbon (DOC) and bicarbonates (HCO_3^-) using standardized methods [16].

Phytoplankton samples were placed into 250 mL volume plastic bottles, preserved with formaldehyde solution (2%) on the field and stored in the dark at 4 °C. Phytoplankton biomass was determined according to the Utermöhl method [17] using a Zeiss AxioVert inverted microscope equipped with an AxioCam MRc camera (Carl Zeiss AG, Jena, Germany). Taxa identification was performed using relevant literature [18–24] and names were assigned according to Algaebase [25]. Images of species were processed using the program AxioVision LE 4.8 (Carl Zeiss AG, Jena, Germany). The species were allocated into appropriate functional groups (FGs or coda) following the relevant literature [5,26]. All sampling and analytical procedures were performed according to following standards: HRN EN ISO 5667-3:2018, HRN EN 15204:2008, HRN EN 16695:2015 [27–29].

2.3. DNA Isolation

Samples for DNA extraction were filtered on Nucleopore track-etched polycarbonate membrane filters (47 mm diameter, 0.2 μm pore size; Whatman International Ltd., Maidstone, UK) in a volume of approximately 300 to 400 mL, depending on the suspended particles in the sampled water. After filtration, the filters were stored at −20 °C until further processing. Filters were cut into smaller pieces for DNA extraction using the DNeasy PowerWater Kit (Qiagen, Hilden, Germany). The manufacturer's instructions were followed for isolation, with a minor change in the final step, where 60 μL of sterile DNA-free PCR-grade water was added instead of Qiagen's C6 Solution. The quality of extracted DNA was assessed with a NanoDrop spectrophotometer (BioSpec—nano, Shimadzu Corporation, Kyoto, Japan).

2.4. PCR and Bioinformatic Processing

The hypervariable V9-region of the SSU rRNA gene (ca. 130 bp) was amplified using the universal eukaryotic primer pair according to the protocol of Stoeck et al. [30]. The primers used were 1391F (5′-GTACACACCGCCCGTC-3′) and EukB (5′-TGATCCTTCTGCAGGTTCACCTAC-3′), designed by Amaral-Zettler et al. [31]. The Polymerase chain reactions (PCR) program included the initial step at 98 °C for 30 s, 30 cycles of 94 °C for 30 s, 57 °C for 45 s and 72 °C for 30 s, with the final elongation step at 72 °C for 5 min [32]. PCR products were assessed by visualizing on a 1% agarose gel. Sequencing libraries were prepared using the NEB Next® Ultra™ DNA Library Prep Kit for Illumina (NEB, USA). Libraries were sequenced on an Illumina MiSeq platform, generating 250-bp paired-end reads (SeqIT GmbH & Co. KG, Kaiserslautern, Germany).

The raw Illumina reads for the V9-region were demultiplexed using Cutadapt v3.0 [33], removing the barcodes in the 5′ to 3′ combination. Subsequently, the quality of the demultiplexed raw sequencing was checked using the FastQC tool [34]. After the initial steps, reads were processed using the QIIME2-2020.11 pipeline using the following steps: importing and demultiplexing raw sequencing data, then quality filtering and denoising using the DADA2 plugin [35] for correcting Illumina sequencing amplicon errors. Reads were trimmed at 5′ end for 20 bp (primer removal) and truncated to 185 nucleotides to remove the last, poor-quality nucleotides. Taxonomic assignment of the resulting amplicon sequencing variants (ASVs) was performed using the Naïve Bayes classifier. The Naïve Bayes classifier was pretrained on the Protist Ribosomal Reference (PR2) database v.4.14.0 [36] with a 99% OTU identity threshold. Metazoa sequences were filtered out from the dataset by using taxa filtering. A phylogenetic tree was created from the filtered taxa tables to support the phylogenetic diversity metrics used in the q2-diversity plugin. The raw sequence reads are deposited in the European Nucleotide Archive (ENA) under the project number PRJEB60049.

Ochrophyta, Dinoflagellata, Cryptophyta and Fungi accounted for the majority of eukaryote reads (Figure S1), and unassigned eukaryotes were excluded from graphical presentation. Only ASVs taxonomically assigned to the phytoplankton community were filtered from the main data and used for all further statistical analyses.

2.5. Statistical Analyses

All statistical multivariate analyses were carried out in PRIMER v7 for Windows (Primer-E Ltd., Plymouth, UK). Principal Component Analysis (PCA) was performed to outline and visualize the relationships between environmental variables. One-way analysis of similarity (ANOSIM) was used to test significant differences between composite and discrete samples and determine if composite samples can be used as a relevant substitute for discrete sampling in the characterization of phytoplankton communities in the euphotic zone. Non-metric Multidimensional Scaling (NMDS) was carried out to determine the spatial patterns in the phytoplankton community structure with respect to the sampling stations. Prior to all analyses, the data were normalized using log-transformation. Graphical charts were created in Microsoft Excel for Microsoft 365 (Microsoft Corporation, Redmond, WA, USA). The map of the study area was created using the Free and Open Source QGIS 3.16 software [37].

3. Results

3.1. Physical and Chemical Parameters

The environmental variables of water measured on nine sampling stations for composite samples are presented in Table 1. Secchi depth ranged from 2.5 m (station V2) to 5 m (station V9). Stations with the highest maximum depths were V7 and V9 (24 m and 23 m, respectively). The shallowest station was V10 (5 m) with the euphotic zone extending along the entire water column. The temperature of water ranged from 20.6 °C on station V7 to 26.2 °C on station V10. The lowest O_2 concentration was recorded on station V4 (9.03 mg L^{-1}) and the highest on station V1 (11.68 mg L^{-1}). Oxygen saturation was in the range from 102.6% (station V4) to 138% (station V1). The lowest pH was recorded on station V7 (7.75), whilst the highest was on station V1 (8). The electrical conductivity of water ranged from 494 µS cm^{-1} on station V10 to 562 µS cm^{-1} on station V6 (Table 1).

Table 1. Environmental variables on 9 sampling stations (V1 to V10, excluding station V8) for composite samples.

Station	Max Depth (m)	SD * (m)	Z_{EU} * (m)	T * (°C)	O_2 * (mg L^{-1})	O_2 (%)*	pH	EC * (µS cm^{-1})
V1	15.2	3.0	7.5	23.8	11.68	138.0	8.00	545
V2	18.0	2.5	6.25	23.3	11.03	129.8	7.99	547
V3	18.0	3.0	7.5	22.0	10.00	115.0	7.84	556
V4	15.5	4.0	10.0	21.5	9.03	102.6	7.79	559
V5	18.0	3.5	8.75	22.1	9.79	113.0	7.94	559
V6	15.0	3.5	8.75	21.7	9.09	103.3	7.77	562
V7	24.0	3.5	8.75	20.6	11.14	124.5	7.75	545
V9	23.0	5.0	12.5	25.9	10.18	125.6	7.84	505
V10	5.0	4.5	5.0	26.2	9.97	123.5	7.98	494

Note: * SD—Secchi depth, Z_{EU}—euphotic zone depth, T—temperature, O_2—oxygen concentration, O_2 (%)—oxygen saturation, EC—electrical conductivity.

The environmental variables of water measured from discrete-depth samples in the euphotic and aphotic zone on nine sampling stations are presented in Table S1. The minimum value of water temperature in the euphotic zone was 22.4 °C, measured on station V2, and the maximum was 26.4 °C on station V10. The lowest concentration of O_2 was recorded on station V10 (10.13 mg L^{-1}), whilst the highest concentration and saturation of O_2 in the euphotic zone were measured on station V7 (12.28 mg L^{-1} and 147.5%, respectively). The lowest saturation of O_2 was observed on station V5 (80.1%). The lowest pH (7.88) was recorded on station V2, whilst the highest (8.20) was on station V5. The electrical conductivity of water in the euphotic zone ranged from a minimum of 495 µS cm^{-1} on station V10 to a maximum of 562 µS cm^{-1} on station V2. In the aphotic zone, water temperature varied from 16.5 °C to 20.8 °C (stations V7 and V1, respectively). The lowest concentration and saturation of O_2 were recorded on station V9 (2.91 mg L^{-1} and 27.4%, respectively) and the highest on station V1 (9.60 mg L^{-1} and 108.2%, respectively). Value of pH in the aphotic zone ranged from 7.05 to 7.80 (stations V4 and V5, respectively), whilst electrical conductivity of water varied from 517 µS cm^{-1} to 587 µS cm^{-1} (stations V9 and V1, respectively).

Chemical parameters of water are presented in Table S2. The concentration of NO_3^--N was very low on most stations (<0.1 mg L^{-1}), with slightly higher values on stations V4 (0.7 mg L^{-1}) and V5 (0.3 mg L^{-1}). Very low concentrations of NO_2^--N and NH_4^+-N were recorded on all stations (<0.001 mg L^{-1} and <0.01 mg L^{-1}, respectively). The concentration of PO_4^{3-}-P varied from the lowest measured on six stations in total (<0.01 mg L^{-1}) to the highest measured on station V9 (0.6 mg L^{-1}). The values of SiO_2 ranged from the lowest on station V9 (1.5 mg L^{-1}) to the highest on station V6 (4.5 mg L^{-1}). The highest concentration of TN was measured on stations V4 and V5 (1 mg L^{-1}), while it was very low on the other eight stations (<1 mg L^{-1}). The lowest concentrations of inorganic carbon compounds were detected on station V3 (TIC and DIC of 11.48 mg L^{-1} and 10.32 mg L^{-1}, respectively), while the highest was present on station V7 (TIC and DIC of 13.98 mg L^{-1} and 13.83 mg L^{-1}, respectively). TOC was in the range between 1.55 mg L^{-1} (station V3) and 2.39 mg L^{-1} (station V7). The values of DOC ranged from 0.80 mg L^{-1} (V2) to 2.37 mg L^{-1} (V1), whilst HCO_3^- ranged from 151 mg L^{-1} (V4) to 212 mg L^{-1} (V7).

Principal component analysis (PCA) performed for the 14 environmental variables explained 59% of the total variance on the first two PC axes (Table S3). NO_2^--N and NH_4^+-N were excluded from PCA because their concentrations did not vary across the stations. The most important parameters for PCA axis 1 were temperature, DIC and electrical conductivity (intra-set correlations: 0.433, −0.361 and −0.338, respectively). Regarding axis 2, O_2 concentration, nitrate and TOC were the variables with the most weight for ordination (intra-set correlations: −0.464, 0.369 and −0.368, respectively). PCA arranged samples (Figure 2) into four groups: the first group consisted of samples from the central stations (V3, V4, V5, V6) and the upper station sample V2, whilst the second group included samples from the lower part of the Lake (V9 and V10). Sample V1 from the uppermost part and sample V7 taken from the lower part of the Lake were singled out.

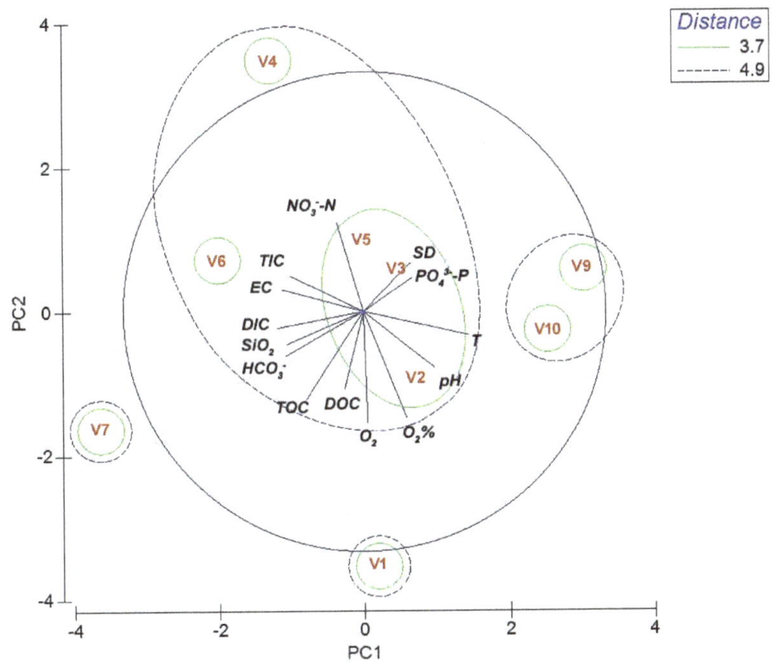

Figure 2. Principal Component Analysis (PCA) ordination diagram performed on the environmental variables and sampling stations (V1 to V10, excluding station V8) in Lake Visovac during August 2018. O_2—oxygen concentration, O_2%—oxygen saturation, T—temperature, SD—Secchi depth, EC—electrical conductivity, NO_3^--N—nitrate, PO_4^{3-}-P—ortophosphate, SiO_2—total silica, TIC—total inorganic carbon, DIC—dissolved inorganic carbon, TOC—total organic carbon, DOC—dissolved organic carbon, HCO_3^-—bicarbonates.

3.2. Characterization of Phytoplankton Community According to Morphological Approach

Based on the morphological approach, a total of 51 phytoplankton taxa were found during the research period. Identified taxa belonged to eight major groups: Chlorophyta (23), Bacillariophyta (8), Ochrophyta (7), Cyanobacteria (4), Charophyta (4), Cryptophyta (2), Miozoa (2), Euglenozoa (1). In total, six taxa contributed more than 5% of the total biomass. The main descriptive phytoplankton species was dinoflagellate *Ceratium hirundinella* (O.F. Müller) Dujardin, followed by centric diatom Pantocsekiella ocellata (Pantocsek) K.T. Kiss & E. Ács, cryptophytes *Cryptomonas* sp. and *Plagioselmis nannoplanctica* (H. Skuja) G. Novarino, I.A.N. Lucas & S. Morrall, chlorophyte *Tetraselmis cordiformis* (N. Carter) Stein, and dinoflagellate *Parvodinium inconspicuum* (Lemmermann) Carty (Table S4).

In terms of biomass share, the most dominant group in the composite samples on stations V1 and V2 (Figure 3) was Miozoa (54% and 48%, respectively), followed by Chlorophyta (24% and 29%, respectively) and Cryptophyta (14% and 12%, respectively). A corresponding situation regarding the phytoplankton assemblage was observed in the euphotic zone samples on the same stations, with a slight difference in shares of major taxonomic groups. Cryptophyta emerged as a dominant group (41%) in the composite sample on station V3, whilst Bacillariophyta and Miozoa were subdominant (25% and 18%, respectively). Conversely, the euphotic zone samples on station V3 were dominated by Miozoa (33%), followed by Bacillariophyta and Cryptophyta (26% and 18%, respectively). Phytoplankton community in the composite sample on station V4, situated close to the littoral zone in the central part of Lake Visovac, was characterized by the dominance of Miozoa with 68% of the total phytoplankton biomass, followed by subdominant groups Bacillariophyta (13%) and Chlorophyta (13%). The most dominant group in the euphotic zone samples on station V4 was Bacillariophyta (30%), accompanied by the subdominant groups Chlorophyta and Cryptophyta (28% and 23%, respectively). Cryptophyta and Bacillariophyta were the main groups characterizing the community in both composite and euphotic zone samples on station V5, positioned in the limnetic zone of the central part of the Lake (36% and 34%; 27% and 29%, respectively). Miozoa was the most dominant group on station V6 in both composite and euphotic zone samples (46% and 43%, respectively), with Bacillariophyta and Cryptophyta as subdominant groups. Bacillariophyta dominated in the composite sample on station V7, followed by Cryptophyta (44% and 30%, respectively), whilst Miozoa dominated the assemblage (50%) in the euphotic zone sample. Station V9, placed in the lower part of the Lake, was characterized by a complete dominance of Miozoa in the composite sample, reaching 80% of the total phytoplankton biomass, whilst in the euphotic zone sample, this share was 49%. Cryptophyta appeared as a dominant group in the composite sample on station V10 (46%). However, in the euphotic zone on the same station, Miozoa outnumbered Cryptophyta in their biomass share (36% and 30%, respectively).

Regarding the aphotic zone samples, stations V1 and V3 were co-dominated by Chlorophyta and Miozoa (39% and 34%; 41% and 43%, respectively). Chlorophyta was also a dominant group on station V2, followed by Miozoa (50% and 29%, respectively). In the aphotic zone samples on stations V4, V5 and V7, the domination of Miozoa was observed (61%, 57% and 64%, respectively). The aphotic zone on station V6 was characterized by Cryptophyta and Miozoa (30% and 28%, respectively), whilst on station V9, Chlorophyta took over domination (51%). Ochrophyta were present in all samples in the range from 1% (composite sample on station V6) to 11% (composite sample on station V10). There was no aphotic zone on station V10 due to the shallow maximum depth of 5 m (see Table 1).

The descriptive Reynolds' functional groups on upper stations V1 and V2 were **Lo** and **X2** in both composite and euphotic zone samples (Figure 4). The composite and euphotic zone samples on station V3 were characterized by functional group **X2** with the highest biomass share (50% and 37%, respectively), followed by associations **C** and **Lo** (25% and 18%; 26% and 33%, respectively). Functional group **Lo** dominated the assemblage in the composite sample on station V4, with 68% of the total phytoplankton biomass, whereas associations **C** and **X2** were both contributing with 12% to the total phytoplankton biomass.

Meanwhile, the functional group **X2** dominated the assemblage (51%), with coda **C** and **Lo** being subdominant (30% and 15%, respectively) in the euphotic zone samples on station V4. The functional group **X2** was dominant, whilst codon **C** appeared as subdominant in both composite and euphotic zone samples on station V5. Group **Lo** again became the most dominant in both samples on station V6, followed by coda **X2** and **C**. Codon **C** prevailed in the composite sample on station V7, together with association **X2** (43% and 39% of the total phytoplankton biomass, respectively), whilst in the euphotic zone samples, codon **Lo** gained dominance (50%). Codon **Lo** dominated the assemblage with 80% of the total phytoplankton biomass in the composite sample on station V9, as was also the case in the euphotic zone samples, although with a lower share (49%). The descriptive functional group in the composite sample on station V10 was **X2** (57% of the total phytoplankton biomass), whereas in the euphotic zones sample, it co-dominated with codon **Lo**. Species belonging to codon **E** (genus *Dinobryon*) were present in a very low biomass share on every station, except for the somewhat higher shares noted in the composite sample on station V10 and the euphotic zone sample on station V9 (11% and 9%, respectively).

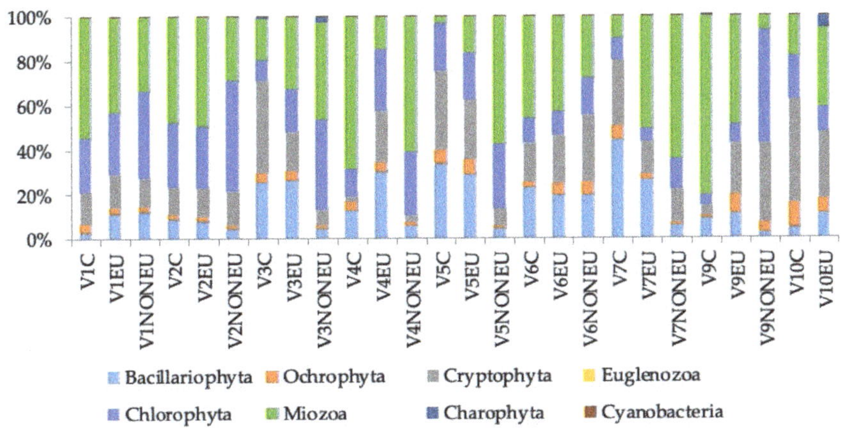

Figure 3. Relative biomass of phytoplankton taxonomic groups (expressed in percentages) in the composite (C), euphotic (EU) and aphotic (NONEU) zone samples from the sampling stations (V1 to V10, excluding station V8) in Lake Visovac during August 2018 according to the morphological approach.

In the aphotic zone samples, stations V1, V2 and V6 were dominated by codon **X2** (51%, 64% and 45% of the total phytoplankton biomass, respectively), followed by codon **Lo** (34%, 29% and 28%, respectively). Functional groups **X2** and **Lo** co-dominated the phytoplankton assemblage in the aphotic zone on station V3 (47% and 43%, respectively). On stations V4, V5 and V7 the descriptive codon was **Lo** (61%, 57% and 64%, respectively), followed by codon **X2** (33%, 37% and 28%, respectively). Functional group **X2** clearly dominated the assemblage on station V9 (77% of the total phytoplankton biomass).

The NMDS analysis of phytoplankton taxonomic composition according to the morphological approach (Figure 5) indicated segregation of almost all discrete-depth samples into two separate groups, the first one comprising the euphotic zone and the second one including the aphotic zone. The samples from station V10 were all counted into the euphotic zone.

Considering the horizontal distribution, NMDS analysis of the composite samples showed grouping of samples from limnetic and littoral stations (Figure 6). *Cryptomonas* sp. and *P. ocellata* dominated on central limnetic stations V3 and V5. Stations from the littoral zone were clustered into two distinct groups as follows: the first group, dominated by *C. hirundinella* (stations V1, V2, V4, V6, V9), and the second group, dominated by *P. ocellata* (station V7) and *Cryptomonas* sp. (station V10).

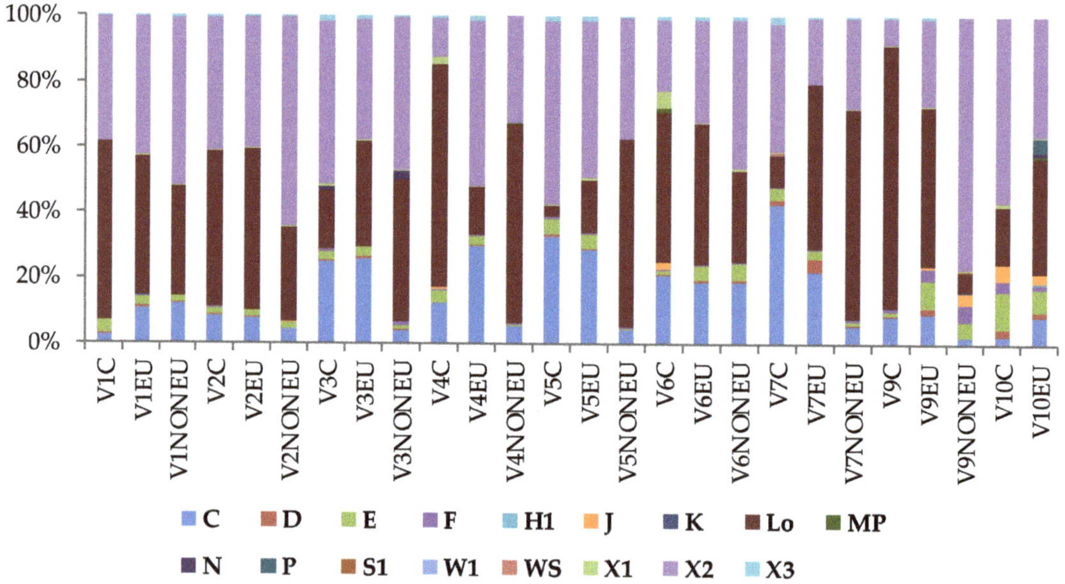

Figure 4. Relative biomass shares of Reynolds' functional groups (expressed in percentages) in the composite (C), euphotic (EU) and aphotic (NONEU) zone samples from the sampling stations (V1 to V10, excluding station V8) in Lake Visovac during August 2018 according to the morphological approach.

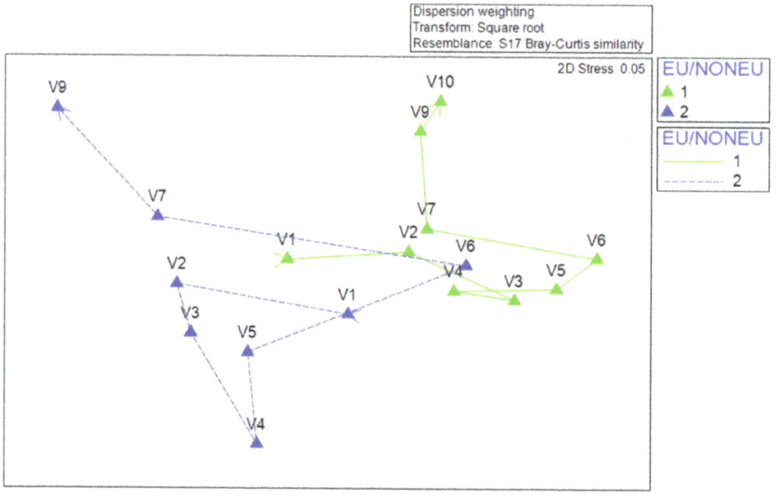

Figure 5. Non-metric multidimensional scaling (NMDS) ordination based on the Bray–Curtis similarity distance in the taxonomic composition of phytoplankton community on sampling stations (V1 to V10, excluding station V8) in Lake Visovac during August 2018 according to morphological approach. 1-EU-euphotic zone samples, 2-NONEU- aphotic zone samples.

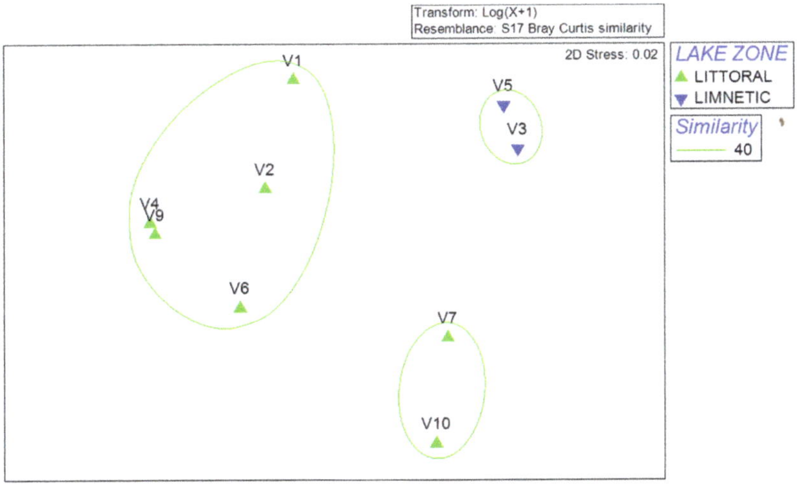

Figure 6. Non-metric multidimensional scaling (NMDS) ordination based on Bray–Curtis similarity distance in the taxonomic composition of phytoplankton community of composite samples on sampling stations (V1 to V10, excluding station V8) in Lake Visovac during August 2018 according to the morphological approach.

The results of one-way ANOSIM pairwise tests (Table 2) have demonstrated significant differences when comparing composite samples vs. aphotic zone samples (p = 0.042), and euphotic vs. aphotic zone samples (p = 0.006). On the other hand, the negative R statistic close to zero and the high significance level indicated very low differences between composite samples and euphotic zone samples.

Table 2. One-way analysis of similarity (ANOSIM) between the composite (C), euphotic (EU) and aphotic (NONEU) zone samples in Lake Visovac. Statistically significant values are marked in bold.

Pairwise Tests					
Comparison of Samples	**R Statistic**	**Significance Level (p)**	**Possible Permutations**	**Actual Permutations**	**Number >= Observed**
C vs. EU	−0.023	0.548	24310	999	547
C vs. NONEU	0.168	**0.042**	24310	999	41
EU vs. NONEU	0.391	**0.006**	24310	999	5

3.3. Characterization of Phytoplankton Community According to Molecular Approach

ANOSIM pairwise testing (Table 2) confirmed that the composite samples can be used as representative of the phytoplankton community in Lake Visovac and further characterized by molecular analyses. Correspondingly, eDNA metabarcoding analyses were performed on composite and aphotic zone samples. A total of 1,010,539 quality reads were yielded in 19 samples for eukaryotes, featuring 7140 ASVs at the 99% similarity level. After the Metazoan sequences were filtered out, a total of 902,798 quality reads with 7002 ASVs were found. The mean frequency per sample was 47,515 (min. 9779; max. 108,716). Only 97 ASVs corresponded to algal community with a total of 150,005 quality reads (Table S5).

Based on the sum of the relative abundance of the family ranks determined using the molecular approach, the taxonomically unassigned phytoplankton ASVs were found to have the highest relative abundance in all recorded composite and aphotic zone samples, with approximately 50% recorded in composite sample V7 (Figure 7). The four most abundant family ranks were polar centric diatoms (Mediophyceae), cryptophyta (Cryptomonadales), chrysophytes (Chrysophyceae_X), and dinoflagellates (Peridiniales). Interestingly, the two most abundant family ranks had the highest relative abundance in aphotic zone samples: 38% in V6 for Mediophyceae and 58% in V9 for Cryptomonadales. Furthermore, two other most abundant family ranks had the highest relative abundance in composite samples in V2 (31%) for Chrysophyceae_X and in V10 (67%) for Peridiniales. The families with total relative abundance between 65% and 20% in all samples were: Chlorodendrales, Suessisales, Pyrenomonadales, Katablepharidales, Cryptophyceae_X, Dictyochophyceae_X and Pseudodendromonadales. Of these family ranks, four had the highest relative abundance in the composite samples as follows: Suessisales (V1), Dictyochophyceae_X (V4), Catablepharidales (V6), and Pseudodendromonadales (V9). In contrast, Cryptophyceae_X (V2), Chlorodendrales (V3), and Pyrenomonadales (V7) had the highest relative abundance in the aphotic zone samples, while their relative abundance in the composite samples was about or less than 1%. The total relative abundance of the two family ranks corresponding to the Sphaeropleales and Synurales was less than 20% in all samples, but their abundance was highest in the aphotic zone samples (V3 for Sphaeropleales and V4 for Synurales). Perkinsida_X, Chrysochromulinaceae and Bicoecales corresponded to families with a total relative abundance in all samples equaling less than 10%. Families with a proportion of less than 1% were Gymnodiniaceae, Dolichomastigales, Cyanidiales, Charophyceae_X, Chlorellales, Distigmidae, Euglenophyceae, Petalomonadida, Choanoflagellida_X, Nucleariida, Araphid-pennate, Bacillariophyta_X, Raphid-pennate, Eustigmatophyceae_X and Xanthophyceae_X.

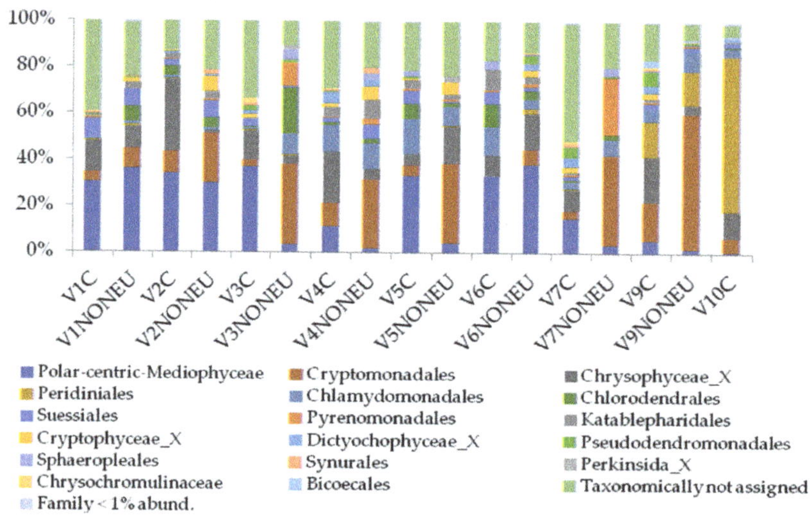

Figure 7. Relative abundance shares of phytoplankton family ranks (expressed in percentages) in the composite (C) and aphotic zone (NONEU) samples of sampling stations (V1 to V10, excluding station V8) in Lake Visovac during August 2018 according to the molecular (eDNA) approach.

The descriptive functional groups in the eDNA composite sample of station V1 were **Lo** (47%) and **C** (30%) (Figure 8). In the composite eDNA sample of station V2, the dominant association was group **C** (with 34%), whereas coda **X3**, **X2** and **Lo** were subdominant (with 29%, 18% and 13%, respectively). Functional groups **C** and **Lo** dominated the assemblage on station V3 (37% and 34%, respectively), whilst **X3** and **X2** were subdominant (14% and 10%, respectively). Codon **X3** and **X2** emerged as dominant on the station V4 (45% and 34%, respectively), whilst codon **C** appeared as a subdominant (11%). Functional groups **C** and **X2** were the most dominant on stations V5 (34% and 32%, respectively) and V6 (33% and 32%, respectively), whilst codon **X3**, **Lo** and **J** were subdominant. Functional group **X3** dominated the assemblage again on station V7 (53%), followed by functional groups **X2**, **C** and **Lo** (16%, 15% and 7%, respectively). Codon **X2**, **X3** and **Lo** prevailed on station V9 (with 31%, 26% and 21%, respectively). At station V10, codon **Lo** showed clear dominance (73%).

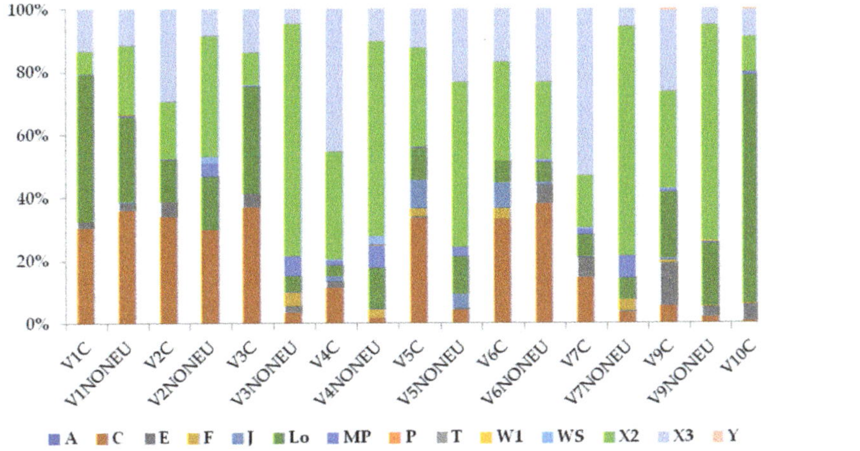

Figure 8. Relative biomass shares of Reynolds' functional groups (expressed in percentages) in the eDNA composite (C) and aphotic zone (NONEU) samples of sampling stations (V1 to V10, excluding station V8) in Lake Visovac during August 2018.

Regarding the eDNA aphotic zone samples, functional group **X2** prevailed with over 50% of total phytoplankton biomass on the majority of studied stations (V3 with 74%, V4 with 62%, V5 with 52%, V7 with 72%, and V9 with 68%). Station V1 was characterized by the dominance of codon **C** (36%) with coda **Lo** and **X2** as subdominant (27% and 22%, respectively), whilst coda **X2** and **C** had the highest biomass share on station V2 (38% and 30%, respectively). Aphotic zone samples from station V6 were characterized by functional groups **C**, **X2** and **X3** (38%, 25% and 23%, respectively).

NMDS analysis of phytoplankton eDNA metabarcoding has disclosed two groups, the first one including composite samples and the second one comprising aphotic zone samples (Figure 9).

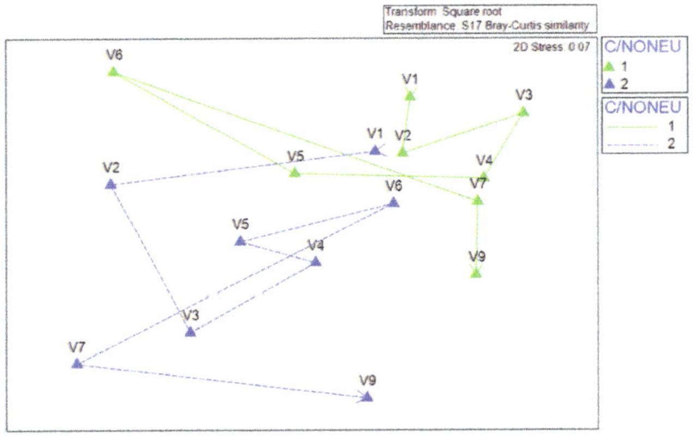

Figure 9. Non-metric multidimensional scaling (NMDS) ordination based on Bray–Curtis similarity distance in the taxonomic composition of phytoplankton community on sampling stations (V1 to V9, excluding station V8) in Lake Visovac during August 2018 according to the molecular eDNA approach. 1-C- composite samples, 2-NONEU-aphotic zone samples.

3.4. Morphological and Molecular Diversity of Phytoplankton in Lake Visovac

The highest number of phytoplankton taxa in the composite samples based on the morphological approach (Figure 10a) was detected on station V10 (26) and the lowest on station V1 (16). The Margalef Richness Index in the composite samples ranged from 0.99 on station V1 to 1.79 on station V10 (Figure 10b). Pielou's Evenness Index (Figure 10c) in the composite samples varied between 0.39 (station V4) and 0.78 (station V9). Concordantly, the lowest Shannon–Wiener Diversity Index value (Figure 10d) was present on station V4 (1.22), while the highest was on station V9 (2.38).

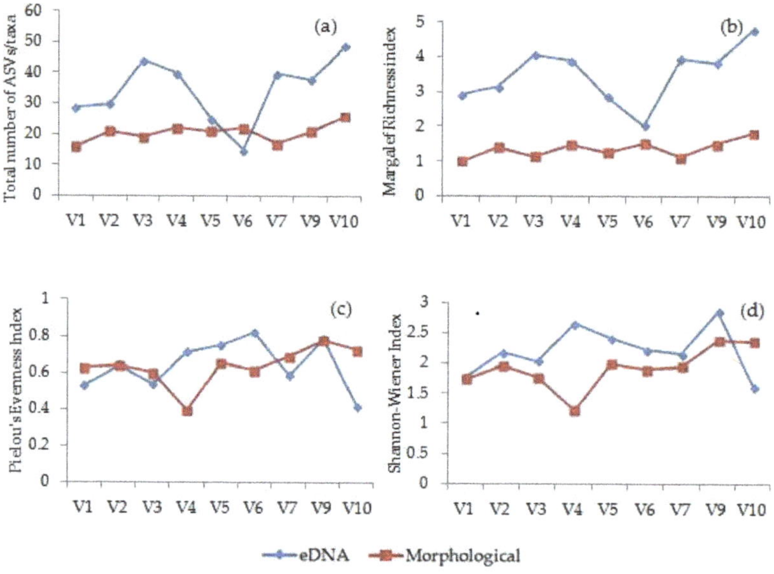

Figure 10. Phytoplankton diversity: (**a**) Total number of phytoplankton ASVs and taxa, (**b**) Margalef Richness Index (*d*), (**c**) Pielou's Evenness Index (*J'*), and (**d**) Shannon–Wiener Diversity Index (*H'*) for eDNA (blue line) and morphological composite samples (red line) on sampling stations (V1 to V10, excluding station V8) in Lake Visovac during August 2018.

The highest number of phytoplankton ASVs provided by eDNA metabarcoding in the composite samples (Figure 10a) was detected on station V10 (49), whilst the lowest on station V6 (15). The lowest Margalef Richness Index in the eDNA composite samples (Figure 10b) was obtained on station V6 (2.04), whilst the highest taxa richness was recorded on station V10 (4.76). Results obtained by eDNA metabarcoding showed higher values for both indices compared to morphological approach data, with the exception of station V6. Pielou's Evenness Index in the eDNA composite samples (Figure 10c) varied between 0.41 (station V10) and 0.82 (station V6). Compared to morphological composite samples, Pielou's Index was higher in the eDNA composite samples on stations V2, V4, V6, V8 and V9 but lower on stations V1, V3, V7 and V10. The Shannon–Wiener Diversity Index (Figure 10d) for eDNA composite samples varied between 1.61 (station V10) and 2.86 (station V9), and provided higher values than morphological approach data, except for station V10.

The total number of taxonomically assigned ASVs detected by the molecular approach was two times higher than the number of taxa revealed by traditional morphological identification using microscopy. Considering the results on each station, the number of ASVs was higher than taxa number on every station except for station V6, which proved to be an outlier just as for morphological samples. The Margalef Richness Index was higher in all eDNA samples compared to morphological samples. Pielou's Evenness index was lower in the eDNA samples on stations V1 and V3 (about 40% of not assigned ASVs and high share domination of Polar centric Mediophyceae), V7 (about 60% of unassigned ASVs) and V10 (clear domination of Peridiniaceae). Results obtained by the eDNA metabarcoding showed higher Shannon–Wiener Diversity Index values, except for station V10 due to the explicit prevalence of Peridiniaceae.

According to the morphological approach, the phytoplankton community was characterized by dinoflagellate taxa (Miozoa) belonging to functional group **Lo**, followed by cryptophyte and chlorophyte taxa (mainly *Cryptomonas* sp. and *Tetraselmis cordiformis*, respectively) from codon **X2** (Figure 4, Table S4). A higher share of centric diatom *Pantocsekiella ocellata* (Bacillariophyta) assorted into codon **C** was apparent on central stations (V3–V6) and station V7. Mixotrophic genus *Dinobryon* (Ochrophyta) from codon **E** was present across all stations in a small share. According to eDNA metabarcoding approach, members of the family Cryptomonadales belonging to functional group **X2** had the highest phytoplankton biomass share on central stations (V3-V6), whilst coda **Lo** (Peridiniales), **C** (Polar centric Mediophyceae) and **X3** (Chrysophyceae_X) appeared either as dominant or with a high biomass share on other stations (Figures 7 and 8).

4. Discussion

Comparison of identified taxa according to morphological approach indicated a high level of similarity between composite samples and euphotic zone samples in the karst Lake Visovac. A similar number of recorded species was found in both types of samples as well as similar descriptive taxa. Especially, a higher level of similarity in both euphotic zone samples and composite samples was found for the following taxa: *C. hirundinella*, *Cryptomonas* sp. and *P. ocellata*. The high level of similarity was further supported by the ANOSIM analysis, which indicated that euphotic zone samples can be reliably exchanged by composite samples to provide an accurate characterization of phytoplankton communities in the euphotic zone. Moreover, apart from some mismatches in samples from stations V4 and V7, in particular codon **Lo**, which consequently affected other relative biomass shares, the results on the phytoplankton community from composite samples and euphotic zone samples were mainly congruent. Gaps in shares of phytoplankton coda can be explained due to deviations in biomass when determining species with a large biovolume value, i.e., a difference of just one or two found cells in the observed sample can influence the results [38]. Nevertheless, the composite sampling of phytoplankton can be used as an optimal sampling approach in Lake Visovac during regular monitoring. Lake phytoplankton communities depend on factors such as nutrient availability, light and temperature, as well as hydrological conditions [39], all of which condition a predictably lower phytoplankton biomass in the aphotic zone of Lake Visovac than in the euphotic zone. Moreover, a significantly lower oxygen concentration (<5 mg L^{-1}) was measured on several stations (V3, V5 and V9) in the aphotic zone of Lake Visovac indicating hypoxic conditions. As for the horizontal distribution, the similarity within the phytoplankton community was more evident with respect to the sampling microlocation than to the sampling station, thus resulting in a strong separation between limnetic and littoral zone (the nearshore environment) samples in Lake Visovac. The relative predominance of centric diatoms in both the limnetic and littoral zones may be associated with their capability to resuspend from the bottom due to short retention time, fast water flow, and a relatively deep mixing depth, which can prevent centrics from sinking to the hypolimnion and allow them to dominate in the water column [40]. On the other hand, the flagella-bearing *Cryptomonas* sp. and *Ceratium hirundinella* can actively swim in the water column to obtain sufficient amounts of light and nutrients [41,42]. The segregation of littoral samples could be attributed to several biotic factors, such as the impact of macrophytic vegetation on development of phytoplankton, grazing by zooplankton, and interspecific competition, and abiotic factors such as variations in water chemistry and differences in nutrient availability [43,44].

The phytoplankton functional groups are based on ecological sensitivities and tolerances of taxa [5,26], so their main advantage over traditional taxonomic division is the possibility of generalizing the results [45]. Moreover, they can also be successfully integrated in ecological assessments where eDNA metabarcoding datasets are present [8]. The euphotic and aphotic zones were both dominated by a large mixotrophic swimming dinoflagellate *C. hirundinella*, known to build up higher biomass during the summer stratification period [46,47]. The flagellar motility enables cells to vertically migrate throughout the water column, thus facilitating effective exploitation of nutrients and boosting photosynthesis [26,48,49]. This conclusively allows them to dominate in thermally stratified mesotrophic lakes during the summer period [5]. Besides *C. hirundinella*, dinoflagellate *Parvodinium inconspicuum* was also a descriptive representative of **Lo**, a functional group characteristic of the summer epilimnion of mesotrophic lakes and tolerant of nutrient deficiency [5,26]. Dominance of the *Ceratium* species has risen significantly with warming caused by climate change [50]. The mixotrophic nutrition strategy is widespread and often dominant in freshwater ecosystems [51,52], enabling phytoplankton to be capable of bacterivory in nutrient-depleted conditions [53,54]. Different mixotrophic species may vary in their ability to regulate the shift between autotrophic and heterotrophic lifestyles, with regards to the changes in the environment, which might cause differences in their temperature response [55]. Changes in the functional role of mixotrophs from primary producers to consumers may cascade through food webs, altering species interactions, as well as the magnitude and direction of the carbon flux [55].

Among other descriptive taxa, larger abundance of cryptophyte *Cryptomonas* sp. was observed on all stations, whilst centric diatom *Pantocsekiella ocellata* showed higher abundance on all central stations and lower station V7. *Cryptomonas* sp. belongs to functional group **X2,** together with another cryptophyte species *Plagioselmis nannoplanctica* and chlorophyte *Tetraselmis cordiformis* [5]. The representatives of codon **X2** are acknowledged as meso-eutrophic indicators [5] and exhibit a wide range of tolerance to changes in ecological conditions in Lake Visovac [56]. *P. ocellata* is sorted into codon **C**, adaptated to high lake stability [57] and low light availability [5,58], and known to dominate in the mesotrophic ecosystems [59]. *P. ocellata* was confirmed to be one of the key descriptors of the phytoplankton community in previous investigations on Lake Visovac [8,46,47,56]. Chrysomonads of the mixotrophic genus *Dinobryon* from codon **E** were present in lower shares across all stations, and usually denote small, shallow, base-poor lakes or heterotrophic ponds [5,26].

The comparability of traditional light microscopy and eDNA metabarcoding approach is evident in the designation of centric diatoms, dinoflagellates and cryptophytes as descriptive taxa of Lake Visovac. Although dinoflagellate *Ceratium hirundinella,* diatom *Pantocsekiella ocellata* and cryptophytes *Cryptomonas* sp. and *Plagioselmis nannoplanctica* were identified to the species level by using traditional microscopy, the eDNA metabarcoding method designated the higher ranks of Polar centric Mediophyceae, Cryptomonadales and Peridiniales as the most dominant algal groups. The possible reason that the descriptive taxa were identified only by microsocpy and not by eDNA to the species level is that taxonomic assignments of short amplicon reads to the species level are still problematic because too many species are missing from the reference database [30,31]. For this reason, eDNA analyses of eukaryotic phytoplankton diversity were based at the family level, because metabarcoding at the V9-region of SSU rRNA genes allows correct identification from the genus to the higher taxonomic level, as recognized in previous studies [30,60]. Although there are several more specific primers for detecting diversity of diatoms, such as ribosomal (*rbcL*) genes [61,62], the small universal hypervariable V9-region of the 18S rRNA gene was chosen in this study because it provides a comprehensive overview of the community and has the ability to capture assemblages of photosynthetic organisms, especially when dealing with phytoplankton consisting of many different algal groups [8,30]. On the other hand, according to the relative abundance, there were a lot of taxonomically unassigned ASVs in the whole molecular dataset. First, the reason for this gap might be the choice of primers, as mentioned above, since they are crucial for species recognition [10]. Second, the method of sampling may also affect the results, as eDNA samples require different volumes of water to be filtered. In this case, water volumes were lower than usual to reduce potential PCR inhibitors from filtering larger volumes of water, but this could potentially limit detection of all taxa present in represented samples [10,63].

When comparing the results of the phytoplankton community defined by functional classification, both approaches proved reliable in detecting functional groups (**Lo**, **C**, **X2**, **X3**) with similar ecological demands and were congruent with previous studies [8]. Differences between the approaches can be attributed to the ability to distinguish indistinct morphological characteristics, as the existence of cryptic species, pico- and concealed phytoplankton can be difficult to ascertain from morphological analyses [7]. In addition, the most frequent coda were used in the assigning of class and family ranks into functional groups, which could also affect the results. The above mentioned could explain a higher share of functional group **X3** to which the family rank Chrysophyceae_X was assigned in the molecular approach.

As for alpha diversity, in most cases, the eDNA metabarcoding results provided higher values than the morphological approach for all indices, except for Pielou's Index which has shown contrasting results for all samples. This may be related to the occurrence of similar morphological features between microscopically recorded species, which may lead to difficulties in species discrimination [8,64]. In addition, certain small phytoplankton can be easily detected by eDNA metabarcoding, whereas they are usually missed by light microscopy, which may affect species diversity [65,66]. Several previous studies also reported that V9-region has potential for broader recognition spectrum when obtaining results for Shannon diversity [67,68]. On the other hand, there was a discrepancy in the Pielou index, which showed very low values mostly for eDNA samples, which was due to a clear dominance of Polar centric Mediophyceae and Peridiniales [67]. Although the total number of taxonomically assigned ASVs identified by the molecular approach was two times higher than the number of taxa identified by traditional morphological identification, these results should be viewed with caution, especially for those ASVs that could not be taxonomically assigned, as they do not reflect the same percentage or number of unidentified taxa [12]. There is still a problem in translating abundance from sequence data to biological abundance because the rDNA copy number varies among taxa. Therefore, caution should always be used when interpreting the most abundant taxa detected by amplicon sequences, because the sequences of Alveolata (dinoflagellates) show variation in rDNA copy numbers [69]. However, in this study the eDNA results were compared, and to some extent confirmed and verified with the results of the morphological approach.

Non-metric multidimensional scaling (NMDS) analysis demonstrated the corresponding grouping of samples in both morphological and molecular approaches. Similarly comparable results between molecular and morphological approaches using beta diversity were also confirmed in several recent studies [70–72]. The results of NMDS analysis indicated a significant dissimilarity between composite samples and aphotic zone samples in both approaches, thus confirming the applicability of eDNA metabarcoding in the routine biomonitoring of Lake Visovac. Significant differences in horizontal and vertical distribution of the phytoplankton in Lake Visovac were found previously by Ciglenečki-Jušić et al. [73].

5. Conclusions

The morphological and eDNA metabarcoding approaches offer comparable results in describing the phytoplankton community and are applicable tools for biodiversity monitoring in Lake Visovac. Moreover, eDNA metabarcoding should be used as a feasible supplement to traditional microscopy in the phytoplankton community assessments, with regards to the drawbacks of each method. It is important to emphasize the essential continual advancement of eDNA metabarcoding in providing more accurate results.

Supplementary Materials: The following supporting information can be downloaded at: https://www.mdpi.com/article/10.3390/w15071379/s1. Supplement file: Table S1. Physical and chemical parameters in the euphotic and aphotic zone samples on the sampling stations (V1 to V10, excluding station V8).; Table S2. Chemical parameters measured in the laboratory for composite samples, sampling stations (V1 to V10, excluding station V8).; Table S3. Relative variance explained and factor coordinates of the variables for the first two principal components (PC1 and PC2) of the Principal Component Analysis (PCA).; Figure S1. Shares of all Eukaryota class groups on the sampling stations (V1 to V10, excluding station V8) in Lake Visovac during August 2018 according to analysis of V9-region.; Table S4. Phytoplankton taxa biomass (mg L^{-1}) on the sampling stations (V1 to V10, excluding station V8) in Lake Visovac during August 2018.; Table S5. Assigned taxonomic amplicon sequencing variants for phytoplankton community on the sampling stations (V1 to V10, excluding station V8) in Lake Visovac during August 2018.

Author Contributions: Formal analysis, investigation, data curation, writing—Original draft preparation, visualization, M.Š.; formal analysis, investigation, writing—Review and editing, visualization, A.K.; formal analysis, investigation, validation, writing—Review and editing, visualization, P.Ž.; conceptualization, investigation, supervision, validation, writing—Review and editing M.G.U. All authors have read and agreed to the published version of the manuscript.

Funding: This study was financed by the Public Institute Krka National Park as part of the project "Istraživanja horizontalne i vertikalne raspodjele fitoplanktona Visovačkog jezera u svrhu ocjene ekološkog stanja primjenom standardnih i molekularnih metoda (FVORUM)".

Institutional Review Board Statement: Not applicable.

Informed Consent Statement: Not applicable.

Data Availability Statement: The datasets generated during and/or analyzed during the current study are available from the corresponding author on reasonable request.

Acknowledgments: We thank Mirela Šušnjara for contributions in the laboratory and field work and Darija Damić for help with data visualization in QGIS.

Conflicts of Interest: The authors declare no conflict of interest.

References

1. Myers, N.; Mittermeier, R.A.; Mittermeier, C.G.; da Fonseca, G.A.B.; Kent, J. Biodiversity Hotspots for Conservation Priorities. *Nature* **2000**, *403*, 853–858. [CrossRef] [PubMed]
2. Mikac, I.; Fiket, Ž.; Terzić, S.; Barešić, J.; Mikac, N.; Ahel, M. Chemical Indicators of Anthropogenic Impacts in Sediments of the Pristine Karst Lakes. *Chemosphere* **2011**, *84*, 1140–1149. [CrossRef] [PubMed]
3. Gaedke, U. Trophic Dynamics in Aquatic Ecosystems. In *Encyclopedia of Inland Waters*; Elsevier: New York, NY, USA, 2009; pp. 499–504; ISBN 978-0-12-370626-3.
4. European Commision Directive of the European Parliament and of the Council. *2000/60/EC. Establishing a Framework for Community Action in the Field of Water Policy (Water Framework Directive)*; The Publications Office of the European Union (L'Office des publications de l'Union européenne: Luxembourg, 2000.
5. Reynolds, C.S.; Huszar, V.; Kruk, C.; Naselli-Flores, L.; Melo, S. Towards a Functional Classification of the Freshwater Phytoplankton. *J. Plankton Res.* **2002**, *24*, 417–428. [CrossRef]
6. Bellinger, E.G.; Sigee, D.C. *Freshwater Algae: Identification, Enumeration and Use as Bioindicators*, 1st ed.; Wiley-Blackwell: New York, NY, USA, 2015; ISBN 978-1-118-91716-9.
7. Huo, S.; Li, X.; Xi, B.; Zhang, H.; Ma, C.; He, Z. Combining Morphological and Metabarcoding Approaches Reveals the Freshwater Eukaryotic Phytoplankton Community. *Environ. Sci. Eur.* **2020**, *32*, 37. [CrossRef]
8. Hanžek, N.; Gligora Udovič, M.; Kajan, K.; Borics, G.; Várbíró, G.; Stoeck, T.; Žutinić, P.; Orlić, S.; Stanković, I. Assessing Ecological Status in Karstic Lakes through the Integration of Phytoplankton Functional Groups, Morphological Approach and Environmental DNA Metabarcoding. *Ecol. Indic.* **2021**, *131*, 108166. [CrossRef]
9. Borics, G.; Abonyi, A.; Salmaso, N.; Ptacnik, R. Freshwater Phytoplankton Diversity: Models, Drivers and Implications for Ecosystem Properties. *Hydrobiologia* **2021**, *848*, 53–75. [CrossRef]
10. Pawlowski, J.; Apothéloz-Perret-Gentil, L.; Mächler, E.; Altermatt, F. *Environmental DNA Applications in Biomonitoring and Bioassessment of Aquatic Ecosystems*; Environmental Studies. no 2010; Federal Office for the Environment: Bern, Switzerland, 2020; p. 71. [CrossRef]

11. Hering, D.; Borja, A.; Jones, J.I.; Pont, D.; Boets, P.; Bouchez, A.; Bruce, K.; Drakare, S.; Hänfling, B.; Kahlert, M.; et al. Implementation Options for DNA-Based Identification into Ecological Status Assessment under the European Water Framework Directive. *Water Res.* **2018**, *138*, 192–205. [CrossRef]
12. Kulaš, A.; Gligora Udovič, M.; Tapolczai, K.; Žutinić, P.; Orlić, S.; Levkov, Z. Diatom EDNA Metabarcoding and Morphological Methods for Bioassessment of Karstic River. *Sci. Total Environ.* **2022**, *829*, 154536. [CrossRef]
13. Bonacci, O.; Andrić, I.; Roje-Bonacci, T. Hydrological Analysis of Skradinski Buk Tufa Waterfall (Krka River, Dinaric Karst, Croatia). *Env. Earth Sci.* **2017**, *76*, 669. [CrossRef]
14. Official Gazette of the Republic of Croatia No. 96/19 Regulation on Water Quality Standards. Zagreb, Croatia, 2019. Available online: https://leap.unep.org/countries/hr/national-legislation/regulation-water-quality-standard (accessed on 6 March 2023).
15. Poikane, S. *Water Framework Directive Intercalibration Technical Report*; Part 2: Lakes; Office for Official Publications of the European Communities (OPOCE): Luxembourg, 2009.
16. APHA Standard Methods for the Examination of Water and Wastewater. American Public Health Association: Washington, DC, USA, 2017. Available online: https://www.scirp.org/(S(351jmbntvnsjt1aadkozje))/reference/referencespapers.aspx?referenceid=2459667 (accessed on 6 March 2023).
17. Utermöhl, H. Zur Vervollkommnung Der Quantitativen Phytoplankton-Methodik. *SIL Commun.* **1958**, *9*, 1–38. [CrossRef]
18. Krammer, K.; Lange-Bertalot, H. Bacillariophyceae 3. Teil: Centrales, Fragilariaceae, Eunotiaceae. In *Süßwasserflora von Mitteleuropa Band 2/3*; Ettl, H., Gerloff, J., Heynig, H., Mollenhauer, D., Eds.; Spektrum Akademischer Verlag: Heidelberg, Germany, 2000.
19. Komárek, J.; Anagnostidis, K. Cyanoprokaryota 1. Teil: Chroococcales. In *Süsswasserflora von Mitteleuropa 19/1*; Ettl, H., Gärtner, G., Heynig, H., Mollenhauer, D., Eds.; Gustav Fischer: Jena, Germany; Stuttgart, Germany; Lübeck, Germany; Ulm, Germany, 1998.
20. Komárek, J.; Anagnostidis, K. Cyanoprokaryota 2. Teil/ 2nd Part: Oscillatoriales. In *Süsswasserflora von Mitteleuropa 19/2*; Büdel, B., Krienitz, L., Gärtner, G., Schagerl, M., Eds.; Elsevier/Spektrum: Heidelberg, Germany, 2005.
21. Krammer, K. *Bacillariophyceae*; Spektrum Akademischer: Heidelberg, Germany, 2000.
22. Kristiansen, J.; Preisig, H.R. Süßwasserflora von Mitteleuropa 1. In *Chrysophyte and Haptophyte Algae: Part 2: Synurophyceae*, 2nd ed.; Spektrum Akademisher Verlag: Berlin, Germany, 2007.
23. Hofmann, G.; Werum, M.; Lange-Bertalot, H. Diatomeen Im Süßwasser—Benthos von Mitteleuropa. In *Bestimmungsflora Kieselalgen Für Die Ökologische Praxis*; Über 700 Der Häufigsten Arten Und Ihre Ökologie; Gantner, A.R.G., Ed.; Koeltz Scientific Books: Ruggell, Liechtenstein, 2011.
24. John, D.M.; Whitton, B.A.; Brook, A.J. *The Freshwater Algal Flora of the British Isles: An Identification Guide to Freshwater and Terrestrial Algae*; Cambridge University Press: Cambridge, UK, 2011.
25. Guiry, M.D.; Guiry, G.M. AlgaeBase. In *World-Wide Electronic Publication*; National University of Ireland: Galway, Ireland, 2022. Available online: https://www.algaebase.org (accessed on 3 August 2022).
26. Padisák, J.; Crossetti, L.O.; Naselli-Flores, L. Use and Misuse in the Application of the Phytoplankton Functional Classification: A Critical Review with Updates. *Hydrobiologia* **2009**, *621*, 1–19. [CrossRef]
27. HRN EN ISO 5667-3:2018 Water Quality—Sampling—Part 3: Preservation and Handling of Water Samples. Available online: https://repozitorij.hzn.hr/norm/HRN+EN+ISO+5667-3%3A2018 (accessed on 16 January 2019).
28. HRN EN 15204:2008; Water Quality—Guidance Standard on the Enumeration of Phytoplankton Using Inverted Microscopy (Utermöhl Technique). Available online: https://repozitorij.hzn.hr/norm/HRN+EN+15204%3A2008 (accessed on 11 June 2020).
29. HRN EN 16695:2015; Water Quality—Guidance on the Estimation of Phytoplankton Biovolume. Available online: https://repozitorij.hzn.hr/norm/HRN+EN+16695%3A2015 (accessed on 11 June 2020).
30. Stoeck, T.; Bass, D.; Nebel, M.; Christen, R.; Jones, M.D.M.; Breiner, H.-W.; Richards, T.A. Multiple Marker Parallel Tag Environmental DNA Sequencing Reveals a Highly Complex Eukaryotic Community in Marine Anoxic Water. *Mol. Ecol.* **2010**, *19*, 21–31. [CrossRef]
31. Amaral-Zettler, L.A.; McCliment, E.A.; Ducklow, H.W.; Huse, S.M. A Method for Studying Protistan Diversity Using Massively Parallel Sequencing of V9 Hypervariable Regions of Small-Subunit Ribosomal RNA Genes. *PLoS ONE* **2009**, *4*, e6372. [CrossRef]
32. Stoeck, T.; Kochems, R.; Forster, D.; Lejzerowicz, F.; Pawlowski, J. Metabarcoding of Benthic Ciliate Communities Shows High Potential for Environmental Monitoring in Salmon Aquaculture. *Ecol. Indic.* **2018**, *85*, 153–164. [CrossRef]
33. Martin, M. Cutadapt Removes Adapter Sequences from High-Throughput Sequencing Reads. *EMBnet. J.* **2011**, *17*, 10. [CrossRef]
34. Andrews, S. *FastQC: A Quality Control Tool for High Throughput Sequence Data*; Babraham Bioinformatics; Babraham Institute: Cambridge, UK, 2010. Available online: http://www.bioinformatics.babraham.ac.uk/projects/fastqc (accessed on 3 May 2019).
35. Callahan, B.J.; McMurdie, P.J.; Rosen, M.J.; Han, A.W.; Johnson, A.J.A.; Holmes, S.P. DADA2: High-Resolution Sample Inference from Illumina Amplicon Data. *Nat. Methods* **2016**, *13*, 581–583. [CrossRef]
36. Guillou, L.; Bachar, D.; Audic, S.; Bass, D.; Berney, C.; Bittner, L.; Boutte, C.; Burgaud, G.; de Vargas, C.; Decelle, J.; et al. The Protist Ribosomal Reference Database (PR2): A Catalog of Unicellular Eukaryote Small Sub-Unit RRNA Sequences with Curated Taxonomy. *Nucleic Acids Res.* **2012**, *41*, D597–D604. [CrossRef]
37. QGIS A Free and Open Source Geographic Information System. Open Source. Geospatial Foundation Project. 2009. Available online: https://qgis.org/en/site/ (accessed on 21 March 2023).
38. Zarauz, L.; Irigoien, X.; Fernandes, J.A. Changes in Plankton Size Structure and Composition, during the Generation of a Phytoplankton Bloom, in the Central Cantabrian Sea. *J. Plankton Res.* **2008**, *31*, 193–207. [CrossRef]

39. Ptacnik, R.; Lepistö, L.; Willén, E.; Brettum, P.; Andersen, T.; Rekolainen, S.; Lyche Solheim, A.; Carvalho, L. Quantitative Responses of Lake Phytoplankton to Eutrophication in Northern Europe. *Aquat. Ecol.* **2008**, *42*, 227–236. [CrossRef]
40. Žutinić, P.; Gligora Udovič, M.; Kralj Borojević, K.; Plenković-Moraj, A.; Padisák, J. Morpho-Functional Classifications of Phytoplankton Assemblages of Two Deep Karstic Lakes. *Hydrobiologia* **2014**, *740*, 147–166. [CrossRef]
41. Clegg, M.R.; Maberly, S.C.; Jones, R.I. Behavioral Response as a Predictor of Seasonal Depth Distribution and Vertical Niche Separation in Freshwater Phytoplanktonic Flagellates. *Limnol. Oceanogr.* **2007**, *52*, 441–455. [CrossRef]
42. Jamal, A.; Yusoff, F.M.; Banerjee, S.; Shariff, M. Littoral and Limnetic Phytoplankton Distribution and Biodiversity in a Tropical Man-Made Lake, Malaysia. *Adv. Stud. Biol.* **2014**, *6*, 149–168. [CrossRef]
43. Lemly, A.D.; Dimmick, J.F. Phytoplankton Communities in the Littoral Zone of Lakes: Observations on Structure and Dynamics in Oligotrophic and Eutrophic Systems. *Oecologia* **1982**, *54*, 359–369. [CrossRef] [PubMed]
44. Mulderij, G.; Van Nes, E.H.; Van Donk, E. Macrophyte–Phytoplankton Interactions: The Relative Importance of Allelopathy versus Other Factors. *Ecol. Model.* **2007**, *204*, 85–92. [CrossRef]
45. Blaum, N.; Mosner, E.; Schwager, M.; Jeltsch, F. How Functional Is Functional? Ecological Groupings in Terrestrial Animal Ecology: Towards an Animal Functional Type Approach. *Biodivers. Conserv.* **2011**, *20*, 2333–2345. [CrossRef]
46. Gligora Udovič, M.; Kralj Borojević, K.; Žutinić, P.; Šipoš, L.; Plenković-Moraj, A. Net-Phytoplankton Species Dominance in a Travertine Riverine Lake Visovac, NP Krka. *Nat. Croat.* **2011**, *20*, 411–424.
47. Gligora Udovič, M.; Žutinić, P.; Kralj Borojević, K.; Plenković-Moraj, A. Co-Occurrence of Functional Groups in Phytoplankton Assemblages Dominated by Diatoms, Chrysophytes and Dinoflagellates. *Fundam. Appl. Limnol.* **2015**, *187*, 101–111. [CrossRef]
48. Sommer, U. Growth and Survival Strategies of Plankton Diatoms. In *Growth and Reproductive Strategies of Freshwater Phytoplankton*; Sandgren, C.D., Ed.; Cambridge University Press: Cambridge, UK, 1988; pp. 227–260; ISBN 0-521-32722-9.
49. Grigorszky, I.; Padisák, J.; Borics, G.; Schitchen, C.; Borbély, G. Deep Chlorophyll Maximum by Ceratium hirundinella (O. F. Müller) Bergh in a Shallow Oxbow in Hungary. *Hydrobiologia* **2003**, *506*, 209–212. [CrossRef]
50. Olrik, K.; Cronberg, G.; Annadotter, H. Lake Phytoplankton Responses to Global Climate Changes. In *Climatic Change and Global Warming of Inland Waters: Impacts and Mitigation for Ecosystems and Societies*; Goldman, C.R., Kumagai, M., Robarts, R.D., Eds.; Wiley: Chichester, UK, 2012; pp. 173–199.
51. Stoecker, D.K.; Johnson, M.D.; deVargas, C.; Not, F. Acquired Phototrophy in Aquatic Protists. *Aquat. Microb. Ecol.* **2009**, *57*, 279–310. [CrossRef]
52. Sanders, R.W. Alternative Nutritional Strategies in Protists: Symposium Introduction and a Review of Freshwater Protists That Combine Photosynthesis and Heterotrophy1. *J. Eukaryot. Microbiol.* **2011**, *58*, 181–184. [CrossRef]
53. Kamjunke, N.; Henrichs, T.; Gaedke, U. Phosphorus Gain by Bacterivory Promotes the *Mixotrophic Flagellate Dinobryon* Spp. during Re-Oligotrophication. *J. Plankton Res.* **2007**, *29*, 39–46. [CrossRef]
54. Hansen, P.J. The Role of Photosynthesis and Food Uptake for the Growth of Marine Mixotrophic Dinoflagellates1. *J. Eukaryot. Microbiol.* **2011**, *58*, 203–214. [CrossRef]
55. Wilken, S.; Huisman, J.; Naus-Wiezer, S.; Van Donk, E. Mixotrophic Organisms Become More Heterotrophic with Rising Temperature. *Ecol. Lett.* **2013**, *16*, 225–233. [CrossRef]
56. Šimunović, M.; Kulaš, A.; Žutinić, P.; Goreta, G.; Gligora Udovič, M. Phytoplankton Metrics for Trophic and Ecological Status Assessment of a Natural Karstic Lake. *Acta Bot. Croat.* **2022**, *81*, 185–196. [CrossRef]
57. Beamud, S.G.; León, J.G.; Kruk, C.; Pedrozo, F.; Diaz, M. Using Trait-Based Approaches to Study Phytoplankton Seasonal Succession in a Subtropical Reservoir in Arid Central Western Argentina. *Environ. Monit. Assess.* **2015**, *187*, 271. [CrossRef]
58. Reynolds, C.S. Vegetation Processes in the Pelagic: A Model for Ecosystem Theory. Xxvii, 371 p. Oldendorf/Luhe, Germany: Ecology Institute, 1997. (Excellence in Ecology No. 9). Price DM 68.00. *J. Mar. Biol. Ass.* **1997**, *77*, 919. [CrossRef]
59. Hu, R.; Han, B.; Naselli-Flores, L. Comparing Biological Classifications of Freshwater Phytoplankton: A Case Study from South China. *Hydrobiologia* **2013**, *701*, 219–233. [CrossRef]
60. Choi, J.; Park, J.S. Comparative Analyses of the V4 and V9 Regions of 18S RDNA for the Extant Eukaryotic Community Using the Illumina Platform. *Sci. Rep.* **2020**, *10*, 6519. [CrossRef]
61. Rimet, F.; Vasselon, V.; A.-Keszte, B.; Bouchez, A. Do We Similarly Assess Diversity with Microscopy and High-Throughput Sequencing? Case of Microalgae in Lakes. *Org. Divers. Evol.* **2018**, *18*, 51–62. [CrossRef]
62. Vasselon, V.; Domaizon, I.; Rimet, F.; Kahlert, M.; Bouchez, A. Application of High-Throughput Sequencing (HTS) Metabarcoding to Diatom Biomonitoring: Do DNA Extraction Methods Matter? *Freshw. Sci.* **2017**, *36*, 162–177. [CrossRef]
63. Bruce, K.; Blackman, R.; Bourlat, S.J.; Hellström, A.M.; Bakker, J.; Bista, I.; Bohmann, K.; Bouchez, A.; Brys, R.; Clark, K.; et al. *A Practical Guide to DNA-Based Methods for Biodiversity Assessment*; Pensoft Publishers: Sofia, Bulgaria, 2021; ISBN 978-619-248-053-0.
64. Wilmotte, A.; Dail Laughinghouse IV, H.; Capelli, C.; Rippka, R.; Salmaso, N. Taxonomic Identification of Cyanobacteria by a Polyphasic Approach. In *Molecular Tools for the Detection and Quantification of Toxigenic Cyanobacteria*; John Wiley & Sons, Ltd.: Chichester, UK, 2017; pp. 79–134.
65. Not, F.; Zapata, M.; Pazos, Y.; Campaña, E.; Doval, M.; Rodríguez, F. Size-Fractionated Phytoplankton Diversity in the NW Iberian Coast: A Combination of Microscopic, Pigment and Molecular Analyses. *Aquat. Microb. Ecol.* **2007**, *49*, 255–265. [CrossRef]
66. Xiao, X.; Sogge, H.; Lagesen, K.; Tooming-Klunderud, A.; Jakobsen, K.S.; Rohrlack, T. Use of High Throughput Sequencing and Light Microscopy Show Contrasting Results in a Study of Phytoplankton Occurrence in a Freshwater Environment. *PLoS ONE* **2014**, *9*, e106510. [CrossRef] [PubMed]

67. Maritz, J.M.; Rogers, K.H.; Rock, T.M.; Liu, N.; Joseph, S.; Land, K.M.; Carlton, J.M. An 18S RRNA Workflow for Characterizing Protists in Sewage, with a Focus on Zoonotic Trichomonads. *Microb. Ecol.* **2017**, *74*, 923–936. [CrossRef] [PubMed]
68. Tragin, M.; Zingone, A.; Vaulot, D. Comparison of Coastal Phytoplankton Composition Estimated from the V4 and V9 Regions of the 18S RRNA Gene with a Focus on Photosynthetic Groups and Especially Chlorophyta. *Environ. Microbiol.* **2018**, *20*, 506–520. [CrossRef]
69. Medinger, R.; Nolte, V.; Pandey, R.V.; Jost, S.; Ottenwalder, B.; Schlotterer, C.; Boenigk, J. Diversity in a Hidden World: Potential and Limitation of next-Generation Sequencing for Surveys of Molecular Diversity of Eukaryotic Microorganisms. *Mol. Ecol.* **2010**, *19*, 32–40. [CrossRef]
70. Lanzén, A.; Lekang, K.; Jonassen, I.; Thompson, E.M.; Troedsson, C. High-Throughput Metabarcoding of Eukaryotic Diversity for Environmental Monitoring of Offshore Oil-Drilling Activities. *Mol. Ecol.* **2016**, *25*, 4392–4406. [CrossRef]
71. Maitland, V.C.; Robinson, C.V.; Porter, T.M.; Hajibabaei, M. Freshwater Diatom Biomonitoring through Benthic Kick-Net Metabarcoding. *PLoS ONE* **2020**, *15*, e0242143. [CrossRef]
72. Kahlert, M.; Bailet, B.; Chonova, T.; Karjalainen, S.M.; Schneider, S.C.; Tapolczai, K. Same Same, but Different: The Response of Diatoms to Environmental Gradients in Fennoscandian Streams and Lakes—Barcodes, Traits and Microscope Data Compared. *Ecol. Indic.* **2021**, *130*, 108088. [CrossRef]
73. Ciglenečki-Jušić, I.; Ahel, M.; Mikac, N.; Omanović, D.; Vdović, N. *Investigation of Natural Characteristics and Assesment of Antropogenic Influences on Water Quality of Visovac Lake*; Ruđer Bošković Institute: Zagreb, Croatia, 2013. Available online: https://www.bib.irb.hr/662477 (accessed on 21 June 2020).

water

MDPI

Article

Effects of Environmental Factors on the Distribution and Diversity of Aquatic Oligochaetes

Ana Atanacković *, Nataša Popović, Nikola Marinković, Jelena Tomović, Jelena Đuknić, Jelena Stanković and Momir Paunović

Department of Hydroecology and Water Protection, Institute for Biological Research "Siniša Stanković"—National Institute of the Republic of Serbia University of Belgrade, University of Belgrade, Bulevar despota Stefana 142, 11060 Belgrade, Serbia; natasa.popovic@ibiss.bg.ac.rs (N.P.); nikola.marinkovic@ibiss.bg.ac.rs (N.M.); jelena.tomovic@ibiss.bg.ac.rs (J.T.); jelena.djuknic@ibiss.bg.ac.rs (J.Đ.); jelena.stankovic@ibiss.bg.ac.rs (J.S.); mpaunovi@ibiss.bg.ac.rs (M.P.)
* Correspondence: adjordjevic@ibiss.bg.ac.rs

Abstract: The aim of our study was to detect the actual distribution of oligochaete species and to identify their ecological differentiation with respect to environmental factors: altitude, temperature, oxygen concentration, conductivity, total organic carbon, and waterbody type. Although widespread, differentiation of oligochaete communities in four waterbody types and altitudinal groups can be observed through alpha and beta diversity. Their differences were analyzed using MANOVA, while the ecological preferences of species were presented with logistic Gaussian regression analyses. The highest number of the species of Oligochaeta was recorded in oligochaete communities in medium and large rivers. Total beta diversity decreased with the decreasing of waterbody size, the increasing of size of the substrate particles, river flow velocity, as well as altitude. Communities from small mountain rivers and streams and large and medium rivers with coarser substrate differed from other oligochaete communities. When coarser substrate was prevalent in smaller and medium rivers, a domination of a certain family was observed: Lumbriculidae (>800 m a.s.l.), Propappidae and Enchytraeidae (500–800 m), and Naididae (<500 m a.s.l.). Common species of Oligochaeta, with significantly overlapping ranges in running waters in Serbia, still show a clear grouping with respect to preference for certain types of waterbodies.

Keywords: aquatic worms; communities; diversity; waterbody types; Serbia

1. Introduction

According to Limnofauna Europea [1], there are around 190 species of aquatic oligochaetes in Europe. Due to their potentially high densities, wide distribution, and indicator value, aquatic oligochaetes may be important for water management [2] but may also be indicative of a variety of environmental conditions other than pollution. The influence of stream hydrology and physical and chemical factors on aquatic Oligochaeta has been studied by many authors [3–6].

The diversity of the Oligochaeta fauna in Serbia is in accordance with research in European countries: the Netherlands, Belgium, Germany [7], Poland [8], the Czech Republic [9], and Estonia [6], as well as in the region: Slovenia [10], Bulgaria [11], Croatia [12], Montenegro [13], and Albania [14]. Generally, in these countries the number of species is around 100–150, depending on the amount of examined material, habitat types, and detailed determination.

Previous investigations of oligochaete fauna in Serbian freshwaters included community structure, composition, and species distribution in different waterbody types [15]. The first comprehensive species list with oligochaete fauna of the large lowland rivers, the Danube and Sava, as well as hilly and mountainous rivers south of the Danube corridor was presented. The high diversity of aquatic oligochaetes in Serbia was noted, with the largest participation of potamal and rhithral species, and running waters were divided into four groups based on the oligochaete community. An obvious distinction based on the dominant taxa in the community was as follows: naidids (naidins and tubificins), enchytraeids, and lumbriculids. The Danube basin was distinguished by a high species diversity and dominance of *Limnodrilus hoffmeisteri*. Lower river stretches of the Danube and Sava that flow through Serbian territory could be compared to lake ecosystems, representing the typical potamal. Generally, the diversity and relative abundance of the macroinvertebrate fauna is significantly influenced by substrate type and river current [16], so the qualitative composition of oligochaete assemblages in Serbian freshwaters had a clear pattern—lower diversity in tributaries, and an increased diversity in the main watercourses. Of course, due to anthropogenic factors, discrepancies in the distribution of oligochaetes can be observed in waterbodies with increased sedimentation, slower river current, and increased organic pollution of tributaries, allowing habitation by cosmopolitan species. Under these conditions, Schenková and Helešic [17] observed that substrate type does not have a crucial role, and the normal distribution of Oligochaeta can change in response to organic pollution, in the way that lower diversity could be observed in the polysaprobic zones of river stretches and higher diversity in oligosaprobic zones. On the other hand, saprobic conditions are important abiotic factors for the distribution of Oligochaeta, but the substrate type can reduce the indicator value of some taxa, which has been shown in previous research [15].

The aim of this work is to better understand the distribution and diversity of aquatic oligochaetes in smaller rivers and mountain watercourses, since they are usually neglected in these types of waters. The results presented should help to make oligochaetes more reliable in the biological validation of waterbody typology according to European best management practice.

2. Materials and Methods

2.1. Field and Labarotory Work

An investigation conducted in 2019 and 2020 (spring/summer) covered 119 watercourses (181 locations/sites) and included a variety of waterbodies from mountain streams, upper river stretches, and downstream to the large lowland rivers. Macroinvertebrate samples were collected using a combination of the kick and sweep and multihabitat sampling technique according to European Standards [18] using a FBA hand net (mesh size 500 μm and 250 μm). The samples were pooled, and the material was preserved in ethyl alcohol (70%). In total, 3600 oligochaeta individuals were collected. For species identification, appropriate keys were used [7,19]. Families Naididae (with subfamilies Naidinae, Tubificinae and Pristininae), Propappidae, Lumbricidae, and Lumbriculidae were identified to the species level, while family Enchytraeidae was identified to the lowest possible taxonomic level.

Water temperature (°C) at the moment of sampling, conductivity (μS/cm), and oxygen concentration (mg/L O_2) were measured with the Horiba W-23XD multiparametric probe (HORIBA Instruments Corporation, Irvine, CA, USA) in the field and total organic carbon (TOC; mg/L C; SRPS ISO 8245:1994) in the laboratory.

2.2. Data Analyses

The frequency of occurrence (F) for each species in oligochaete assemblages was calculated using the formula:

$$F = n/N,$$

where n is the number of samples in which a taxon was found, and N is the total number of samples.

Oligochaeta were analyzed according to waterbody type by classifying each locality according to its characteristics (hydromorphological properties) into the following: Type 1—large rivers with fine substrate (silt, clay mud, and sand); Type 2—mix of large and medium rivers with coarser substrate (gravel, stones, and rocks); Type 3—small watercourses (up to 500 m a.s.l.); Type 4—small mountain rivers and streams (above 500 m a.s.l.);

and Type 5—slow flowing/stagnant waters (artificial canals and reservoirs), and in relation to the elevation gradient: 1—localities up to 500 m a.s.l.; 2—localities from 500 to 800 m a.s.l.; 3—localities above 800 m a.s.l.

A range of waterbody types represents a complex hydromorphological gradient. Environmental factors (flow velocity, bottom properties, temperature, oxygen concentrations) change predictably through waterbody types (5→1→2→3→4) and with increasing altitude. The logistic Gaussian regression [20,21] was used to detect the response and ecological preferences of the analyzed species, along the altitudinal gradient and gradient of waterbody types. The logistic Gaussian regression was performed using FLORA software version 2013 [22].

Within each group of the Oligochaeta community, components of alpha and beta diversity were investigated. Alpha diversity was assessed using Species Richness, the Shannon Index, and the Equitability Index. Components of beta diversity were analyzed using the procedures described by Baselga [23].

MANOVA [24] was utilized to find a combination of species that maximally discriminates groups of communities.

3. Results

Of the total number of macroinvertebrate samples, the Oligochaeta were found in 70 samples from 56 waterbodies. The Oligochaeta were represented by 34 taxa, belonging to 21 genera within 5 families. The distribution of the taxa recorded and the frequency of their occurrence are presented in Table 1.

Table 1. Distribution of taxa recorded with name abbreviations and the frequency of their occurrence.

		Waterbody Type					**Altitude**		
Taxon	Abb.	Type 1	Type 2	Type 3	Type 4	Type 5	Alt 1	Alt 2	Alt 3
f. Naididae									
subf. Naidinae									
Branhiodrilus hortensis (Stephenson, 1910)	Bra hor	0.2				0.09	0.04		
Chaetogaster diaphanus (Gruithuisen, 1828)	Cha dia		0.1				0.02		
Dero digitata Müller, 1773	Der dig					0.09	0.02		
Dero dorsalis Ferronière, 1899	Der dor					0.09	0.02		
Dero obtusa d'Udekem, 1835	Der obt					0.18	0.04		
Nais alpina Sperber, 1948	Nai alp				0.17			0.14	
Nais barbata Müller, 1773	Nai bar			0.03					0.1
Nais bretscheri Michaelsen,1899	Nai bre		0.2	0.37	0.17	0.27	0.3	0.29	0.2
Nais communis Piguet, 1906	Nai com			0.03	0.17		0.02		0.1
Nais elinguis Müller, 1774	Nai eli		0.4	0.16	0.17		0.13	0.29	0.2
Nais pseudobtusa Piguet, 1906	Nai pse		0.2				0.02	0.14	
Nais variabilis Piguet, 1906	Nai var		0.1				0.02		
Ophidonais serpentina (Müller, 1773)	Oph ser	0.2		0.05		0.27	0.11		
Uncinais uncinata (Ørsted, 1842)	Unc unc					0.09	0.02		
Stylaria lacustris (Linnaeus, 1767)	Sty lac	0.4		0.05		0.64	0.21		
subf. Pristininae									
Pristina aequiseta Bourne, 1891	Pri aeq			0.03			0.02		
subf. Tubificinae									
Bothrioneurum vejdovskyanum Štolc, 1888	Bot vej		0.1	0.03			0.04		

Table 1. *Cont.*

		Waterbody Type					Altitude		
Branchiura sowerbyi Beddard, 1892	Bra sow	0.2	0.1	0.05		0.09	0.09		
Ilyodrilus templetoni (Southern, 1909)	Ily tem		0.1				0.02		
Limnodrilus claparedeanus Ratzel, 1868	Lim cla	0.4	0.2	0.11			0.13	0.14	
Limnodrilus hoffmeisteri Claparède, 1862	Lim hof	0.8	0.6	0.55	0.33	0.55	0.62	0.43	0.3
Limnodrilus udekemianus Claparède, 1862	Lim ude	0.2	0.1	0.13		0.09	0.15		
Potamothrix hammoniensis (Michaelsen, 1901)	Pot ham	0.2	0.6	0.39	0.33	0.09	0.36	0.43	0.3
Potamothrix vejdovskyi (Hrabě, 1941)	Pot vej				0.17				0.1
Psammoryctides albicola (Michaelsen, 1901)	Psa alb			0.08	0.17		0.04		0.2
Psammoryctides barbatus (Grube, 1861)	Psa bar		0.1	0.05			0.04	0.14	
Lophochaeta ignota (Štolc, 1886)	Lop ign			0.03			0.02		
Tubifex tubifex (Müller, 1774)	Tub tub	0.2		0.08			0.08		
f. Propappidae									
Propappus volki (Michaelsen, 1916)	Pro vol		0.1	0.08			0.04	0.29	
f. Enchytraeidae									
Enchytraeidae gen. sp.	Enc	0.2	0.1	0.18	0.17	0.18	0.15	0.29	0.2
Fridericia sp.	Fri sp.			0.03		0.09	0.04		
Henlea ventriculosa (d'Udekem, 1854)	Hen ven			0.08			0.06		
Cernosvitoviella sp.	Cer sp.		0.2				0.02	0.14	
f. Lumbriculidae									
Stylodrilus heringianus Claparède,1862	Sty her		0.2	0.26	0.5		0.15	0.14	0.6
f. Lumbricidae									
Eiseniella tetraedra (Savigny, 1826)	Eis tet		0.2	0.05	0.17		0.04	0.14	0.2

Five localities stood out due to the high participation of oligochaetes in the macroinvertebrate community: Vrla—Vladičin Han (57.14% of individuals), Bjelica—Lučani (87.5%), Jablanica—Leće (88.46%), Krivaja—Bačka Topola (96.76%), and Tamnava—Koceljeva (98.60%).

In the oligochaete assemblages, *Limnodrilus hoffmeisteri* (Claparède, 1862) was the dominant species in most waterbody types (68.70% in Type 1, 59.48% in Type 2, 45.40% in Type 3, and 31.18% in Type 5), with exception of Type 4, where *Stylodrilus heringianus* (Claparède, 1862) had the highest participation (75.55%). Regarding altitudes, *L. hoffmeisteri* had the highest percentage participation in localities below 500 m a.s.l. (52.03%). In localities from 500 to 800 m a.s.l. the dominant species was *P. hammoniensis* (27%), and at altitudes above 800 m it was *S. heringianus* (67.9%).

The most frequent species in our investigation was *L. hoffmeisteri* (F = 0.55), followed by *Potamothrix hammoniensis* (Michaelsen, 1901) (F = 0.36), *Nais bretscheri* (Michaelsen, 1899) (F = 0.29), and *S. heringianus* (F = 0.21). Other species were recorded with frequency of occurrence less than 0.2. *Limnodrilus hoffmeisteri* was the most frequent in waterbodies Type 1, 2, and 3; *S. heringianus* was the most frequent in waterbodies Type 4, while *Stylaria lacustris* (Linnaeus, 1758) was the most frequent in waterbodies Type 5. Regarding altitudes, the most frequent species below 500 m a.s.l. was *L. hoffmeisteri* (F = 0.62), followed by *P. hammoniensis* (F = 0.36) and *S. lacustris* (F = 0.21). At altitudes from 500 to 800 m a.s.l., the most frequent species were still *L. hoffmeisteri* and *P. hammoniensis* (F = 0.43), but also naidins (*N. bretscheri*, *N. elinguis*), propappids (*Propappus volki*), and enchytraeids were frequent; F = 0.29 each. At altitudes above 800 m, the most frequent species was *S. heringianus* (F = 0.6).

A few species were recorded only in certain types of waterbodies with low frequencies (Table 1): *Chaetogaster diaphanus* (Gruithuisen, 1828), *Nais pseudobtusa* Piguet, 1906, *N. variabilis* Piguet, 1906, *Ilyodrilus templetoni* (Southern, 1909) (Type 2), *N. barbata*, *Pristina aequiseta* Bourne, 1891, *Lophochaeta ignota* (Štolc, 1886), *Henlea ventriculosa* (d'Udekem, 1854), *Cernosvitoviella* sp. (Type 3), *Nais alpina* Sperber, 1948, *Potamothrix vejdovskyi* (Hrabě, 1941) (Type 4), *Dero* sp., *Uncinais uncinata* (Ørsted, 1842) (Type 5). *N. alpina* was found only at one locality at an altitude of 750 m a.s.l.

Tubificines, naidines, and enchytraeids were recorded in all waterbody types, but the tubificines were the most diverse in Type 3 and the naidines in Type 5 (Figure 1). Most of the families were recorded in all altitude groups of localities, except Propappidae and Pristininae (Figure 2). In altitudes below 500 m, the highest number of species was detected for Naidinae and Tubificinae.

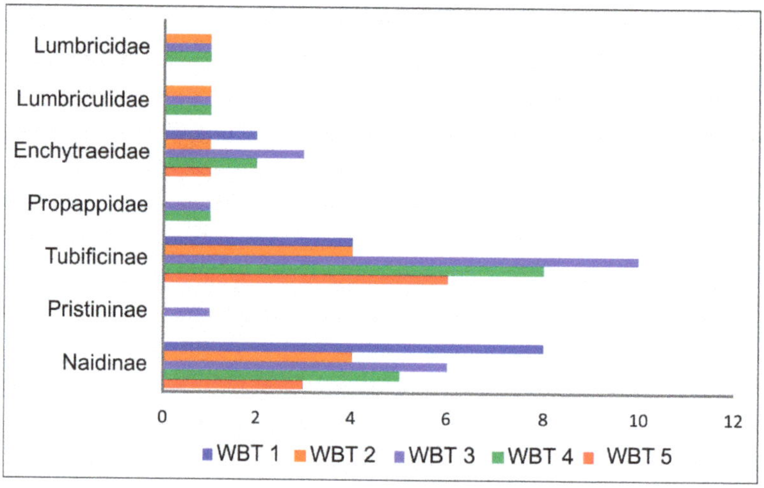

Figure 1. Number of species in different waterbody types (WBT).

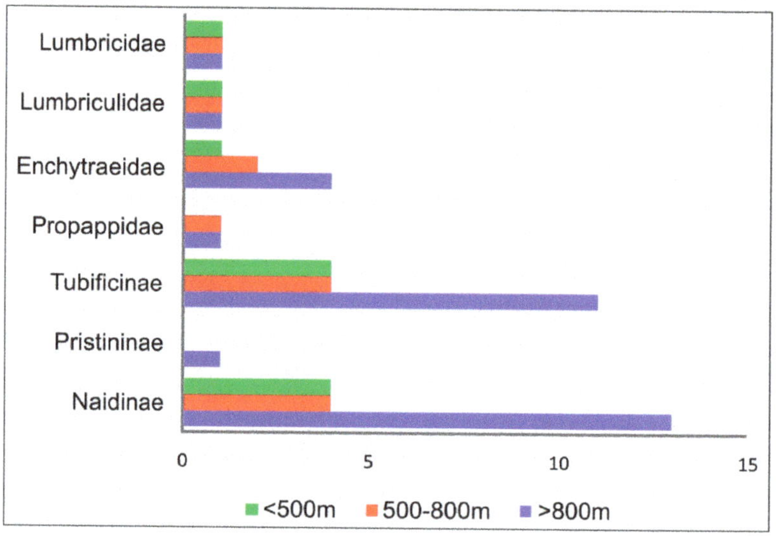

Figure 2. Number of species in different altitudes (Alt).

The components of alpha diversity within the oligochaete community were analyzed in relation to altitude (Figure 3) and waterbody types (Figure 4). Three groups of oligochaete communities differed significantly in relation to altitudes, where all three components of alpha diversity (Shannon entropy, Species richness, and Equitability) showed similar trends. Communities at altitudes of 500–800 m showed the highest values of alpha diversity components, while communities at the highest altitudes showed the lowest values. Differentiation of oligochaete groups in relation to different waterbody types is not so obvious. Species richness and equitability showed similar trends. The highest species richness was observed in WBT 2.

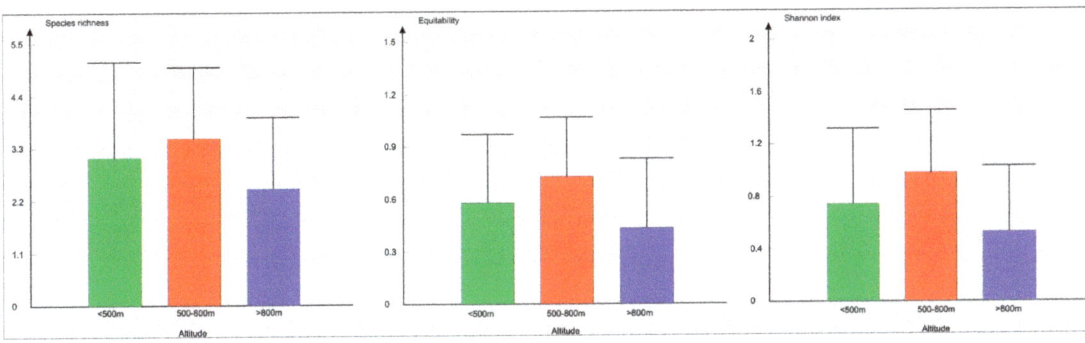

Figure 3. Alpha diversity components within oligochaetes communities in relation to altitude.

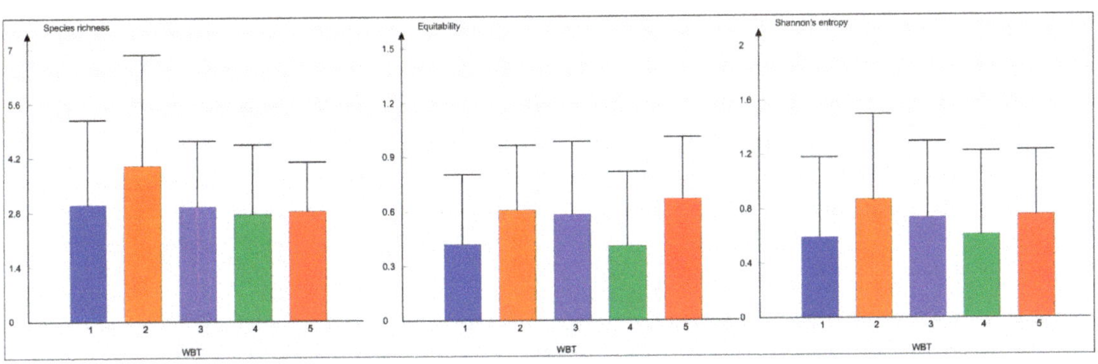

Figure 4. Alpha diversity components within oligochaetes communities in relation to the waterbody types (WBT).

The highest number of oligochaete species was recorded in the group of oligochaete communities in medium and large rivers (WBT 2) and at altitudes of 500 to 800 m.

The total beta diversity, and its components, was analyzed in relation to different waterbody types and altitudes (Figure 5a,b).

The lowest values of nestedness with the highest species turnover were in WBT 1, with the highest total beta diversity as well. Total beta diversity decreases with decreasing waterbody size, increasing substrate size, and higher river flow velocity, as well as with increasing altitude. The oligochaete communities in small mountain rivers and streams, at altitudes above 800 m, showed the highest values for nestedness with the lowest species turnover.

A comparison of oligochaete communities' composition using MANOVA (Figure 6) showed differences in relation to altitudes. Three groups of communities were separated: those at altitudes above 800 m were characterized by a representative of the family Lumbriculidae, *S. heringianus* (followed by Lumbricidae, *E. tetraedra* and Naididae, *N. barbata*, *N.*

communis); communities at altitudes of 500–800 m were characterized by family Propappidae, *P. volki* (followed by Enchytraeidae, *Cernosvitoviella* sp. and Naididae, *Psammoryctides barbatus*); and communities at altitudes below 500 m were distinguished by the species *L. hoffmeisteri* (family Naididae).

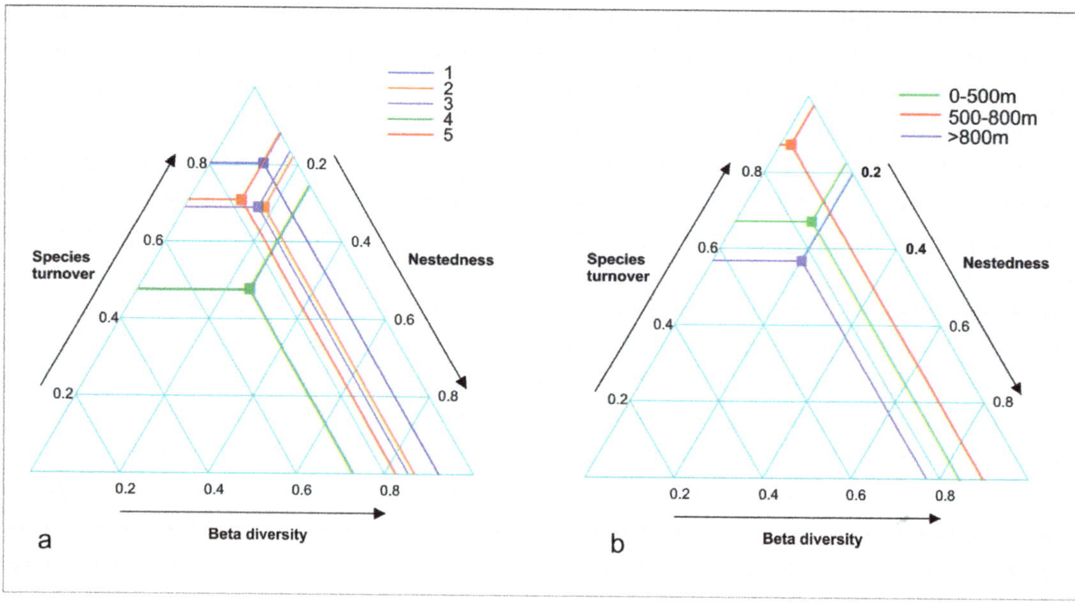

Figure 5. Beta diversity components in (**a**) different waterbody types (1–5) and (**b**) altitudes.

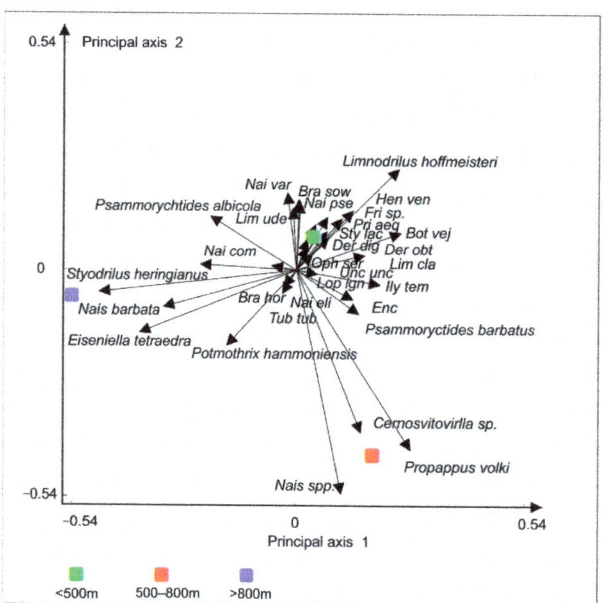

Figure 6. MANOVA analysis of oligochaete communities with respect to different altitudes.

MANOVA analyses with respect to WBT showed differences of oligochaete communities (Figure 7). Small mountain rivers and streams (WBT 4) and large and medium rivers with a larger substrate type (WBT 2) were distinguished from other communities. In WBT 2, enchytraeids (*Cernosvitoviella* sp.) and propappids (*P. volki*) dominated, while in WBT 4 lumbriculids (*S. heringianus*) and naidids (*N. communis, N. elinguis, P. hammoniensis*) were dominant. The communities in other waterbody types showed fewer differences (Figure 7).

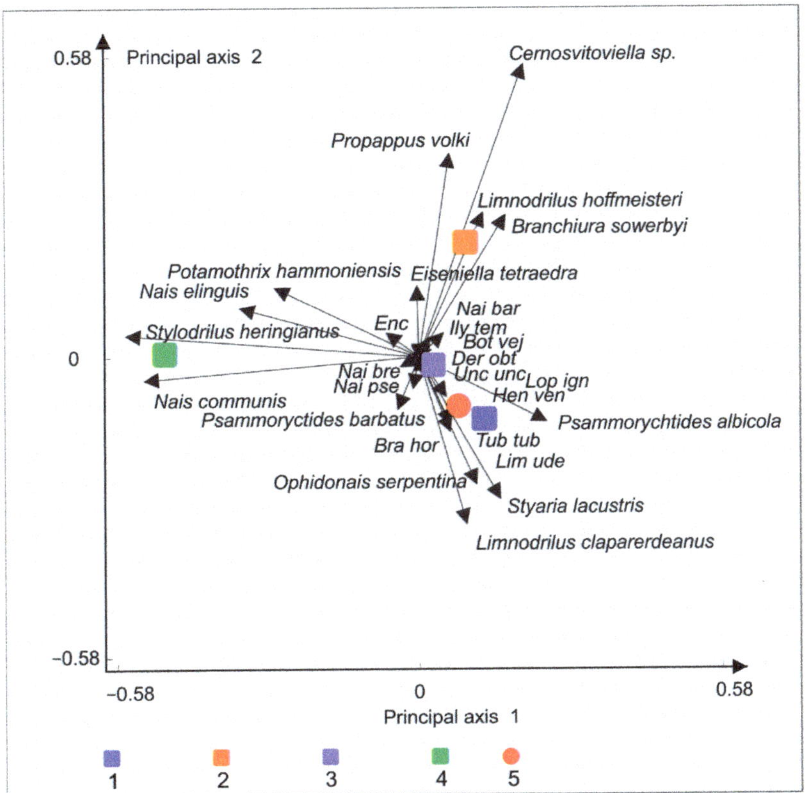

Figure 7. MANOVA analysis of oligochaete communities with respect to different waterbody types.

Response curves of selected species with respect to several environmental factors are shown in Figure 8. Gaussian logistic regression showed the ecological differentiation of the species which characterized oligochaete communities along the oxygen concentration, altitudinal gradient, and with respect to the waterbody type. The greatest ecological tolerance with respect to altitude was observed for *S. heringianus* and *E. tetraedra* (over 1200 m a.s.l.), while *B. sowerbyi, O. serpentina,* and *S. lacustris* were distinguished by their narrow range of distribution along the altitudinal gradient, with the optimal altitude up to 300 m and below.

The ecological tolerance of analyzed species with respect to waterbody type was wide (most of the species occurred in all waterbody types). The ecological differentiation was obvious for *P. volki, P. barbatus,* and *S. heringianus.* These species avoid large rivers with fine substrate, while *Stylaria lacustris* preferred slow and stagnant waters with fine sediment (Types 1 and 5). Most species had a wide response curve when this gradient was observed. *Eiseniella tetraedra* showed the narrowest tolerance curve with respect to oxygen concentration. A high concentration of oxygen is preferred by this species, but also by *P. volki* and *N. elinguis,* which showed a wider range of tolerances. With respect to the temperature gradient, the narrow ecological tolerance was recorded only for *E. tetraedra* and *B. sowerbyi*. With respect to other analyzed ecological preferences, conductivity and total organic carbon, the ecological tolerances of the analyzed species were wide. The ecological optimum of all species ranged from 6 to 9 mg/L, except for *E. tetraedra*, which was recorded in assemblages with oxygen concentrations higher than 9 mg/L.

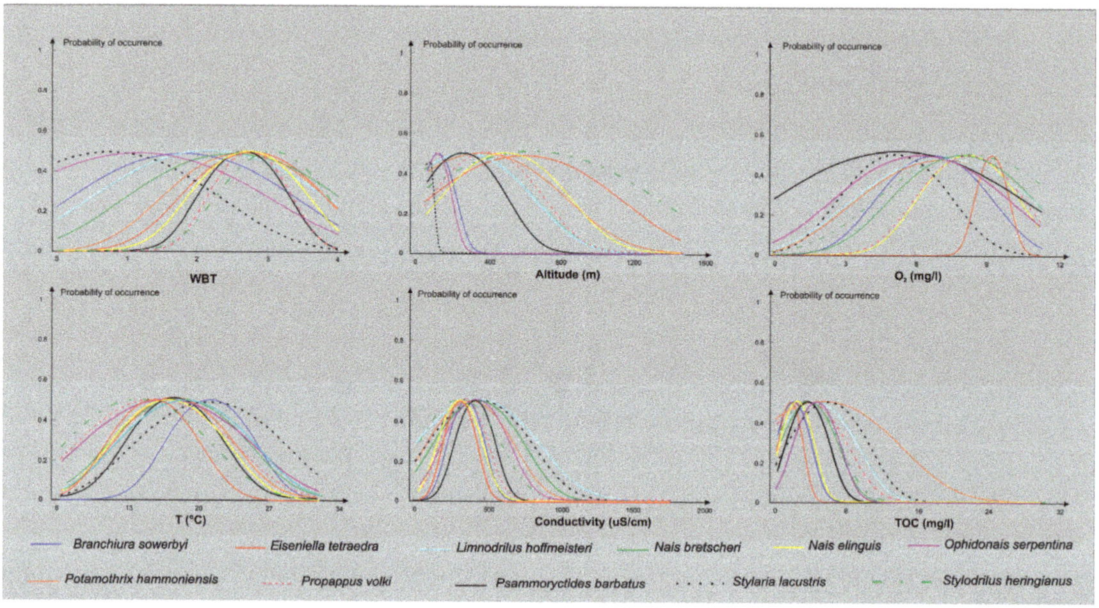

Figure 8. The ecological differentiation of the analyzed species with respect to selected factors.

The highest range of ecological tolerance with respect to conductivity and TOC gradient showed *P. hammoniensis* and *L. hoffmeisteri*.

4. Discussion

In the fauna of freshwaters of Serbia, a total of 97 species (45 genera from 8 families) have been recorded so far [8]. A third of the species (37) were recorded only in the main flow of the Danube River. The Danube is considered as the center of biodiversity and the main corridor for the spread of Ponto-Caspian oligochaetes species [25]. Since the Danube was not included in this study, many species typically found in this catchment area were missing from the results, especially the species from the genera *Potamothrix*, *Psammoryctides*, and *Isochaetides*. Oligochaete communities from this investigation include 34 taxa (21 genera from 5 families), which is comparable with previous investigations in Serbia and in the surrounding countries that belong to the same biogeographical territory. Oligochaete fauna of running waters in Serbia has been enriched by a new species. The record of *N. alpina* is interesting because it is a rare species; some consider it endemic to Europe [7]. It is a rheophil, stenotherm species, inhabits stony bottoms in the upper and middle courses of brooks and rivers, and prefers cooler waters [7]. The exotic, tropical species *B. hortensis*, which could be invasive, is rare in Europe. So far, it has been recorded only in seven European countries [26]. In Serbia, it has expanded from the main course of the Danube to the main tributaries and canals, and it has successfully established stable populations in the new environment and is regularly recorded during annual monitoring. The species is adapted to a wide range of substrate types, from small wetland pools to large ponds [27], and it seems to be tolerant to organic load and pollution. In Serbia, it is successfully expanding its range to the area north of the Danube (Pannonian Plain), while it has not yet been observed in the hilly-mountainous region south of the Danube. As van Haaren and Soors [7] pointed out, the specific ecological demands for this species remain to be determined.

This study confirmed previous conclusions [15] that *Limnodrilus hoffmeisteri* is the most important edificator of the oligochaete fauna in the Serbian waters, as it was recorded in all waterbody types, with the highest frequency of occurrence and percentage participation in oligochaete communities in waterbody Types 1, 2, and 3. It was typical for large lowland rivers, but also present in some small- and medium-sized watercourses at altitudes above 500 m a.s.l. Like in the research of [15], the characteristic species of hilly and mountainous types of watercourses (especially above 800 m) was *Stylodrilus heringianus*, the dominant species in waterbody Type 4. This species typically inhabits springs, preferring a lower water temperature and level of eutrophication, harder substrates, and faster currents [28]. Phytophilous *Stylaria lacustris* was dominant in waterbody Type 5, a typical lentic habitat type, with slow current or stagnant water and with the presence of detritus and macrophytes. In periphyton, the domination of naidines is expected, particularly the domination of *S. lacustris* [29].

Microhabitat complexity in large rivers enables the presence of euryvalent, a cosmopolitan species, such as most of the species from the Tubificinae subfamily, with the dominance of *L. hoffmeisteri*. Microhabitat complexity with altitude decreases and an increase in the number of taxa that prefer low organic load was observed. The same patterns are noted by Atanacković et al. [15]. Such significant reduction in microhabitat complexity is a selective pressure, which reduces the number of species, as concluded by Marinković et al. [30] for leeches, another group of Annelids. High beta diversity in communities dominated by *L. hoffmeisteri* (WBT 1, 2, and 3) is attributable to the difference in species composition (species turnover) and not to species richness (nestedness). These waterbodies differed in their habitat characteristics, providing a variety of microhabitats that offer a range of suitable conditions for different species. The opposite was observed in communities dominated by *S. heringianus* (WBT 4), with the highest values for nestedness and the lowest for species turnover. A plausible explanation for this phenomenon could be that these streams at higher altitudes have characteristics that are a limiting factor for most other oligochaete species, such as higher flow velocity and less organic matter and silt, resulting in each successive community being a subset of the previous community, while species substitution does not occur [30,31]. With harder substrate, sand and rocks, and a fast current, the lowest species turnover was expected due to the unfavorable environmental conditions for oligochaetes. In regard to oligochaete communities, the biodiversity of these habitats is generally low, the majority of species belong to the families Enchytraeidae and Lumbriculide, and the frequent occurrence of *Stylodrilus heringianus* is evident [28].

Environmental predictors can explain a relatively small part of total variability of oligochaete distribution. The altitudinal gradient affects substrate particle size and river current [32], so it had an effect on the community structure. *Eiseniella tetraedra*, *N. elinguis*, *S. heringianus*, *P. volki*, and *P. barbatus* preferred rivers with harder substrate and faster current and were absent in localities with fine substrate. This distinguishes them from *O. serpentina* and *S. lacustris*, which occurred only in rivers covered with fine substrate and with a slow-to-medium water current.

A higher abundance of enchytraeids and propappids is characteristic of rheo- and helocrene watercourses [33]. Communities in rivers with harder substrate (WBT 3 and 4) were distinguished by the presence of edificatory species from these families, while communities in small hilly and mountainous rivers (WBT 4) were characterized by edificatory species of lumbriculids. Representatives of the family Naididae showed a distribution pattern characterized by different preferences for flow velocity and substrate composition. *Nais bretscheri*, *N. barbata*, *N. pseudobtusa*, and *Bothrioneurum vejdovskyanum* preferred higher flow velocity and coarser substrate. When the water flow slowed down and the substrate was finer, these species were replaced by naidids that prefer almost stagnant waters (*S. lacustris*, *Chaetogaster* sp.). Martínez-Ansemil and Collado [34] reported that substrate and water velocity are the most important factors influencing the distribution of oligochaetes, while Marchand [35] noted that DO and organic matter affect the distribution of oligochaetes. Large rivers and lower river stretches could be compared to lake ecosystems (with high depth, slow flow, lower oxygen concentration) representing a typical potamal type, and as Atanacković et al. [15] showed, the slowing down of the river current contributes to more intensive sedimentation and in that way could significantly influencethe diversity and relative abundance of the oligochaete fauna. Thus, the oligochaetes in the tributaries are less diverse and abundant than in the main stream of the river [15]. The pelophilous group, which consists of the genera *Limnodrilus, Branchiura, Tubifex,* and *Pothamothrix,* was characteristic of a slow river current and fine substrate, and the psammophilous group (*Stylodrilus, Henlea, Nais* spp., and *Eiseniella*) was characteristic of habitats with harder substrates (sand, pebbles, and stones) and faster currents. Also, a lentic environment influenced the distribution of phytophilous *Ophidonais serpentina* and *Stylaria lacustris*. According to the present results, due to heterogeneous microhabitats, higher species richness was observed in oligochaete assemblages in these river stretches.

5. Conclusions

This study analyzed oligochaete communities with a focus on smaller rivers and mountain watercourses. It revealed a high diversity of oligochaetes at higher altitudes (500–800 m) and in rivers with coarser substrate. Also, it shows that substrate particle size and current velocity have a significant influence on the distribution and diversity of oligochaetes. Although the majority of aquatic oligochaetes prefer silt, clay, and slower water currents, some species such as *Stylodrilus heringianus* are typical inhabitants of mountain rivers and streams. The results also indicate that *Limnodrilus hoffmeisteri* is the most prominent edificator of the oligochaete fauna in the waters of Serbia. Further investigations that encompass all ecosystems, booth small and large rivers, as well as reservoirs, could give us the larger datasets necessary to answer the question of where the highest diversity of oligochaetes is. This study is an important step for using oligochaetes more reliably and effectively, as they are one of the necessary BQEs (biological quality elements) in the biological validation of waterbody typology in routine monitoring practice.

Author Contributions: Conceptualization, A.A., N.P., N.M. and M.P.; methodology, N.M., A.A., J.T., J.S. and J.Đ.; validation, N.P., N.M. and M.P., formal analysis, A.A., N.P. and N.M.; investigation, A.A., N.P., J.Đ., J.T., N.M., J.S. and M.P.; writing—original draft preparation, A.A., N.P., J.T. and N.M.; writing—review and editing, A.A., N.P., J.T., J.Đ. and M.P.; visualization, N.P., N.M. and A.A.; supervision, A.A. and M.P.; project administration, A.A.; funding acquisition, M.P. All authors have read and agreed to the published version of the manuscript.

Funding: This research was funded by the Ministry of Science, Technological Development and Innovations of the Republic of Serbia, Contract No. 451-03-47/2023-01/200007.

Data Availability Statement: The data presented in this study are available on request from the corresponding author.

Conflicts of Interest: The authors declare no conflict of interest.

References

1. Ilies, J. *Limnofauna Europaea;* Gustav Fischer Verlag: Stuttgart, Germany, 1978.
2. Verdonschot, P.F.M. The Role of Oligochaetes in the Management of Waters. *Hydrobiologia* **1989**, *180*, 213–227. [CrossRef]
3. Elexová, E.; Némethová, D. The Effect of Abiotic Environmental Variables on the Danube Macrozoobenthic Communities. *Limnologica* **2003**, *33*, 340–354. [CrossRef]
4. Šporka, F.; Nagy, Š. The Macrozoobenthos of Parapotamon-Type Side Arms of the Danube River in Slovakia and Its Response to Flowing Conditions. *Biologia* **1998**, *53*, 633–643.
5. Moog, O. *Fauna Aquatica Austriaca Katalog zur Autökologischen Einstufung Aquatischer Organismen Österreichs;* Wasserwirtschaftskataster, Federal Ministry for Agriculture and Forestry: Vienna, Austria, 2002; p. 670.
6. Timm, T.; Seire, A.; Pall, P. Half a Century of Oligochaeta Research in Estonian Running Waters. *Hydrobiologia* **2001**, *463*, 223–234. [CrossRef]
7. Van Haaren, T.; Soors, J. *Aquatic Oligochaetes of The Netherlands and Belgium;* KNNV Publishing Zeist: Zeist, The Netherlands, 2012.

8. Dumnicka, E. New for Poland Tubificid (Oligochaeta) Species from Karstic Springs. *Pol. J. Ecol.* **2009**, *57*, 395–401.
9. Schenková, J.; Pařil, P.; Petřivalská, K.; Bojková, J. Aquatic Oligochaetes (Annelida: Clitellata) of the Czech Republic: Check-List, New Records, and Ecological Remarks. *Zootaxa* **2010**, *44*, 29–44. [CrossRef]
10. Giani, N.; Sambugar, B.; Martínez-Ansemil, E.; Martin, P.; Schmelz, R.M. The Groundwater Oligochaetes (Annelida, Clitellata) of Slovenia. *Subterr. Biol.* **2011**, *9*, 85–102. [CrossRef]
11. Uzunov, Y. *Aquatic Oligochaets (Oligochaeta Limicola), Annelida: Aphanoneura, Oligochaeta, Branchiobdellea—Catalogus Faunae Bulgaricae 7*; Professor Marin Drinov Academic Publishing House: Sofia, Bulgaria, 2008.
12. Kerovec, M.; Kerovec, M.; Brigić, A. Croatian Freshwater Oligochaetes: Species Diversity, Distribution and Relationship to Surrounding Countries. *Zootaxa* **2016**, *4193*, 73–101. [CrossRef]
13. Šundić, D.; Radujković, B.M.; Krpo-Ćetković, J. Catalogue of Aquatic Oligochaeta (Annelida: Clitellata) of Montenegro, Exclusive of Naidinae and Pristininae. *Zootaxa* **2011**, *18*, 1–25. [CrossRef]
14. Dhora, D. *Regjistër i Specieve të Faunës së Shqipërisë*; Camaj-Pipa, Botimet: Shkodrës, Albania, 2009; ISBN 9789995602956.
15. Atanacković, A.; Zorić, K.; Tomović, J.; Vasiljević, B.; Paunović, M. Distributional Patterns of Aquatic Oligochaeta Communities (Annelida: Clitellata) in Running Waters in Serbia. *Arch. Biol. Sci.* **2020**, *72*, 359–372. [CrossRef]
16. Kang, H.; Bae, M.-J.; Lee, D.-S.; Hwang, S.-J.; Moon, J.-S.; Park, Y.-S. Distribution Patterns of the Freshwater Oligochaete *Limnodrilus hoffmeisteri* Influenced by Environmental Factors in Streams on a Korean Nationwide Scale. *Water* **2017**, *9*, 921. [CrossRef]
17. Schenková, J.; Helešic, J. Habitat Preferences of Aquatic Oligochaeta (Annelida) in the Rokytná River, Czech Republic—A Small Highland Stream. *Hydrobiologia* **2006**, *564*, 117–126. [CrossRef]
18. *EN 27828:1994*; Water Quality—Methods of Biological Sampling—Guidance on Handnet Sampling of Aquatic Benthic Macro-Invertebrates (ISO 7828:1985). European Committee for Standardisation: Brussels, Belgium, 1994.
19. Timm, T. A Guide to the Freshwater Oligochaeta and Polychaeta of Northern and Central Europe. *Lauterbornia* **2009**, *66*, 1–235.
20. Ter Braak, C.J.F. Canonical Correspondence Analysis: A New Eigenvector Technique for Multivariate Direct Gradient Analysis. *Ecology* **1986**, *67*, 1167–1179. [CrossRef]
21. James, G.; Witten, D.; Hastie, T.; Tibshirani, R. *An Introduction to Statistical Learning*; Springer: New York, NY, USA, 2013; p. 112.
22. Karadžić, B. FLORA: A Software Package for Statistical Analysis of Ecological Data. *Water Res. Manag.* **2013**, *3*, 45–54.
23. Baselga, A. Partitioning the Turnover and Nestedness Components of Beta Diversity. *Glob. Ecol. Biogeogr.* **2010**, *19*, 134–143. [CrossRef]
24. Bray, J.H.; Maxwell, S.E.; Maxwell, S.E. *Multivariate Analysis of Variance*; Sage: Thousand Oaks, CA, USA, 1985; ISBN 0803923104.
25. Timm, T. The Genus *Potamothrix* (Annelida, Oligochaeta, Tubificidae): A Literature Review. *Est. J. Ecol.* **2013**, *62*, 121–136. [CrossRef]
26. Atanacković, A.; Zorić, K.; Paunović, M. Invading Europe: The Tropical Aquatic Worm *Branchiodrilus hortensis* (Stephenson, 1910) (Clitellata, Naididae) Extends Its Range. *BioInvasions Rec.* **2021**, *10*, 598–604. [CrossRef]
27. Nesemann, H.; Sharma, G.; Sinha, R.K. Aquatic Annelida (Polychaeta, Oligochaeta, Hirudinea) of the Ganga River and Adjacent Water Bodies in Patna (India: Bihar), with Description of a New Leech Species (Family Salifidae). *Ann. Naturhist. Museum Wien* **2004**, *105*, 139–187.
28. Dumnicka, E. Composition and Abundance of Oligochaetes (Annelida: Oligochaeta) in Springs of Kraków-Częstochowa Upland (Southern Poland): Effect of Spring Encasing and Environmental Factors. *Pol. J. Ecol.* **2006**, *54*, 231–242.
29. Dumnicka, E. Distribution of Oligochaeta in Various Littoral Habitats in the Anthropogenic Reservoirs. *Int. J. Oceanogr. Hydrobiol.* **2007**, *36*, 13–19.
30. Marinković, N.; Karadžić, B.; Pešić, V.; Gligorović, B.; Grosser, C.; Paunović, M.; Nikolić, V.; Raković, M. Faunistic Patterns and Diversity Components of Leech Assemblages in Karst Springs of Montenegro. *Knowl. Manag. Aquat. Ecosyst.* **2019**, *420*, 26. [CrossRef]
31. Petsch, D.K.; Ragonha, F.H.; Gimenez, B.C.G.; Barboza, L.G.A.; Takeda, A.M. Partitioning Beta Diversity of Aquatic Oligochaeta in Different Environments of a Neotropical Floodplain. *Acta Sci.—Biol. Sci.* **2015**, *37*, 41–49. [CrossRef]
32. Vannote, R.L.; Minshall, G.W.; Cummins, K.W.; Sedell, J.R.; Cushing, C.E. The River Continuum Concept. *Can. J. Fish. Aquat. Sci.* **1980**, *37*, 130–137. [CrossRef]
33. Schenková, J.; Kroča, J. Seasonal Changes of an Oligochaetous Clitellata (Annelida) Community in a Mountain Stream. *Acta Univ. Carol. Environ.* **2007**, *21*, 143–150.
34. Von Schiller, D.; Acuña, V.; Aristi, I.; Arroita, M.; Basaguren, A.; Bellin, A.; Boyero, L.; Butturini, A.; Ginebreda, A.; Kalogianni, E.; et al. River Ecosystem Processes: A Synthesis of Approaches, Criteria of Use and Sensitivity to Environmental Stressors. *Sci. Total Environ.* **2017**, *596–597*, 465–480. [CrossRef]
35. Marchand, J. The Influence of Seasonal Salinity and Turbidity Maximum Variations on the Nursery Function of the Loire Estuary (France). *Netherl. J. Aquat. Ecol.* **1993**, *27*, 427–436. [CrossRef]

water

MDPI

Article

Distribution Range of the Endangered Species *Unio crassus* Philipsson, 1788 in Serbia (Western Balkans Region), Historical and Recent Data

Jelena Tomović [1,*], Vladica Simić [2], Ana Petrović [2], Ana Atanacković [1], Katarina Zorić [1], Momir Paunović [1] and Maja Raković [1]

[1] Department of Hydroecology and Water Protection, Institute for Biological Research "Siniša Stanković" National Institute of Republic of Serbia, University of Belgrade, Bulevar Despota Stefana 142, 11060 Belgrade, Serbia; adjordjevic@ibiss.bg.ac.rs (A.A.); katarinas@ibiss.bg.ac.rs (K.Z.); mpaunovi@ibiss.bg.ac.rs (M.P.); rakovic.maja@ibiss.bg.ac.rs (M.R.)

[2] Department of Biology and Ecology, Faculty of Science, University of Kragujevac, Radoja Domanovića 12, 34000 Kragujevac, Serbia; vladica.simic@pmf.kg.ac.rs (V.S.); ana.petrovic@pmf.kg.ac.rs (A.P.)

* Correspondence: jelena.tomovic@ibiss.bg.ac.rs

Abstract: The thick-shelled river mussel, *Unio crassus* Philipson, 1788, is considered to be one of the species with the highest conservation priority in Serbia. The study represents the first comprehensive research of the distribution of *U. crassus* in Serbian waters. The research covered a variety of waterbody types throughout Serbia, and distribution data were considered over three time periods from 1953 to 2019. The paper summarizes all the available literature data, field research and information obtained during the review of the collection of malacological material of the Natural History Museum in Belgrade. The results show a positive population trend, which is reflected in an extension of the distribution area and an increase in population density. After reviewing the museum collection, 13 synonyms for *U. crassus* were identified. The study also revealed a better insight into the habitat requirements and the limiting factors of the species. Substrate characteristics, waterbody types, altitude, and nitrate content of the water seem to be of great importance for the occurrence of the species. The results presented here can improve further measures for the conservation of *U. crassus*, not only in Serbia, but also in the Western Balkans.

Keywords: freshwater mussels; distribution history; re-identification; conservation; Serbia

1. Introduction

Freshwater mussels (Unionida: Unionidae) are one of the most important and widespread groups of aquatic organisms, found in a variety of freshwater habitats, throughout the world. These bivalves are an essential component of freshwater [1] and contribute to sediment stabilization, nutrient cycling and water purification with positive effects on freshwater biodiversity [1–3]. The Unionidae (also known as bivalves, naiads and unionids) are among the most threatened faunistic groups at a global level [4–10]. Of the 16 European species of the order Unionida, nine have the status of near-threatened, endangered, or critically endangered according to the IUCN Red List [10,11]. For this reason, comprehensive environmental and population studies are very important from a conservation perspective. Mussels also have economic importance as a food source and in the ornamental industry. Their over-exploitation for industrial purposes has led to the population decline of some species in many regions [12,13] and even to their local disappearance.

The thick-shelled river mussel, *Unio crassus* Philipsson, 1788 is currently listed by the IUCN classification as endangered-EN at a global level [14]. It is listed in Annexes II and IV of the European Commission Habitats Directive [15], and in Resolution 6 of the Bern Convention [16]) but is also covered by Serbian legislation [17]. These legislations promote the conservation of unionids, including habitat restoration and the reintroduction of mussels and host fish [18,19].

The native distribution area of *U. crassus* extends from France in the west to western Russia in the east, and from Scandinavia in the north to Asia Minor in the southeast. It was also recorded in the basins of the Baltic Sea, the Black Sea, the Azov Sea and the Caspian Sea, up to the Ural River basin in Russia and Kazakhstan [10]. The species is widespread in Europe, with the exception of Great Britain, the Apennines and the Iberian Peninsula, where its occurrence has not been recorded [10,14].

Until the first half of the 20th century, *U. crassus* was the most abundant unionid species in Europe [20]. Recently, declining population densities and the endangered status of *U. crassus* have been observed in most European countries, especially in Western and Central Europe [19]. According to the latest evidence, the decline of *U. crassus* in Europe is estimated to be more than 50% [19]. The species is listed in the national Red List as critically endangered in Switzerland, Austria and Germany with only a few intact populations remaining [21,22], endangered in the Czech Republic, Poland and Sweden, and vulnerable in Albania, Belarus, Finland and Latvia [14].

Insufficient knowledge of the unionids, their current status and the quantification of population changes over time are all problems for further research on population trends and for determining effective conservation measures not only in Serbia but also in other European countries. *U. crassus* is a strictly protected species in Serbia according to national legislation. Taking into account the new population data, we assumed that the population trend in Serbia is continuously changing. To confirm these assumptions, population changes over time need to be documented and quantified, including data on the distribution history of the species. It is also important to understand which factors are potentially responsible for the changes over time.

The aim of this study was to use a large amount of distribution data to gain a better insight into the distribution range of the species *U. crassus* in different time periods; to identify habitat preferences; to discuss anthropogenic factors affecting the distribution of the species; and to solve the problem of using many unaccepted synonyms and not confidently identifying *U. crassus* in the past in Serbia.

To achieve these goals, this manuscript compiles all known records of *U. crassus* in Serbia from the literature and unpublished sampling data up to 2019. The information presented here will significantly improve our knowledge of the current situation of *U. crassus* in Serbia and support the conservation of unionids and their habitats.

2. Materials and Methods

2.1. Data Collection

The data used for the analysis of the distribution of *U. crassus* in Serbia cover the period 1953–2019. The distribution was estimated based on all available data: (1) peer-reviewed articles, monographs, dissertations and reports [23–32]; (2) unpublished data on samples collected during field research of several national projects in Serbia (material deposited in the malacological collection of the Institute for Biological research "Siniša Stanković", University of Belgrade—further referred as the IBISS); (3) material collected during the realization of four international projects (material deposited in the malacological collection of IBISS) [33–36]; (4) BAES database—biodiversity in aquatic ecosystems in Serbia, ex situ conservation [37]; and (5) collection of unionids of the Natural History Museum in Belgrade (collector Ante Tadić)—referred to as historical data in remaining text.

2.2. Historical Data (1953–1973)

In order to determine the historical distribution of *U. crassus* in Serbia, the museum collection of unionids was reviewed (collector of Ante Tadić). The analyzed historical material consisted of 244 individuals from 36 sites in Serbia, collected in the Danube and its main tributaries in the Serbian stretch (the Sava, Tisa, Karaš, Tamiš, Nera and MlavaRivers, as well as the Velika Morava and Timok basins)—Table 1. The museum collection was inspected and the identification of each specimen was checked and verified. A re-identification was carried out and the presence of potential synonyms for the *U. crassus* was considered. Each specimen had an inventory number and a label with sampling site, the name of the collector, the date of collection and the identification. A description of the respective sampling location is given in Tadić [38]. Key features used in identification were external morphology (shell outline, color, umbo sculpture, hinge characteristics as well as the three measured linear shell distances—shell length, height and width). Review of the status and taxonomic history of the species was carried out according to databases: MolluscaBase [39], MUSSELpdb [40] and WoRMS [41].

2.3. Current Data (1990–2019)

Recently, the study of aquatic ecosystems has been intensified and covered the entire territory of Serbia. A total of 540 sites were studied, covering different types of running water—from small and medium-sized streams to large lowland rivers (Figure 1). Various techniques were used to collect mussel samples—kick and sweep sampling and the multi-habitat approach (EN 27828:1994) with the FBA benthic hand net (aperture: 25 × 25 cm, mesh size of 500 and 250 μm) according to European Standards [42], benthic dredging and in some cases visual inspection and snorkeling. To obtain comparable data, abundance was expressed as the number of individuals per sample (relative abundance). For the graphical presentation, abundance per watercourse was pooled.

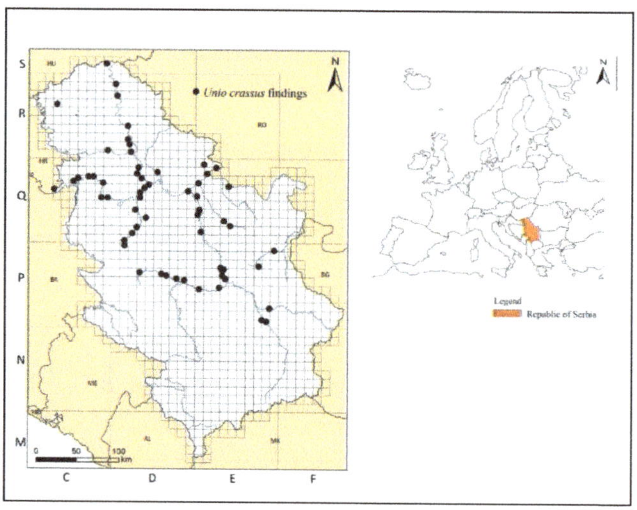

Figure 1. Map of all observed locations of *U. crassus* in watercourses in Serbia in the period (1973–2019).

The distribution data for *U. crassus* are considered over two time periods (1990–2008 and 2009–2019). The number of detections per study period and per river kilometer was carried out to examine the distribution of the species in Serbian waters.

2.4. Environmental Variables and Data Processing

The GPS position and elevation of each site in recent research period were recorded using a GarmineTrex 20× handheld GPS receiver (Garmin ltd). The chemical parameter(NO_3-N) considered in the study was provided by the Serbian Environmental Protection Agency (SEPA), as an official accredited institution (SRPS ISO/IEC 17025:/2017) [43] for the national water-monitoring programs. All water samples were analyzed in the accredited SEPA laboratory according to the following method: nitrates (NO_3-N): UP 1.98/PC 12.

Water parameters were collected once a month. This parameter was selected based on the literature data as potentially one of the main elements affecting the *U. crassus* community.

In each watercourse the substrate type was categorized according to the AQEM protocol [44], which included: 1—megalithal (>40 cm); 2—macrolithal (20–40 cm); 3—mesolithal (6–20 cm); 4—microlithal (2–6 cm); 5—akal (2 mm–6 cm); 6—psammal/psammopelal (6 µm–2 mm); 7—argyllal (<6 µm) and other (organic mud, Xylal, living parts of terrestrial plants, debris) and were categorized into classes based on the percentage of cover (1–7).

According to the modified national typology [45], all surface waters in Serbia are classified into six categories: Type 1—large lowland rivers; Type 2—large rivers; Type 3—small to medium rivers with elevation below 500 m; Type 4—small to medium rivers and streams with elevation above 500 m; Type 5—watercourses of the Pannonian Plain; Type 6—small waterbodies including springs and upper stretches of streams. For this study, the material was collected in three categories of waterbodies (Types 1, 2 and 3).

2.5. Statistical and Graphical Analysis

The study of the ecological preferences of species in terms of the elevation gradient, waterbody types and the gradient of substrate types with the response curve was performed using the STATISTICA 8 software (StatSoft, Inc., Tulsa, OK, USA) [46]. The nitrate–nitrogen content in different watercourses was analyzed using General Discriminant Analysis (GDA). The values of abundance and nitrogen content are graphically represented by mean, maximum and minimum values. The maps were created using Adobe Illustrator CC15 (Adobe Inc., 2015) [47].

3. Results

3.1. Historical Data

By analyzing historical data, 94 individuals of the 244 examined specimens were identified as *U. crassus*. A re-identification of each specimen from the museum collection was carried out and then the scientific names were validated in the database. After re-identification, 13 synonyms for *U. crassus* were identified (Table 1).

Table 1. Re-idetification of the collection from the Natural History Museum in Belgrade (collector Ante Tadić).

Collection Number	Label	Re-Identification	Collection Data	Collection Site
63	*Unio crassus crassus* Philipson 1788	*Unio crassus*	1963	Veliki Bački channel, Sombor
62, 65	*Unio crassus crassus* Philipson 1788	*Unio crassus*	1955	Sava, Stara Bežanija
55	*Unio crassus crassus* Philipson 1788	*Unio crassus*	1967	Tamiš, Pančevo
53	*Unio crassus crassus* Philipson 1788	*Unio crassus*	1967	Dunav, Smederevo
51	*Unio crassus crassus* Philipson 1788	*Unio crassus*	1966	Tisa, Senta
52	*Unio crassus crassus* Philipson 1788	*Unio crassus*	1965	Karaš channel
42	*Unio crassus crassus* Philipson 1788	*Unio crassus*	1958	Z. Morava, Rudinci
43	*Unio crassus crassus* Philipson 1788	*Unio crassus*	1961	Dunav, Zemun
60	*Unio crassus cytherea* Kuster 1833	*Unio crassus*	/	Sava (76 rkm)
56	*Unio crassus cytherea* Kuster 1833	*Unio crassus*	1958	Z. Morava, Trstenik
46	*Unio crassus cytherea* Kuster 1833	*Unio crassus*	1961	Danube, Zemun
94	*Unio crassus batavus* (Maton and Rackett, 1807)	*Unio crassus*	1967	Danube, Golubac
97	*Unio crassus batavus* (Maton and Rackett, 1807)	*Unio crassus*	1953	Sava (35.5 rkm)
93	*Unio crassus batavus* (Maton and Rackett, 1807)	*Unio crassus*	1973	Mlava, Gornjak
99	*Unio crassus batavus* (Maton and Rackett, 1807)	*Unio crassus*	1955	Sava, Stara Bežanija
49	*Unio crassus f. Grandis*	*Unio crassus*	1967	Danube, Zemun
44	*Unio crassus crassus f. Grandis*	*Unio crassus*	1967	Tamiš, Pančevo
45	*Unio crassus crassus f. Grandis*	*Unio crassus*	1958	Z. Morava
47	*Unio crassus crassus f. Grandis*	*Unio crassus*	1958	Z. Morava, Klenjak
48	*Unio crassus crassus f. Grandis*	*Unio crassus*	1958	Z. Morava, Rudinci
124	*Unio amnicus* Rossmässler, 1836	*Unio crassus*	6 August 1958	Z. Morava, Rudinci
123	*Unio amnicus* Rossmässler, 1836	*Unio crassus*	/	Z. Morava, Trstenik
122	*Unio consentaneus* 'Zigel' Rossmässler, 1836	*Unio crassus*	6 September 1958	Z. Morava, Rudinci
120, 121	*Unio serbicus* Drouët, 1884	*Unio crassus*	6 August 1958; 6 September 1958	Z. Morava, Klenjak

Table 1. *Cont.*

Collection Number	Label	Re-Identification	Collection Data	Collection Site
119, 120	*Unio serbicus* Drouët, 1884	*Unio crassus*	1965; 1966	Nera, Bela Crkva
85	*Unio reniformis* 'Schmidt' Rossmässler, 1836	*Unio crassus*	1961	Danube, Zemun
82	*Unio reniformis* 'Schmidt' Rossmässler, 1836	*Unio crassus*	1967	Danube, Medornica confluence
87	*Unio rivalis* Drouët, 1884	*Unio crassus*	1965	Bela Crkva
68	*Unio bosnensis* Möllendorff, 1874	*Unio crassus*	1961	Danube, Zemun
69	*Unio bosnensis* Möllendorff, 1874	*Unio crassus*	1973	Mlava, Gornjak
81	*Unio savensis* Drouët, 1882	*Unio crassus*	1973	Mlava, Petrovac
73	*Unio pančići* Drouët, 1882	*Unio crassus*	1972	Crni Timok, Zaječar

The majority of species names are not considered valid based on current knowledge of the freshwater mussel diversity. Many of these synonyms were introduced into Europe by the French Nouvelle École in the late 19th century [48]. According to the historical data, considering the period from 1953 to 1973, it can be assumed that *U. crassus* was a common species in Serbia with a continuous range. The distribution range of the species based on historical data is shown in Figure 2.

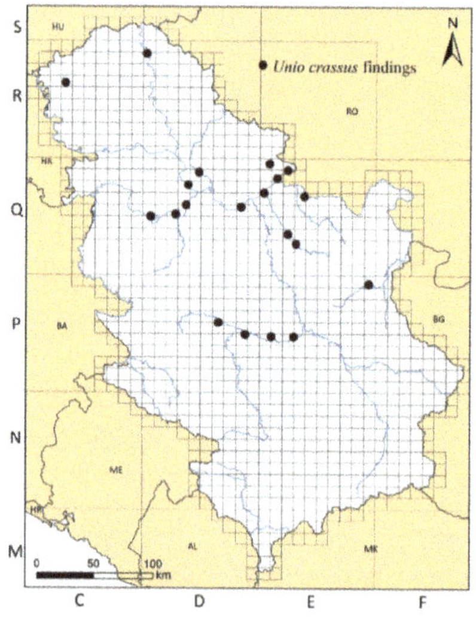

Figure 2. Map of the distribution range of *U. crassus* in studied grid squares 10 × 10 km in the period 1953–1973.

In the archive material, the species was detected at 20 of the 36 examined sites (55%). The species was found to be widespread in the Serbian stretch of the Danube. It was recorded at five out of nine examined sites (55%), from Apatin (1402 rkm) to Golubac (1040 rkm). A total of 16 specimens of the species were collected in the studied section of the Danube. The species was also detected in the Sava River with moderate occurrence, being recorded at three out of the eight examined sites (37%). Five individuals of *U. crassus* were collected in the Serbian stretch of the Sava (from 76 rkm to the mouth of the Danube). Thirty-nine specimens were collected at four sites. A dense population and the highest frequency of occurrence of *U. crassus* was observed in the Zapadna Morava. Even 39 specimens were collected at four sites in Z. Morava. *U. crassus* was detected in the Tamiš (in 33% of the examined samples), Mlava (in 50% of the examined samples) and Nera (in all samples) Rivers, but also detected in the Tisa and Crni Timok Rivers (at only one site). The occurrence of the species is observed in the Veliki Bački and Karaš channels.

Between the 1970s and 1990s, a period of intensive industrialization, *U. crassus* became locally and even regionally extinct. After historical data, there were no records of the species in Serbia until the early 1990s (Figure 3).

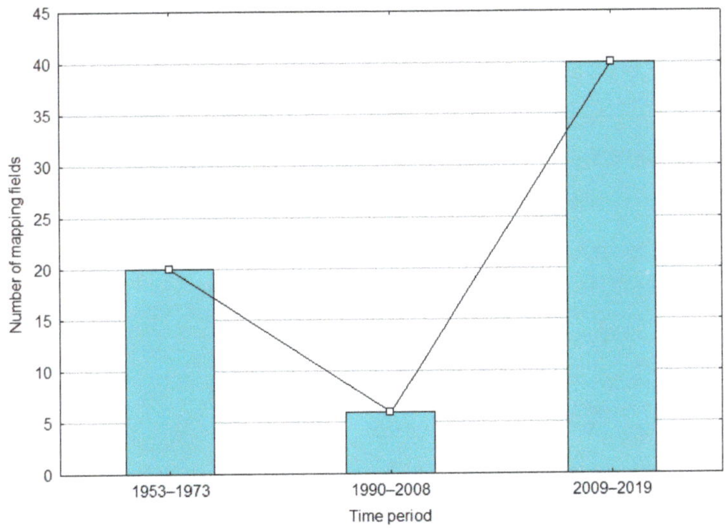

Figure 3. The number of mapping fields in different time periods.

3.2. Current Data

Of the 540 sites surveyed, mussels were detected at 46 sites. The current distribution of the species is shown in Figure 4A,B. The number of detections in the watercourses per study period and the river kilometers are shown in Table 2 and Figure 3. *U. crassus* was detected in the Kolubara [24], Pusta reka [25], Tisa (site Novi Bečej-63rkm) Rivers in 2001 [23], Crni Timok upstream in 2004 [37] and at two sites on the Danube (Stari Banovci and Smederevo) in the period 1990–2008 [26] (Table 2, Figure 4A).

In the period 2009–2019, *U. crassus* was detected in 40 of the 120 examined watercourse sites (33.3%) in Serbia (Figure 4B).

The species was detected in the Danube, Tisa, Sava, Velika and Zapadna Morava Rivers, as well as in the Kolubara River basin (three sites on the main course of the Kolubara and in the Peštan and Ljig Rivers), and according to the literature data, it was also detected in the Južna Morava [39] and Nišava Rivers [40] (Table 2).

During this period, the species was sporadically detected along the Danube, with a low frequency of occurrence and abundance (up to 0.48% of the total mussel community). The species was detected in the Danube only in 2013 at two sites (Čerević-1273 rkm and Tekija-956.2 rkm). It more frequently occurred in the Tisa River. The species was detected at the sites Titel-11 km upstream of the Danube confluence with the Danube (2010), Ada-130 rkm (2013), Martonoš-155 rkm and Tisa, confluence-2 rkm (2019) with low relative abundance (up to 5.78% of the total mussel community). *U. crassus* was also detected in the Sava and Velika Morava Rivers along almost the entire stretch with a higher relative abundance (with a percentage participation of 25.42% and 11.59%, respectively) and in repeated sampling occasions. The occurrence of *U. crassus* in the Kolubara River basin was also confirmed in repeated sampling in the period 2009–2019, but with a low abundance.

The mean value of the population abundance of *U. crassus* is shown in Figure 5, with the minimum and maximum deviation of abundance in the different watercourses, with the highest population abundance recorded in the Sava River.

Table 2. Findings of *U. crassus* in the period 1990–2019, according to the literature data and field investigation.

River/Site	Latitude	Longitude	Period/Year	rkm	Reference
Kolubara (downstream from the Jablanice and Obnice confluence)	44.26163	19.87572	1991–1994	No data	Marković et al., 1999 [24]
Pusta reka	43.08852	21.79819	1998–1999	No data	Živić et al., 2001 [25]
Pusta reka	43.08852	21.79819	1998–1999	No data	Živić et al., 2001 [25]
Tisa, Novi Bečej	20.13447	45.58948	2001	63	JDS-ITR Report 2002 [23]
Dunav, Stari Banovci	44.97855	20.28433	2003–2008	No data	Martinović-Vitanović et al., 2013 [26]
Dunav, Smederevo	44.65945	20.87647	2003–2008	No data	Martinović-Vitanović et al., 2013 [26]
Crni Timok-upstream	43.81826	21.74558	2004	No data	BAES database, Simić et al., 2006 [37]
Dunav, Ćerević	45.22246	19.67268	2013	1273	IBISS
Dunav, Tekija	44.68893	22.41312	2013	956	IBISS
Tisa, Titel	45.21199	20.3188	2010	11	IBISS
Tisa, Ada	45.79409	20.14725	2013	130	IBISS
Tisa, mouth	45.18785	20.31182	2019	11	IBISS
Tisa, Martonoš	46.17644	20.09552	2019	155	IBISS
Velika Morava, Brežane	44.64795	21.07092	2009	9	IBISS
Velika Morava, Varvarin	43.73424	21.37135	13 May 2010	179	IBISS
Velika Morava, Varvarin	43.73424	21.37135	20 September 2010	179	IBISS
Velika Morava, Varvarin	43.73424	21.37135	19 October 2010	179	IBISS
Velika Morava, Varvarin	43.73424	21.37135	16 November 2010	179	IBISS
Velika Morava, Ćuprija	43.94506	21.37101	20 September 2010	146	IBISS
Velika Morava, Markovački most	44.22582	21.15245	20 September 2010	93	IBISS
Velika Morava, Varvarin	43.73424	21.37135	18 January 2011	179	IBISS
Velika Morava, Markovački Most	44.22582	21.15245	31 March 2011	93	IBISS
Veliki Morava, Varvarin	43.73332	21.37018	2019	179	IBISS
Velika Morava, mouth	44.69536	21.03545	2019	2	IBISS
Zapadna Morava-upstream of the Kraljevo and upstream of the Ibar mouth	43.74022	20.73047	2009	4	IBISS
Zapadna Morava, Miločaj	43.77612	20.62904	2012	106	IBISS
Zapadna Morava, Gugaljski Most	43.86874	20.10663	2013	172	IBISS
Južna Morava	42.92038	22.03482	2011		Novaković et al., 2012 [28]
Nišava	43.30647	22.00474	2011		Savić 2012 [32]
Sava, marina	44.80639	20.4438	2010	3	IBISS
Sava, Ostružnica	44.73867	20.31975	2010	16	IBISS
Sava, Šabac	44.7924	19.69151	2011	108	IBISS
Sava, Sremska Mitrovica	44.96211	19.6088	2011	139	IBISS
Sava, Bosut confluence	44.94073	19.36989	2011	162	IBISS
Sava, Bosut confluence	44.94073	19.36989	2012	162	IBISS
Sava, Sremska Mitrovica	44.96211	19.6088	2012	139	IBISS
Sava, Jarak	44.91293	19.75402	2012	124	IBISS
Sava, Umka	44.68449	20.30589	2012	22	IBISS
Sava, Sremska Mitrovica	44.91358	19.7525	2015	139	IBISS
Sava, Šabac	44.76524	19.70304	2015	105	IBISS
Sava, Jamena	44.87813	19.08448	2019	204	IBISS
Sava, mouth	44.79289	20.39587	2019	7	IBISS
Kolubara, Draževac	44.56896	20.21381	2011	14	IBISS
Kolubara, Ćelije	44.37226	20.19992	2012	48	IBISS
Kolubara, Beli Brod	44.37083	20.19956	2013	49	IBISS
Peštan	44.42845	20.25699	2013	1	IBISS
Ljig	44.331578	20.203179	2019	No data	IBISS

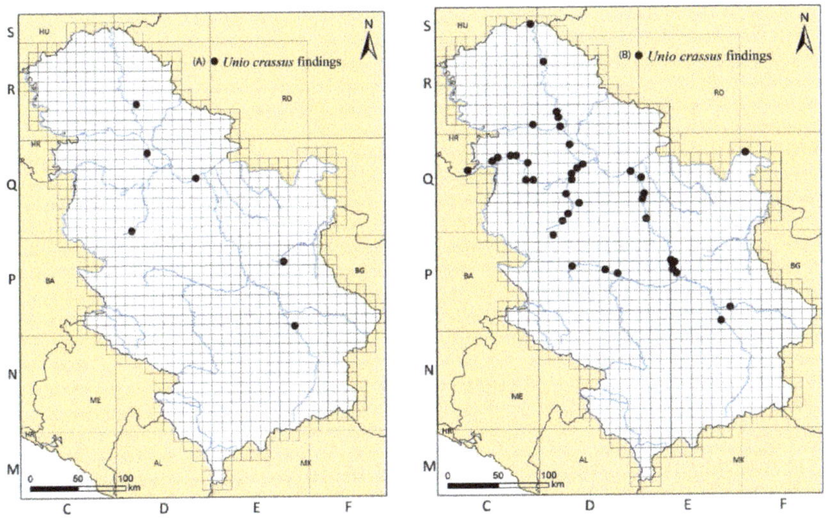

Figure 4. Maps of the distribution range of *U. crassus* in studied grid squares 10 × 10 km in the periods: (**A**) 1990–2008 (according to the literature data only); (**B**) 2009–2019.

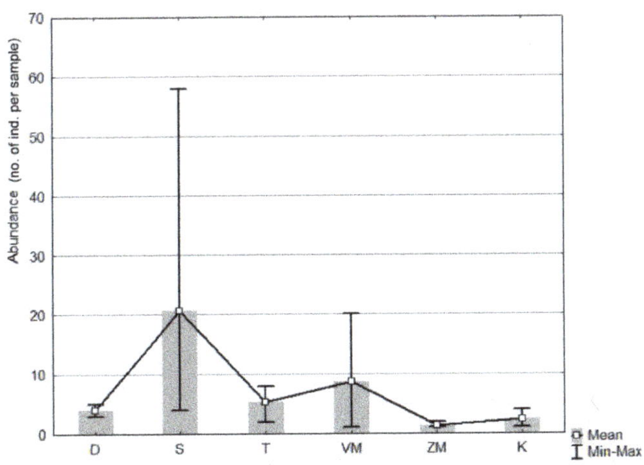

Figure 5. Population abundance of *U. crassus* in different watercourses in Serbia represented by mean, maximum and minimum abundance levels (D—Danube; S—Sava; T—Tisa; VM—Velika Morava; ZM—Zapadna Morava; K—Kolubara).

3.3. Ecological Preferences

The distribution of *U. crassus* is observed predominantly in the littoral reaches of large lowland rivers (waterbody Types 1 and 2), where fine substrate predominates (psammal/psammopelal (6 µm–2 mm) and in small to medium watercourses (Type 3), where coarse substrate (mesolithal 6–20 cm and microlithal 2–6 cm) predominates, at elevations of up to 500 m (Figure 6A–C). It can be characterized as a rheo- to limnophilous species, preferring habitats with slow to moderate water flow.

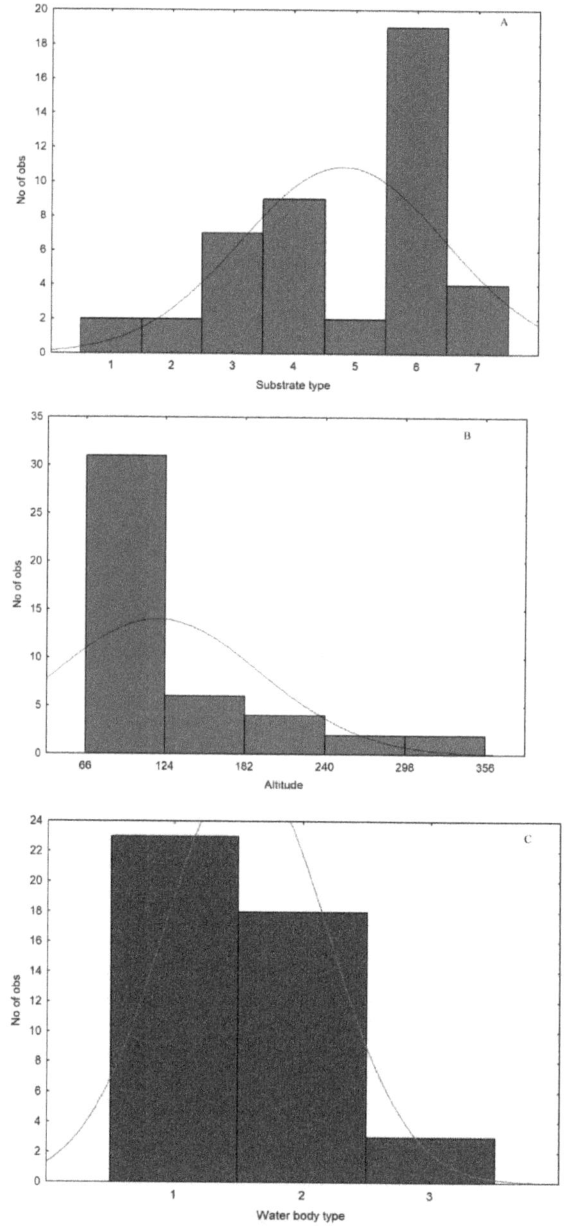

Figure 6. Preference of the mussel assemblages on the (**A**)—substrate type (1—megalithal (>40 cm); 2—macrolithal (20–40 cm); 3—mesolithal (6–20 cm); 4—microlithal (2–6 cm); 5—akal (2 mm–6 cm); 6—psammal/psammopelal (6 µm–2 mm); 7—argyllal (<6 µm); (**B**) altitude and (**C**) waterbody type.

Comparing nitrate–nitrogen levels for the same water bodies between the different monitoring years shows that nitrate–nitrogen levels were higher in the period 1999–2007 than in the most recent monitoring period (2011–2019) (Figures 7 and 8).

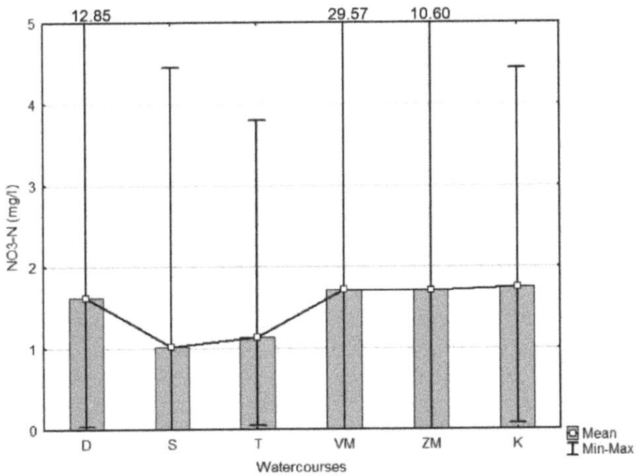

Figure 7. Nitrate–nitrogen concentration in rivers currently or formerly populated by *U. crassus* in period 1999–2007. Mean values are shown as columns with minimum and maximum deviation indicated by lines. D—Danube; S—Sava; T—Tisa; VM—Velika Morava; ZM—Zapadna Morava; K—Kolubara Rivers.

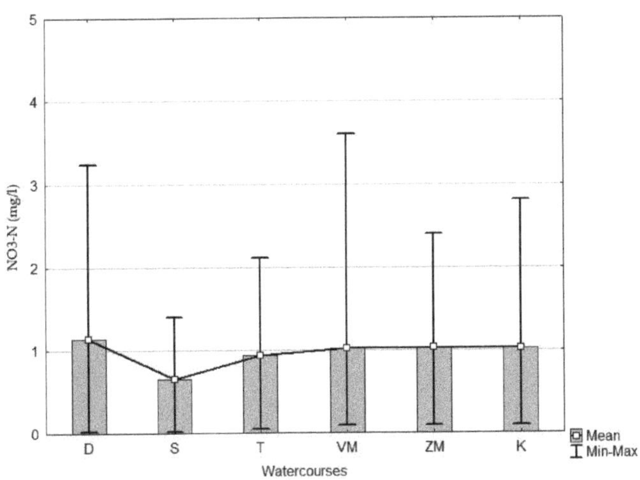

Figure 8. Nitrate–nitrogen concentration in rivers currently or formerly populated by *U. crassus* in period 2011–2019. Mean values are shown as columns with minimum and maximum deviation indicated by lines. D—Danube; S—Sava; T—Tisa; VM—Velika Morava; ZM—Zapadna Morava; K—KolubaraRivers.

4. Discussion

This study represents the first comprehensive research on the distribution of *U. crassus* in Serbian waters, based on historical, literature and field data.

Considering all the collected data, it can be observed that the distribution and abundance of *U. crassus* varies in the different study periods (Figures 2 and 4A,B) and in different watercourses (Figure 5). Re-identification of the archive samples from the Serbian Natural History Museum (period from 1953 to 1973) revealed that *U. crassus* was a common species with continuous distribution throughout Serbia until the mid-1970s. After re-identification (Table 1) of the museum collection, it was observed that there are many synonyms for

U. crassus. Many of these taxa were first described by Henri Drouët (French 'Ecole Nouvelle) in his conchological study of the unionids of Serbia, which also contains an overview of the systematics of only the unionids of Serbia [48]. The great intra-species morphological variability led to an expansion of species' descriptions in the XIX century. Based on all available data on unionids in Serbia and considering the species names according to the valid taxonomy, it can be concluded that the largest number of synonyms exists for the species *U. crassus*. European mussel diversity was also overestimated in the early 1900s due to unreliable taxonomic identification and numerous synonyms, mainly due to the influence of the French 'Ecole Nouvelle [19]. The number of described species in Europe was up to 1500 in the XIX century, but currently 16 species of Unionida are recognized after many synonymies were resolved [19].

Subsequent studies (after the 1970s) showed a decline in population density and a restriction of the distribution range, as well as sporadic findings of the species. In fact, until the early 1990s, there was no data on *U. crassus* in Serbia. Later, the occurrence of *U. crassus* was reported for the Kolubara River in the period 1991–1994 [24], for the Pustareka in the period 1998–1999 [25] and for the Crni Timok River [37]. *U. crassus* was detected in Serbian waters in the Crni Timok in 2004 [37]. Martinović-Vitanović et al. [26] reported findings at two sites in the Serbian stretch of the Danube (Stari Banovci and Smederevo) in the period 2003–2008. All of the above-mentioned findings could be characterized as rare and/or individual findings, indicating that the species was present, but with low population density.

More recent investigations (2009–2019) confirmed the presence of the species in the Danube, Tisa, Sava, Velika and Zapadna Morava Rivers, and in the Kolubara River basin (three sites on the main course of the Kolubara and in the Peštan and Ljig Rivers) (IBISS database), as well as in the Južna Morava [39] and Nišava Rivers [40] (Table 2, Figure 4B). According to the results of the survey of the Sava River in 2012 [28,29], and especially in 2019, a stable population of *U. crassus* was found in the upper and middle stretches of the river. The species was detected at all investigated sites from the site Jamena (204 rkm) to the mouth of the Danube (3 rkm). During the 2019 survey, a high abundance of the species and an almost uniform population of *U. crassus* was detected at the Jamena site. A similar distribution pattern of the species was observed in the KolubaraRiver basin and in the Velika Morava (from 179 rkm to 2 rkm) and Zapadna Morava (from 172 rkm to 2 rkm) Rivers. According to recently published data, an extension of the known range of the species in Serbia and its occurrence in the Južna Morava [28] and Nišava [32] Rivers was also detected. During the investigation of the Tisa River in 2001 [23], the presence of *U. crassus* was detected at one (Novi Bečej-63 rkm) of four investigated sites in the Serbian stretch (lower Tisa), while subsequent surveys from 2010 to 2019 showed an increasing population trend with detection of the species along almost the entire Serbian river section, from the Martonoš (155 rkm) to the Titel (11 rkm) River. Considering that the presence of the species has been confirmed in repeated sampling with significant abundance in the Velika Morava and especially in the Sava River (Figure 5), it could be assumed that the population is recovering, but stable populations are still localized. Furthermore, the permanent finding in the Kolubara River basin could indicate either a recovery of the population or that the population has reached its optimal density for that particular river type. The decreasing trend of the population and the fragmented distribution of *U. crassus* were also confirmed for Europe during studies in the second half of the XX century, with the exception of the northern part (the Baltic basin area), where the species is still considered to be relatively widespread [14]. In contrast to the current data, *U. crassus* was also formerly widespread and the most common unionid in Europe [14].

Knowledge of the habitat requirements of endangered species is of great importance for the implementation of effective conservation strategies, which usually include habitat restoration [49]. In this study, the species was registered in different waterbody types (Types 1, 2 and 3) in areas up to 350 m a.s.l. (Figure 6), in the littoral part of rivers, mostly in fine substrate but also on larger sediment fractions. Most European unionids are lowland species, whereas *U. crassus* can inhabit higher elevations than other unionids [19,30] and can even reach very high densities in mountainous rivers [50], which supports the hypothesis of a wider niche for habitat variables than expected [51].

The population decline and local extinction may be related to the general environmental degradation due to pollution and habitat degradation in the second half of the 20thcentury. It was observed that this species is generally vulnerable to environmental degradation, especially to changes in water chemistry [14]. The high level of eutrophication caused by agricultural drainage is considered to be the main reason for the decline of *U. crassus* [10,22,30,52–54]. Our data show that the mean nitrate–nitrogen concentration varies between the study periods (Figures 7 and 8). The maximum variation in nitrogen concentration indicates highly polluted rivers and poor water quality conditions in almost all the studied rivers in the period between 1999 and 2007 (Figure 7). In the recent period (2011–2019), an improvement in water quality in terms of nitrogen concentration was observed (Figure 8). The most favorable conditions are in the Sava River where the maximum values do not exceed 1.5 NO_3-N mg/L (Figure 8). A significant improvement in water quality was observed in all the studied rivers, which is consistent with our most recent investigations. According to the latest studies, the distribution, the number of detections in selected watercourses and the localities of the first findings, clearly indicate an increasing population trend and an expansion of the distribution range in recent years in Serbia, with a focus on the Sava River basin (Figures 4B and 5). According to research by Zettler and Jueg [22], the increased nitrate–nitrogen caused by eutrophication is one of the main factors in the decline of *U. crassus*. In particular, it is a limiting factor for the growth and maturation of juveniles. A prevailing concentration below 2 mg/L throughout the year and between years indicates successful growth [22]. According to the same authors, limited recruitment of juveniles was observed in moderately polluted streams with nitrogen concentrations between 2 and 10 mg. When the nitrogen concentration exceeds 20 mg/L, the mortality of the mature *U. crassus* population strongly increases [22]. Increased mortality was observed in juveniles above concentrations of 2.3 mg NO_3-N/l [51].

U. crassus was common in the Danube River during the period 1953 to 1973 [37,38], but according to our recent data, its presence in the river was detected only in 2013, with low abundance at only two sites (Čerević-1273 rkm and Tekija-956 rkm) (Table 2). In addition to pollution, the disappearance of this species from the Danube in recent decades could also be related to the hydromorphological changes caused by the construction of dams (Iron Gate) and their impact on the river. Dam construction is probably one of the major threats to the mussel community with direct (damage or removal) or indirect effects on mussels (loss of suitable mussel substrate and decline of host fish) [19,22]. The construction of dams creates barriers to the migration of fish that are potential obligate hosts for the unionid larvae. A lack of suitable host fish can lead to a lack of juvenile recruitment, reducing population density and can potentially lead to species disappearance from habitats or even to extinction [22,55]. The construction of the dam and the forming of a large accumulation lake on the Danube River in Serbia has led to changes in the natural river regime, i.e., the slowing down of the river flow and permanent sediment deposition [56]. Although the dam was built in the lower section, the changes in the river character are noticeable over a long distance downstream and also upstream of the dam. The change in the flow velocity of the river has led to an increased sedimentation rate in the Danube [57]. The increase in sedimentation rate and the change in substrate as a result of the dam [22,58] indirectly affects the mussel community by affecting the potential microhabitats of the species. Changes in river flow due to dam construction and their impact on mussel fauna have already been confirmed for streams and rivers in Europe [14,19,22,59].

On the territory of Serbia, the beginning of mussel exploitation dates back to the 1930s. In the 1950s, organized mussel collection for industrial purposes was performed [38]. This long-term overexploitation has certainly significantly contributed to the decline of mussel populations in our rivers, which can still be observed today. Since the 1850s, freshwater mussels have been exploited for the extraction of pearls and nacre for button making [60].

At the peak of this exploitation, up to 50,000 tons of shells were harvested from North American rivers [61]. Strict laws now prohibit these activities, but poaching continues in some countries [19]. According to Ferreira-Rodríguez et al. [62], overexploitation is only locally significant and is often of secondary importance compared to other pressures that currently exist.

Among other factors, the introduction of exotic species is a possible contributing factor to the decline of freshwater mussels [10]. Over the past 20 years, research on allochthonous species has intensified in Serbia [61]. According to Zorić et al. [63], the Danube is the main corridor for the introduction and spread of alien species in Serbia and their spread to the other major rivers, i.e., the Tisa, Sava and Velika Morava [63]. The invasive bivalve species in Serbian freshwater ecosystems include the zebra mussel *Dreissena polymorpha* (Pallas 1771), the quagga mussel *Dreissena bugensis* Andrusov, 1897, the Asian clam *Corbicula fluminea* (O. F. Müller 1774) and the Chinese pond mussel *Sinanodonta woodiana* (Lea, 1834) [63]. Invasive mussels are widely recognized as an important threat to native biodiversity [64]. The ecological impact of invasive species on native communities is not well documented in Serbia but there is evidence of widespread distribution in Serbian waters, dense populations and coexistence with native fauna [63]. They can cause direct biotic interactions with the native community (e.g., predation and competition) and also indirect changes in habitat conditions (e.g., habitat structure and turbidity) [65,66].Evidence of the negative impact of invasive species on native unionids has already been observed in many European countries as well as in North America [19].

At global, regional and local levels, species important for nature conservation are selected, protected areas are designed and an ecological network is established to link protected areas important for biodiversity conservation and the remaining priority habitat types [64].

The NATURA 2000 network is the main tool for biodiversity protection in the European Union. It is now considered to be the world's largest network of protected areas, covering 30,000 sites that occupy 20% of the EU territory [67].

Nature conservation efforts in the Republic of Serbia are aimed at fulfilling obligations in the framework of preparations for accession to the European Union (EU), which mainly refers to the establishment of the NATURA 2000 ecological network. When the conditions for EU accession are met, biodiversity and habitat diversity in Serbia will become part of the European ecological network NATURA 2000, with the obligation to implement the Directive. Serbia will propose areas important for the conservation of endangered plant and animal species for the ecological network NATURA 2000 and habitat types, as well as other EU member states.

The Balkan Peninsula served as a glacial refuge for several species of freshwater macroinvertebrates [68,69]. The establishment of an ecological network of protected areas will make it possible to ensure the survival of the most valuable species and habitats, promote the protection of numerous ecosystems and ensure that the natural system of Europe, and the Balkan Peninsula in particular, remains healthy and resilient.

5. Conclusions

The results presented show considerable progress in the restoration of the former distribution range of *U. crassus* in Serbian waters. Based on a dataset that includes historical and current data, population trends of this mussel over time were identified and a better understanding of the basic ecological requirements of the species was gained. The interaction of eutrophication, hydrological changes, overexploitation as well as the introduction of invasive species may be possible factors that influenced the local disappearance of *U. crassus* in some sections or the decrease in population density in Serbia. The results of this study can be used for the further development of effective and sustainable conservation strategies for endangered *U. crassus* populations, which usually include habitat restoration. Despite the high conservation status of this species, knowledge about its biology and ecology is insufficient. To improve conservation strategies for *U. crassus*, a systematic understanding of the limiting factors in the species' life cycle is crucial. Further studies on *U. crassus* should include more comprehensive ecological, biological and genetic investigations, as well as detection of new populations on a larger geographical scale.

Author Contributions: Conceptualization, J.T., M.R. and M.P.; methodology, J.T., M.R., V.S., A.P. and M.P.; software, J.T., A.A. and K.Z.; validation, M.R., M.P., V.S. and A.P.; formal analysis, J.T., K.Z., A.A. and A.P.; investigation, J.T., M.R., A.A., K.Z. and M.P.; writing—original draft preparation, J.T., M.R. and M.P.; writing—review and editing, K.Z., A.A., A.P., V.S. and J.T.; visualization, A.A., K.Z., A.P. and J.T.; supervision, M.R., V.S., M.P. and J.T.; project administration, J.T.; funding acquisition, M.P. All authors have read and agreed to the published version of the manuscript.

Funding: This research was funded by the Ministry of Science, Technological Development and Innovations of the Republic of Serbia, Contract No. 451-03-47/2023-01/200007.

Data Availability Statement: The data presented in this study are available upon request from the corresponding author.

Acknowledgments: We would like to thank the participants of the JDS 1-4 expeditions and the International Commission for the Protection of the Danube River (ICPDR). Details of the JDS investigations, including sampling site data are available at: http://www.danubesurvey.org/jds4/, and at: https://jds4.icpdr.org/portal/ (on request from the ICPDR Secretariat, Vienna International Centre, Room D0412, WagramerStrasse 5, A-1220 Vienna, Austria, e-mail: secretariat@icpdr.org; Tel.: +431-260-60-5738).

Conflicts of Interest: The authors declare no conflict of interest.

References

1. Vaughn, C.C.; Hakenkamp, C.C. The Functional Role of Burrowing Bivalves in Freshwater Ecosystems. *Freshw. Biol.* **2001**, *46*, 1431–1446. [CrossRef]
2. Howard, J.K.; Cuffey, K.M. The Functional Role of Native Freshwater Mussels in the Fluvial Benthic Environment. *Freshw. Biol.* **2006**, *51*, 460–474. [CrossRef]
3. Vaughn, C.C.; Nichols, S.J.; Spooner, D.E. Community and Food web Ecology of Freshwater Mussels. *J. N. Am. Benthol. Soc.* **2008**, *27*, 409–423. [CrossRef]
4. Bogan, A.E. Freshwater Bivalve Extinctions (Mollusca: Unionoida): A Search for Causes. *Am. Zool.* **1993**, *33*, 599–609. [CrossRef]
5. Araujo, R.; Ramos, M.A. Status and Conservation of the Giant European Freshwater Pearl Mussel (*Margaritifera auricularia*) (Spengler, 1793) (Bivalvia: Unionoidea). *Biol.Conserv.* **2000**, *96*, 233–239. [CrossRef]
6. Young, M.R.; Cosgrove, P.J.; Hastie, L.C. The Extent of, and Causes for, the Decline of a Highly Threatened Naiad: *Margaritifera margaritifera*. In *Ecology and Evolution of the Freshwater Mussels Unionoida*; Springer: Berlin/Heidelberg, Germany, 2001; pp. 337–357.
7. Lydeard, C.; Cowie, R.H.; Ponder, W.F.; Bogan, A.E.; Bouchet, P.; Clark, S.A.; Cummings, K.S.; Frest, T.J.; Gargominy, O.; Herbert, D.G. The Global Decline of Non marine Mollusks. *Bioscience.* **2004**, *54*, 321–330. [CrossRef]
8. Strayer, D.L.; Downing, J.A.; Haag, W.R.; King, T.L.; Layzer, J.B.; Newton, T.J.; Nichols, J.S. Changing Perspectives on PearlyMussels, North America's Most Imperiled Animals. *Bioscience* **2004**, *54*, 429–439. [CrossRef]
9. Strayer, D.L. *Freshwater Mussel Ecology: A Multifactor Approach to Distribution and Abundance*; University of California Press: Oakland, CA, USA, 2008; Volume 1, ISBN 0520942523.
10. Soroka, M.; Wasowicz, B.; Zając, K. Conservation Status and a Novel Restoration of the Endangered Freshwater Mussel *Uniocrassus* Philipsson, 1788: PolandCase. *Knowl. Manag. Aquat. Ecosyst.* **2021**, *422*, 3. [CrossRef]
11. IUCN. *The IUCN Red List of Threatened Species, version 2017-1*; International Union for Conservation of Nature (IUCN): Gland, Switzerland, 2017.
12. Zieritz, A.; Lopes-Lima, M.; Bogan, A.E.; Sousa, R.; Walton, S.; Rahim, K.A.A.; Wilson, J.-J.; Ng, P.-Y.; Froufe, E.; Mc Gowan, S. Factors Driving Changes in Freshwater Mussel (Bivalvia, Unionida) Diversity and Distribution in Peninsular Malaysia. *Sci. Total Environ.* **2016**, *571*, 1069–1078. [CrossRef]
13. Zieritz, A.; Bogan, A.E.; Froufe, E.; Klishko, O.; Kondo, T.; Kovitvadhi, U.; Kovitvadhi, S.; Lee, J.H.; Lopes-Lima, M.; Pfeiffer, J.M. Diversity, Biogeography and Conservation of Freshwater Mussels (Bivalvia: Unionida) in East and Southeast Asia. *Hydrobiologia.* **2018**, *810*, 29–44. [CrossRef]
14. Lopes-Lima, M.; Kebapçı, U.; Van Damme, D. Unio crassus. In *The IUCN Red List of Threatened Species*; International Union for Conservation of Nature (IUCN): Gland, Switzerland, 2014. [CrossRef]
15. Directive, H. Council of Europe Directive 92/43/EEC on the Conservation of Natural Habitats and of Wild Fauna and Flora. *Brussels* **1992**, *206*, 50.
16. European Council. *Revised Annex I of Resolution 6 of the Bern Convention Listing the Species Requiring Specific Habitat Conservation Measures*; European Council: Brussels, Belgium, 1998.

17. Official Gazette of RS 98/2016. *The Regulation on Proclamation of Strictly Protected and Protected Wild Species of Plants, Animals and Fungi*; Ministry Agriculture, Forestry and Water Management: Belgrade, Serbia, 2016.
18. Collier, K.J.; Probert, P.K.; Jeffries, M. Conservation of Aquatic Invertebrates: Concerns, Challenges and Conundrums. *Aquat. Conserv. Mar. Freshw. Ecosyst.* **2016**, *26*, 817–837. [CrossRef]
19. Lopes-Lima, M.; Sousa, R.; Geist, J.; Aldridge, D.C.; Araujo, R.; Bergengren, J.; Bespalaya, Y.; Bódis, E.; Burlakova, L.; VanDamme, D.; et al. Conservation Status of Freshwater Mussels in Europe: State of the Art and Future Challenges. *Biol. Rev.* **2017**, *92*, 572–607. [CrossRef]
20. Taeubert, J.; Gum, B.; Geist, J. Host-Specificity of the Endangered Thick-Shelled River Mussel (*Unio crassus*, Philipsson 1788) and Implications for Conservation. *Aquat. Conserv. Mar. Freshw. Ecosyst.* **2012**, *22*, 36–46. [CrossRef]
21. Engel, H.; Wächtler, K. SomePeculiaritiesin Development Biology of Two Forms of the Freshwater Bivalve *Unio crassus* inNorthern Germany. *Arch. Hydrobiol.* **1989**, *115*, 441–450. [CrossRef]
22. Zettler, M.L.; Jueg, U. The Situation of the Freshwater Mussel *Unio crassus* (Philipsson, 1788) inNorth-East Germany and ItsMonitoring in Terms of the EC Habitats Directive. *Mollusca.* **2007**, *25*, 165–174.
23. *Report on theJoint Investigation of the Tisza River on the Territory of the FR Yugoslavia within the International JDS-ITR Program*; JDS-ITR National Tisza Report; Republic of Serbia, Ministry for Protection of Natural Resources and Environment, Federal Hydro-Meteorological Institute: Belgrade, Serbia, 2002; pp. 1–52.
24. Marković, Z.; Miljanović, B.; Mitrović-Tutundžić, V. Macrozoobenthos as an Indicator of the River Kolubara Water Quality. [Serbia, Yugoslavia] water quality. In Proceedings of the 28 Konferencija o Aktuelnim Problemima Zastite voda. Zastita voda'99, Soko Banja, Yugoslavia, 12–15 October 1999; Jugoslovensko Drustvo za Zastitu Voda: Belgrade, Serbia, 1999; Volume 28, pp. 261–266.
25. Živić, I.; Marković, Z.; Brajković, M. Macrozoobenthos in the Pusta Reka River, Left Tributary of the South Morava River. *Arch. Biol. Sci.* **2001**, *53*, 109–122.
26. Martinovic-Vitanovic, V.M.; Rakovic, M.J.; Popovic, N.Z.; Kalafatic, V.I. Qualitative Study of Mollusca Communities in the Serbian Danube Stretch (RiverKm1260–863.4). *Biologia* **2013**, *68*, 112–130. [CrossRef]
27. Tomović, J.; Paunović, M.; Atanacković, A.; Marković, V.; Gačić, Z.; Csányi, B.; Simić, V. Biotic Typology of the Danube RiverBased on Distribution of Mollusc Fauna as Revealed by the Second Joint Danube Survey (2007). *Acta Zool. Bulg.* **2014**, *66*, 527–537.
28. Novaković, B. Indicative Ecological Status Assessment of the Južna Morava River Based on Aquatic Macroinvertebrates. *Water Res. Manag.* **2012**, *2*, 45–50.
29. Paunović, M.; Tomović, J.; Kovačević, S.; Zorić, K.; Žganec, K.; Simić, V.; Atanacković, A.; Marković, V.; Kračun-Kolarević, M.; Hudina, S. Macroinvertebrates of the Natural Substrate of the Sava River–Preliminary Results. *Water Res. Manag.* **2012**, *2*, 33–39.
30. Marković, V.; Gojšina, V.; Novaković, B.; Božanić, M.; Stojanović, K.; Karan-Žnidaršič, T.; Živić, I. The freshwater mollusks of Serbia: Annotated checklist with remarks on distribution and protection status. *Zootaxa* **2021**, *5003*, 1–64. [CrossRef]
31. Lucić, A.; Paunović, M.; Tomović, J.; Kovačević, S.; Zorić, K.; Simić, V.; Atanacković, A.; Marković, V.; Kračun-Kolarević, M.; Hudina, S.; et al. Aquatic Macroinvertebrates of the Sava River. In *The Sava River*; The Handbook of Environmental Chemistry; Milačić, R., Ščančar, J., Paunović, M., Eds.; Springer: Berlin/Heidelberg, Germany; NewYork, NY, USA; Dordrecht, The Netherlands; London, UK, 2015; Volume 31, pp. 335–359.
32. Savić, A.V. Ecological Analysis of the Macrozoobenthos Community of the Nišava River. Ph.D. Thesis, Faculty of Biology, University of Belgrade, Belgrade, Serbia, 2012; p. 174.
33. Literáthy, P.; Koller-Kreiml, V.; Liška, I. *Final Report of the Joined Danube Survey*; International Commission for the Protection of the Danube River—ICPDR: Vienna, Austria, 2002; p. 262.
34. Liška, I.; Wagner, F.; Slobodnik, J. *Joint Danube Survey—Final Report ICPDR*; International Commission for the Protection of the Danube River—ICPDR: Vienna, Austria, 2008; p. 235.
35. Liška, I.; Wagner, F.; Sengl, M.; Deutsch, K.; Slobodník, J. *Joint Danube Survey 3—A Comprehensive Analyses of Danube Water Quality*; International Commission for the Protection of the Danube River—ICPDR: Vienna, Austria, 2015; p. 369.
36. Liska, I.; Wagner, F.; Sengl, M.; Deutsch, K.; Slobodnik, J.; Paunovic, M. *Joint Danube Survey 4 Scientific Report. A Shared Analysis of the Danube River*; International Commission for the Protection of the Danube River: Vienna, Austria, 2021; ISBN 978-3-200-07450-7.
37. Simić, V.; Simić, S.; Petrović, A.; Paunović, M.; Šorić, V.; Dimitrijević, V. *Biodiversity in Aquatic Ecosystems in Serbia, Ex Situ Conservation (BAES Ex Situ)*; Faculty of Science: Kragujevac, Serbia, 2006.
38. Tadić, A. *Najvažniji Predstavnici Školjaka Roda Unio iz Save, Dunava I Kopačkog Jezera*; Posebna izdanja; Prirodnjački Muzej: Belgrad, Serbia, 1960; Volume 28, pp. 1–46.
39. MolluscaBase. 2023. Available online: https://www.molluscabase.org (accessed on 6 November 2023).
40. Graf, K.S.; Cummings, D.L. The Freshwater Mussels (Unionoida) of the World (and Other Less Consequential Bivalves); Updated 9 August 2018. Available online: http://www.mussel-project.net/ (accessed on 6 November 2023).
41. WoRMSEditorialBoard. *World Register of Marine Species*. 2023. Available online: https://www.marinespecies.orgatVLIZ (accessed on 6 November 2023).
42. *ISO 7828:1985*; Water Quality—Methods of Biological Sampling—Guidance on Hand net Sampling of Aquatic Benthic Macro-Invertebrates. European Committee for Standardization: Brussels, Belgium, 1994.
43. *SRPS ISO/IEC 17025: 2017*; General Requirements for the Competence of Testing and Calibration Laboratories. Institute of Standardization of Serbia: Belgrade, Serbia, 2017.

44. AQEM. *Manual for the Application of the AQEM System. A Comprehensive Method to Assess European Streams Using Benthic Macroinvertebrates, Developed for the Purpose of the Water Framework Directive, Version 1*; The AQEM Consortium: Duisburg-Essen, Germany, 2002; pp. 619–630.
45. Regulative on Parameters of Ecological and Chemical Status of Surface Waters and Parameters of Chemical and Quantitative Status of Ground Waters. *Off. Gaz.* Republic of Serbia. 2011, No. 67/2011. Available online: https://www.pravno-informacioni-sistem.rs/SlGlasnikPortal/eli/rep/sgrs/ministarstva/pravilnik/2011/74/4/reg (accessed on 6 November 2023).
46. "StatSoft, Inc., Tulsa, OK.: STATISTICA, Version 8," AStA Advances in Statistical Analysis, Springer; German Statistical Society, Volume 91, pp. 339–341. Available online: https://link.springer.com/journal/10182 (accessed on 6 November 2023).
47. 2015. Available online: https://adobe.com/products/illustrator (accessed on 6 November 2023).
48. Drouët, H. *Suppl. Unionidae de Serbie*; Paris, France, 1884; pp. 159–174.
49. Stoeckl, K.; Geist, J. Hydrological and Substrate Requirements of the Thick-Shelled River Mussel *Unio crassus* (Philipsson 1788). *Aquat. Conserv. Mar. Freshw. Ecosyst.* **2016**, *26*, 456–469. [CrossRef]
50. Zając, K. Unio crassus Philipsson, 1788. In *Skójka Gruboskorupowa. Pol. czerwona Księga Zwierząt, Bezkręgowce*; Instytut Ochrony Przyrody PAN: Kraków, Poland, 2004; pp. 353–355.
51. Tatoj, K.; Ćmiel, A.M.; Kwaśna, D.; Lipińska, A.M.; Zając, K.; Zając, T. The Endangered Thick-Shelled River Mussel (*Unio crassus*): A New Host Species for the European Bitterling (*Rhodeus amarus*). *Biodivers. Conserv.* **2017**, *26*, 1217–1224. [CrossRef]
52. Patzner, R.A.; Müller, D. Effects of Eutrophication on Unionids. In *Ecology and Evolution of the Freshwater Mussels Unionoida*; Springer: Berlin/Heidelberg, Germany, 2001; pp. 327–335.
53. Köhler, R. Observations on Impaired Vitality of *Unio crassus* (Bivalvia: Najadae) Populations in Conjunction with Elevated Nitrate Concentrationin Running Waters. *ActaHydrochim. Hydrobiol.* **2006**, *34*, 46–348. [CrossRef]
54. Lyubas, A.A.; Kondakov, A.V.; Tomilova, A.A.; Gofarov, M.Y.; Eliseeva, T.A.; Konopleva, E.S.; Vikhrev, I.V.; Yunitsyna, O.A.; Pešić, V.; Bolotov, I.N. Taxonomic Reassessment of Freshwater Mussels from the Western Balkans Reveals an Overlooked butCritical Refugium and Defines Conservation Priorities. *Diversity* **2022**, *14*, 935. [CrossRef]
55. Zale, A.V.; Neves, R.J. Reproductive Biology of Four Freshwater Mussel Species (Mollusca: Unionidae) in Virginia. *Freshw. Invertebr. Biol.* **1982**, *1*, 17–28. [CrossRef]
56. Mladenović, M.B.; Kolarov, V.; Damjanovic, V. Sediment Regime of the Danube River in Serbia. *Int. J. Sediment Res.* **2013**, *28*, 470–485. [CrossRef]
57. Nedeljkovic, R. Zoobenthos of the Danube in Period after the Construction the Dam in Iron Gate Sector. In Proceedings of the 2nd Congress of the Ecologists of Yugoslavia, Zagreb, Croatia , 1979; pp. 1881–1888.
58. Bril, J. *Measuring Mussel Behavior and Analyzing High Frequency Nitrate Data to Explore New Phenomena in Dynamic Nutrient Cycling*; University of Iowa: Iowa, IA, USA, 2010.
59. Beran, L. Distribution and Recent Status of Freshwater Mussels of Family Unionidae (Bivalvia) in the Czech Republic. *Knowl. Manag. Aquat. Ecosyst.* **2019**, *420*, 45. [CrossRef]
60. Humphries, P.; Winemiller, K.O. Historical Impacts on River Fauna, Shifting Baselines, and Challenges for Restoration. *Bioscience.* **2009**, *59*, 673–684. [CrossRef]
61. Haag, W.R. *North American Freshwater Mussels: Natural History, Ecology, and Conservation*; Cambridge University Press: New York, NY, USA, 2012; ISBN 1139560190.
62. Ferreira-Rodriguez, N.; Akiyama, Y.B.; Aksenova, O.V.; Araujo, R.; Barnhart, M.C.; Bespalaya, Y.V.; Bogan, A.E.; Bolotov, I.N.; Budha, P.B.; Clavijo, C.; et al. Research priorities for freshwater mussel conservation assessment. *Biol. Conserv.* **2019**, *231*, 77–87. [CrossRef]
63. Zorić, K.; Atanacković, A.; Tomović, J.; Vasiljević, B.; Tubić, B.; Paunović, M. Diversity of Alien Macroinvertebrate Species inSerbian Waters. *Water.* **2020**, *12*, 3521. [CrossRef]
64. Ricciardi, A.; Neves, R.J.; Rasmussen, J.B. Impending Extinctions of North American Freshwater Mussels (Unionoida) Following the Zebra Mussel (*Dreissena polymorpha*) Invasion. *J. Anim. Ecol.* **2002**, *67*, 613–619. [CrossRef]
65. Sousa, R.; Novais, A.; Costa, R.; Strayer, D.L. Invasive Bivalves in Fresh Waters: Impacts from Individuals to Ecosystems and Possible Control Strategies. *Hydrobiologia* **2014**, *735*, 233–251. [CrossRef]
66. Crooks, J.A. Characterizing ecosystem-level consequences of biological invasions: The role of ecosystem engineers. *Oikos* **2002**, *97*, 153–166. [CrossRef]
67. Popović, M.; Rajkov, S.; Ružić, M.; Savić, D.; Šćiban, M.; Simović, A.; Mesaroš, G.; Radišić, D. *Vrste Biljaka I Životinja Značajne za Zaštitu Prirode u Evropskoj Uniji*; Mesaroš, G., Ed.; Protego: Subotica, Serbia, 2014.
68. Pešić, V.; Glöer, P. The Handbook of Environmental Chemistry. In *The Skadar/Shkodra Lake Environment*; Pešić, V., Karaman, G., Kostianoy, A., Eds.; Springer: Cham, Switzerland, 2018; pp. 295–310.
69. Raković, M.; Tomović, J.; Popović, N.; Pešić, V.; Dmitrović, D.; Stamenković, V.S.; Hinić, J.; Stefanovska, N.; Lajtner, J.; Paunović, M. Gastropods in Small Water Bodies of the Western Balkans—Endangerments and Threats. *Small Water Bodies West. Balk.* **2022**, 227–249.

water

MDPI

Article

Aquatic Insects (Ephemeroptera, Plecoptera and Trichoptera) Metric as an Important Tool in Water Quality Assessment in Hilly and Mountain Streams

Bojana Tubić *, Stefan Andjus *, Katarina Zorić, Božica Vasiljević, Katarina Jovičić, Jelena Čanak Atlagić and Momir Paunović

Department for Hydroecology and Water Protection, Institute for Biological Research "Siniša Stanković", National Institute of the Republic of Serbia, University of Belgrade, Bulevar despota Stefana 142, 11108 Belgrade, Serbia; katarinas@ibiss.bg.ac.rs (K.Z.); bozica@ibiss.bg.ac.rs (B.V.); katarina.jovicic@ibiss.bg.ac.rs (K.J.); jelena.canak@ibiss.bg.ac.rs (J.Č.A.); mpaunovi@ibiss.bg.ac.rs (M.P.)

* Correspondence: bojana@ibiss.bg.ac.rs (B.T.); stefan.andjus@ibiss.bg.ac.rs (S.A.)

Abstract: The aim of the study was to test the significance of the EPT index in the water quality assessment of three types of water bodies in hilly and mountainous region of Serbia. The aquatic macroinvertebrate community was dominated by the group of insects, of which 95 taxa represent the EPT group. We compared the obtained values of biological indices used for the assessment of water quality according to the national legislation with the overall status assessment represented by the ecological quality classes (EQC). The results of the Spearman correlation test showed a negative correlation of EQC with the EPT index, BMWP score, H′, total number of taxa and number of sensitive taxa, while a positive correlation was observed for the values of SI and Tubificinae %. The values of EQC and biological indices were subjected to principal component analysis (PCA). The results showed that the parameters that contributed most to the differences were the EPT index, the BMWP score and the number of sensitive taxa. The results indicate that the EPT index is an excellent indicator of changes in water quality and an important tool for the ecological categorization of water bodies in mountain regions.

Keywords: aquatic insects; EPT group; diversity; biological indices; water quality assessmant; hilly and mountain streams; Serbia

1. Introduction

The benthic macroinvertebrate fauna is an effective tool for documenting changes in overall ecological status, and also one of the most prominent biological quality elements (BQEs) used for the ecological assessment of rivers under the European Water Framework Directive (WFD) [1]. Aims of WFD are to prevent further deterioration and to protect and improve the status of aquatic ecosystems, with the explicit goal of achieving at least "good ecological status" for all surface waters by 2027 [2].

Macroinvertebrate offer numerous advantages for biomonitoring. Sampling is relatively simple and has minimal adverse effects on the resident biota [3–5]. Macroinvertebrate taxa generally occur in characteristic and limited habitats within their geographic range and are usually most abundant near their respective ecological optimum [6,7]. The benthic macroinvertebrates are good indicators of local conditions and are particularly suitable for assessing site-specific impacts, as many of them have restricted migration patterns or a sessile lifestyle [8]. Larval stages will respond quickly to stress. Many of these groups are relatively easy to identify to the lower taxonomic levels such as genus and species [9,10].

Many aquatic insects are intolerant to various types of pressures, therefore along with increased pollution, lower diversity can be expected. Many authors compared a large number of macroinvertebrate indices and found that simple species counts, especially sensitive taxa, were most effective in determining water body impairment [10–12].

Some of the macroinvertebrate-based indices of river health have been proven to be particularly useful and effective, such as the EPT (Ephemeroptera + Plecoptera + Trichoptera) index. This index was named after three orders of aquatic insects common in the benthic macroinvertebrate community: Ephemeroptera (mayflies), Plecoptera (stoneflies) and Trichoptera (caddisflies). It represents the sum of the taxa richness of these three orders [13].

Publications from the 1950s stated that the species of these three orders are generally intolerant of pollution [14]. The EPT larvae are easy to sort and identify and are often used as an indicator of water quality. The EPT index has been recommended as possibly the most efficient of the macroinvertebrate indices, especially in lotic ecosystems where they are the dominant component of the representative community [15,16].

A complex river network covers most of the territory of Serbia and belongs to the catchment area of the Black, Adriatic and Aegean Seas. The Serbian territory belongs to South-East Europe, covering the central part of the Balkan Peninsula and the southern part of the Pannonian Plain. It can be clearly divided into two regions—the Pannonian plain and the hilly and mountainous region south of the Danube and Sava rivers [17]. The distribution of aquatic organisms in this area is therefore a complex issue. The diversity of benthic macroinvertebrates in this area is significant and it is considered the main diversity hotspot of aquatic insects in Europe, especially the EPT groups [18]. Paunović et al. [19] used the distribution of macroinvertebrates to delineate the boundaries of ecoregions on the territory of Serbia with regard to the original concept of Illies [17], which was accepted by the WFD. According to these authors [19], the hilly and mountainous region of the country belongs to the ecoregion 5 (Dinaric Western Balkan) and ecoregion 7 (Eastern Balkan).

The aims of the study were: to assess the significance of the EPT index in the evaluation of the water quality assessment of three types of water bodies in the mountainous regions of Serbia; to test the relationship between the obtained values of biological indices used for the assessment of water quality of these types of water bodies, according to the national legislation and the overall ecological status assessment represented by the ecological quality classes (EQC); and to test whether the EPT index is sufficiently meaningful and self-sufficient for the assessment of water quality of hilly and mountainous streams.

2. Materials and Methods

2.1. Study Area

The collection of benthic macroinvertebrate samples in the spring and fall 2019 was conducted to supplement the data for the revision of the Water Management Plan for the territory of the Republic of Serbia. It included 119 watercourses that have not previously been part of the routine water quality monitoring led by the Environmental Protection Agency of the Republic of Serbia. Large lowland rivers and rivers with a predominance of fine and medium sediment (akal, psammal/psammopelal, argyllal) [20], and artificial water bodies, canals and reservoirs, were excluded from this study. According to the national legislation, the EPT index is not used as a biological metric to assess water quality in these types of water bodies [21].

The analyzed dataset in this paper included the group of hilly and mountainous small- to medium-sized streams with a predominantly hard bottom substrate, classified according to the Serbian typology of watercourses [21]. The focus was on 44 sites located on three selected stream types (Figure 1). The selected types of watercourses were: type 3—small and medium streams, altitude up to 500 m a.s.l., dominance of larger substrate (mesolithal, macrolithal, megalithal) (24 sites); type 4—small and medium streams, altitude above 500 m a.s.l. and dominance of larger substrate (mesolithal, macrolithal, megalithal) (6 sites) and type 6—small watercourses outside the area of the Pannonian Plain that do not fall under types 3 and 4 and are not covered by the regulation on the establishment of surface and groundwater bodies (combination of different types of substrate) (14 sites) [22] (Table 1).

Most of the investigated sites have microhabitat substrate characterized by boulders and cobbles and fast velocity of flow. The size of the microhabitat substrates were defined according to Hering et al. [20].

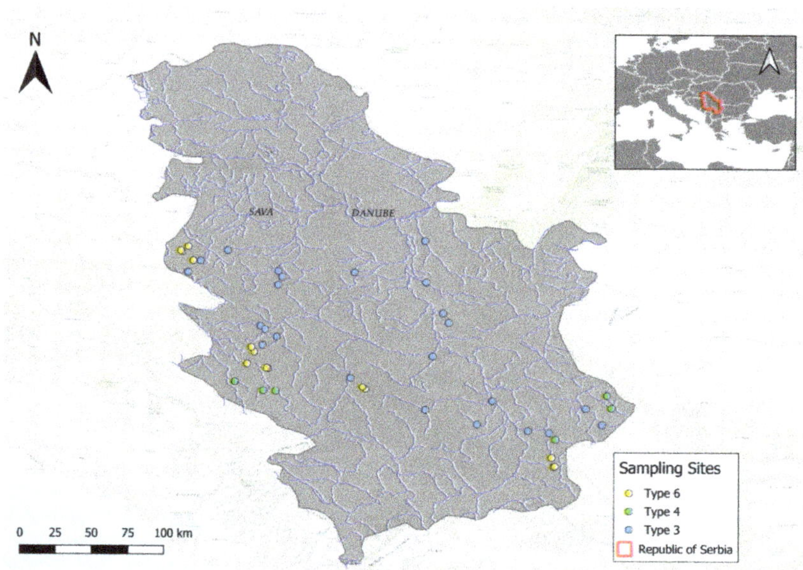

Figure 1. Map of sampling localities.

Table 1. Sampling localities by watercourse type with GPS coordinates.

No.	Type of Watercourse	Localities	N	E	Altitude (m)
1	Type 3	Kamenica	44.27879	20.69594	203
2		Grza	43.86491	21.47602	210
3		Ravanica	43.94375	21.42883	176
4		Likodra	44.37108	19.41078	390
5		Krupinska reka	44.27692	19.30736	313
6		Veliki Rzav 1	43.67234	19.93187	498
7		Veliki Rzav 2	43.74391	20.05248	480
8		Nišava 1	43.01863	22.73374	413
9		Nišava 2	43.15188	22.59950	381
10		Pusta reka	43.01950	21.71199	246
11		Toplica 1	43.21316	21.83449	272
12		Toplica 2	43.13895	21.28398	425
13		Vlasina 1	42.96810	22.12787	329
14		Vlasina 2	42.94951	22.30421	410
15		Rasina	43.58722	21.34405	152
16		Mlava	44.54519	21.27974	93
17		Jošanica	43.40491	20.66400	463
18		Resava	44.19887	21.28998	148
19		Ribnica 1	44.28707	20.06403	163
20		Ribnica 2	44.23879	20.09669	260
21		Ribnica 3	44.17276	20.06269	375
22		Tamnava 2	44.45617	19.63632	236
23		Đetinja 1	43.83131	19.91463	642
24		Đetinja 2	43.80448	19.95518	458
25	Type 4	Mileševka	43.37128	19.69763	726
26		Uvac	43.29290	19.93810	1134
27		Vapa	43.29000	20.04355	1023
28		Dojkinačka reka 1	43.26172	22.77364	1223
29		Dojkinačka reka 2	43.15595	22.81343	1009
30		Gradska reka	42.89735	22.35245	846

Table 1. *Cont.*

No.	Type of Watercourse	Localities	N	E	Altitude (m)
31	Type 6	Tisovica	43.48065	19.97928	1167
32		Trudovačka reka	43.48476	19.96600	1167
33		Čađavica	44.37178	19.34831	491
34		Korenita	44.48985	19.30783	261
35		Štira 1	44.44366	19.25502	526
36		Štira 2	44.45272	19.24496	391
37		Pritoka Uvca	43.52039	19.80133	948
38		Ljubišnica	43.61730	19.86022	1018
39		Katušnica	43.65501	19.83772	896
40		Cvetkova reka	42.74631	22.32128	1380
41		Jarčev potok	42.66968	22.33745	1258
42		Simonova reka	42.67491	22.34782	1258
43		Samokovska reka 1	43.30976	20.78675	1506
44		Samokovska reka 2	43.32760	20.76281	1506

2.2. Macroinvertebrate Sampling and Processing

Samples were collected using the kick and sweep sampling method from all microhabitat types according to European standards [23] with an FBA hand net (25 × 25 cm, mesh size 500 µm). According to Tubić et al. [24] the kick and sweep sampling method is more effective compared to quantitative Surber net sampling in terms of general taxa richness and taxa richness within the main components of the benthic communities in the water body type of small- to medium-sized streams with a predominantly coarse bottom substrate.

The biological material was pooled and transferred to sample containers (250 mL) and preserved with 70% ethanol. The benthic macroinvertebrates were identified based on their morphological characteristics using stereomicroscope Zeiss Stemi 2000C (×50) (Carl Zeiss Microscopy, LLC, White Plains, NY, USA) and Nikon SMZ 800N (×75) (Nikon Instruments Inc., Melville, NY, USA), Zeiss Axio Lab. A1 (×630) (Carl Zeiss Microscopy, LLC, White Plains, NY, USA) at the lowest possible taxonomic level using appropriate identification keys [25–28].

2.3. Biological Metrics

The following biological metrics were used for the analysis and comparison: EPT index, total number of taxa per sample (No. of taxa), Saprobic Index (SI) [29], Biological Monitoring Working Party (BMWP score) [30], Diversity Index/Shannon-Wiener Index (H′) [31], number of sensitive taxa (No. of sensitive taxa), percentage participation of subfamily Tubificinae (Oligochaeta) (Tubificinae %). Average Score Per Taxon (ASPT) is a slightly modified version of BMWP score. It is calculated by dividing the BMWP values by the sum of the relative abundances of the present families. As it is based on the values of the BMWP score, it was not taken into account.

In accordance with the national legislation and the established class boundaries for the metrics, the relevant parameters were used to assess the ecological status of each river type based on macroinvertebrate metrics [22]. This is represented with the ecological quality classes (EQC). The ecological status assessed as high corresponds to class I, good corresponds to class II, moderate corresponds to class III, poor corresponds to class IV and bad corresponds to class V. The EQC obtained for selected types of watercourses were taken into account in order to compare them with the values of the biological indices used for the assessment of water quality according to the national legislation.

2.4. Data Analyses

All metrics calculations, based on macroinvertebrate taxa lists, were performed using the ASTERICS 4.04 software package [32]. This software is commonly used in similar studies [24,33] as a tool for assessing ecological quality in European streams with benthic macroinvertebrates.

The obtained metric values were tested for normality by the Kolmogorov–Smirnov test. Since the variables lacked normality of distribution, nonparametric tests were applied. Spearman's nonparametric correlation test ($p < 0.05$) was used to assess the relationship between the EQC and EPT index, No. of taxa, SI, BMWP score, H′ and No. of sensitive taxa and Tubificinae % for three types of watercourses. The relationship between EQC and biological indices was analyzed using principal component analysis (PCA). PCA was performed for all ecological indices with respect to EQC, aiming to understand the ordination. IBM SPSS Statistics for Windows Software (Version 22.0; IBM Corp, Armonk, NY, USA) was used for the data processing.

3. Results

The ecological characteristics of the community recorded at the 44 studied sites correspond to the communities typical of hilly and mountainous watercourses. Insects were the most dominant group in the community. Mayflies, stoneflies and caddisflies were recorded at all sampling sites as an important component of the community. Together they represented 28 families and 95 taxa (8/30 mayflies, 6/18 stoneflies, 14/47 caddisflies families/taxa). Species *Taeniopteryx nebulosa* (Linnaeus, 1758), *Baetis (Baetis) pavidus* Grandi, 1949 and *Epeorus (Ironopsis) yougoslavicus* (Samal, 1935) classified as strictly protected according the national legislation were recorded) [34]. The number of recorded representatives of the EPT group and their presence or absence in the three types of analyzed water bodies are shown in Table 2 below.

Table 2. List of recorded EPT taxa in three types of watercourses [21].

No.	Order	Family	Taxon	Type 3	Type 4	Type 6
1	Ephemeroptera	Baetidae	*Baetis (Acentrella)* sp.	x		x
2			*Baetis (Baetis) alpinus* (Pictet, 1843)	x	x	x
3			*Baetis (Baetis) fuscatus* (Linnaeus, 1761)	x	x	x
4			*Baetis (Nigrobaetis) muticus* (Linnaeus, 1758)	x	x	x
5			*Baetis (Baetis) lutheri* Müller-Liebenau, 1967	x		x
6			*Baetis (Baetis) meridionalis* Ikonomov, 1954	x		
7			*Baetis (Rhodobaetis) rhodani* (Pictet, 1843)	x	x	x
8			*Baetis (Baetis) pavidus* Grandi, 1949 *		x	x
9			*Baetis (Baetis) scambus* Eaton, 1870			x
10			*Baetis (Baetis) vernus* Curtis, 1834	x		
11			*Baetis* sp.	x	x	x
12			*Cloeon (Cloeon) dipterum* (Linnaeus, 1761)	x	x	
13		Ephemeridae	*Ephemera (Ephemera) danica* Müller, 1764	x	x	x
14		Heptageniidae	*Ecdyonurus (Helvetoraeticus) subalpinus* Klapalek, 1907		x	
15			*Ecdyonurus (Ecdyonurus) aurantiacus* (Burmeister, 1839)		x	
16			*Ecdyonurus (Helvetoraeticus) helveticus* Eaton, 1883.		x	
17			*Epeorus (Epeorus) sylvicola* (A.E Pictet, 1865)	x	x	x
18			*Epeorus (Ironopsis) yougoslavicus* (Samal, 1935) *	x	x	x
19			*Ecdyonurus (Ecdyonurus) venosus* (Fabricius, 1775)	x	x	x
20			*Ecdyonurus* sp.			
21			*Heptagenia (Heptagenia) sulphurea* (Müller, 1776)	x	x	x
22			*Rhithrogena gr. semicolorata* (Curtis, 1834)	x	x	x
23		Caenidae	*Caenis macrura* Stephens, 1835	x		
24			*Caenis luctuosa* (Burmeister, 1839)	x	x	x
25		Potamanthidae	*Potamanthus luteus* (Linnaeus, 1767)	x		
26		Oligoneuriidae	*Oligoneuriella rhenana* (Imhoff, 1852)	x	x	x
27		Leptophlebiidae	*Habrophlebia fusca* (Curtis, 1834)		x	x
28			*Paraleptophlebia submarginata* (Stephens, 1835)	x	x	x
29		Ephemerellidae	*Ephemerella ignita* (Poda, 1761)	x	x	x
30			*Torleya major* (Klapalek, 1905)	x	x	x
31	Plecoptera	Chloroperlidae	*Chloroperla* sp.			x
32			*Siphonoperla torrentium* (Pictet, 1841)			x

Table 2. *Cont.*

No.	Order	Family	Taxon	Type 3	Type 4	Type 6
33		Leuctridae	*Leuctra gr. hippopus* Kempny, 1899	x	x	x
34			*Leuctra fusca* (Linnaeus, 1758)		x	
35		Nemouridae	*Amphinemura sulcicollis* (Stephens, 1836)		x	x
36			*Nemoura* sp.	x	x	x
37			*Protonemura montana* Kimmins, 1941		x	x
38			*Protonemura praecox* (Morton, 1894)		x	
39			*Protonemura* sp.			x
40		Perlidae	*Perla marginata* (Panzer, 1799)	x	x	x
41			*Dinocras megacephala* (Klapálek, 1907)	x	x	x
42			*Dinocras* sp.		x	
43		*Perlodidae*	*Isoperla gramatica* (Poda, 1761)	*x*	*x*	*x*
44			*Isoperla obcura* (Zetterstedt, 1840			x
45			*Isogenus nubecula* Newman, 1833			x
46			*Perlodes microcephalus* (Pictet, 1833)			x
47		Taeniopterygidae	*Taeniopteryx nebulosa* (Linnaeus, 1758) *	x	x	x
48			*Rhabdiopteryx acuminata* Klapálek, 1905.		x	x
49	Trichoptera	Rhyacophilidae	*Rhyacophila pubescens* Pictet 1834			x
50			*Rhyacophila fasciata* Hagen, 1859	x		x
51			*Rhyacophila dorsalis* (Curtis, 1834)	x		x
52			*Rhyacophila tristis* Pictet, 1834	x	x	x
53			*Rhyacophila torrentium* Pictet, 1834			x
54			*Rhyacophila obliterata* McLachlan, 1863	x		x
55			*Rhyacophila vulgaris* Pictet, 1834.		x	
56			*Rhyacophila* sp.	x		x
57		Glossosomatidae	*Glossosoma* sp.		x	x
58		Hydroptilidae	*Hydroptila occulta* (Eaton, 1873)	x		
59		Philopotamidae	*Philopotamus montanus* (Donovan, 1813)		x	
60		Hydropsychidae	*Hydropsyche incognita* Pitsch, 1993	x		
61			*Hydropsyche instabilis* (Curtis, 1834)	x		
62			*Hydropsyche fulvipes* (Curtis, 1834)	x		x
63			*Hydropsyche pellucidula* (Curtis, 1834)	x	x	x
64			*Hydropsyche tabacarui* Botosaneanu, 1960	x		
65			*Hydropsyche* sp.	x	x	x
66			*Cheumatopsyche lepida* (Pictet, 1834)	x	x	
67		Polycentropodidae	*Plectrocnemia conspersa* (Curtis, 1834)	x		x
68			*Polycentropus flavomaculatus* (Pictet, 1834)		x	
69			*Cyrnus trimaculatus* (Curtis, 1834)		x	x
70		Psychomydae	*Psychomyia pusilla* (Fabricius, 1781)	x	x	
71			*Tinodes* sp.			x
72		Brachycentridae	*Brachycentrus montanus* Klapalek, 1892.	x	x	x
73			*Brachycentrus subnubilis* Curtis, 1834	x		
74			*Micrasema morosum* (McLachlan, 1868)			x
75			*Micrasema setiferum* (Pictet, 1834)		x	
76		Limnephilidae	*Ecclisopteryx madida* (McLachlan, 1867)		x	x
77			*Halesus digitatus* (Schrank, 1781)			x
78			*Micropterna lateralis* (Stephens, 1837)			x
79			*Potamophylax cingulatus* (Stephens, 1837)	x	x	x
80			*Limnephilus auricula* Curtis, 1834	x	x	
81			*Limnephilus sparsus* Curtis, 1834			x
82			*Melampophylax melampus* (McLachlan, 1876)		x	x
83			*Anabolia furcata* Brauer, 1857.	x		
84			*Allogamus uncatus* (Brauer, 1857)			x
85			*Chaetopteryx villosa* (Fabricius, 1798)			x
86			*Glyphotaelius pellucidus* (Retzius, 1783)		x	
87		Goeridae	*Goera pilosa* (Fabricius, 1775)		x	x
88			*Silo pallipes* (Fabricius, 1781)		x	
89			*Silo nigricornis* (Pictet, 1834)		x	

Table 2. *Cont.*

No.	Order	Family	Taxon	Type 3	Type 4	Type 6
90		Lepidostomatidae	*Lepidostoma hirtum* (Fabricius, 1775)	x		
91		Leptoceridae	*Athripsodes aterrimus* Stephens, 1836	x		
92			*Athripsodes cinereus* (Curtis, 1834)	x		
93			*Adicella* sp.		x	
94		Sericostomatidae	*Sericostoma personatum* (Kirby and Spence, 1826)	x	x	x
95		Odontoceridae	*Odontocerum albicorne* (Scopoli, 1763)		x	x

Notes: x—taxa recorded at certain types of watercourses; *—strictly protected species according to Official Gazette [34].

In type 3 watercourses, 194 taxa of benthic macroinvertebrates were identified. Insects were the principal components of the macroinvertebrate communities with 142 taxa. The EPT groups (53 taxa) were one of the main components of macroinvertebrate communities. A significant number of recorded taxa belongs to the orders Ephemeroptera (23.8% of the total community, 23 taxa), Plecoptera (1.03%, 6 taxa) and Trichoptera (7.21%, 24 taxa) (Figure 2).

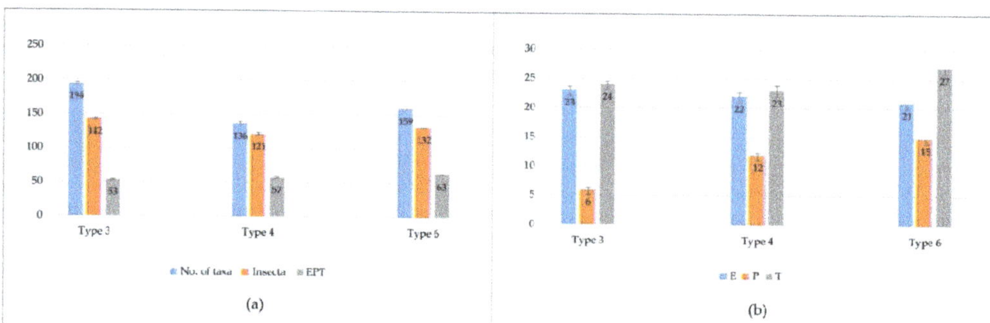

Figure 2. (**a**) Total number of taxa, number of Insecta taxa, and number of EPT taxa at three types of watercourses; (**b**) Number of taxa of EPT orders, Ephemeroptera (E), Plecoptera (P), Trichoptera (T) at three types of watercourses.

In type 4 watercourses, a total of 136 taxa were identified in the macroinvertebrate community at the analyzed sites. In terms of taxa richness and number of individuals, insects were the dominant group in the community (78.16%, 121 taxa). Of the identified insects, almost half of the community (57 identified taxa) belonged to the target group (EPT) organisms in terms of taxa richness and number of individuals with 89.64%/22 taxa, 7.21%/12 taxa and 3.14%/19 taxa (Figure 2).

Of the 14 sites analyzed that belonged to a type 6 watercourse, 159 benthic macroinvertebrate taxa were identified. In terms of the number of taxa, the most diverse group was the Insecta (132 taxa). The EPT groups represented an important component of the community at the sites surveyed and were represented by a total of 63 taxa. Regarding the percentage of these three taxonomic groups, Ephemeroptera were the most represented with 22.12% of the total macroinvertebrate community recorded, while Plecoptera and Trichoptera were represented with 7.61% and 5.4% of the total number of taxa recorded, respectively. Ephemeroptera with 21 taxa and Plecoptera with 15 taxa recorded, had a high diversity in the total community. Trichoptera (27 taxa) had the highest number of taxa in the total community (Figure 2).

Differences in the distribution of target groups of insects in analyzed types of watercourses were evident. The structure of the Ephemeroptera community in respect to three different types of watercourses is shown in Figure 3. The family Baetidae was present with the largest number of species (12) and the species *Baetis rhodani* (Pictet, 1843) was the representative of the Ephemeroptera group with a significant abundance in all three types of studied water bodies. The species *Ecdyonurus (Helvetoraeticus) subalpinus* Klapalek, 1907 was only found in type 4, but it was the most abundant species of the Ephemeroptera group, along with *Baetis (Baetis) alpinus* (Pictet, 1843).

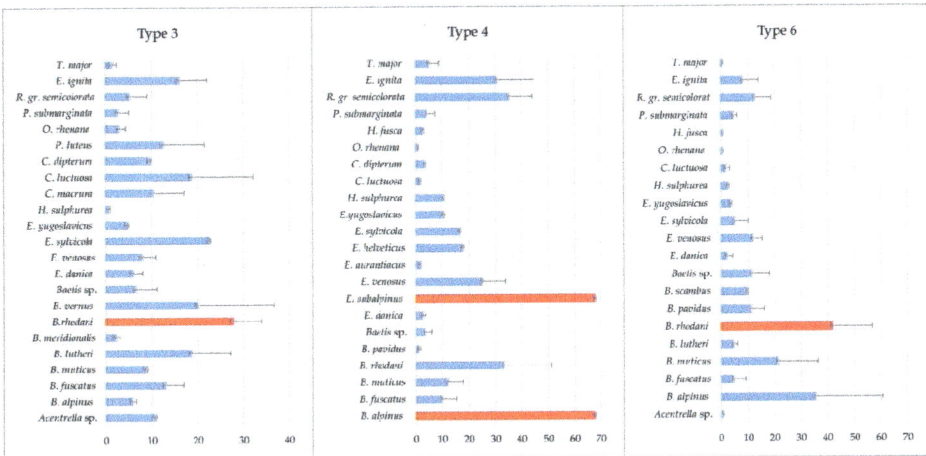

Figure 3. Average abundance of Ephemeroptera taxa at three types of watercourses (individuals/m^2). The most abundant species are indicated by red bars.

The species *Leuctra gr. hippopus* Kempny, 1899, was the representative of the Plecoptera group, and was recorded with a significant abundance in all three types of studied water bodies, with type 3 showing the highest values. In water types 4 and 6, species *Protonemura montana* Kimmins, 1941, and *Siphonoperla torrentium* (Pictet, 1841), which favor watercourses at higher altitudes, with a domination of larger substrates, were the most numerous among the Plecoptera. (Figure 4).

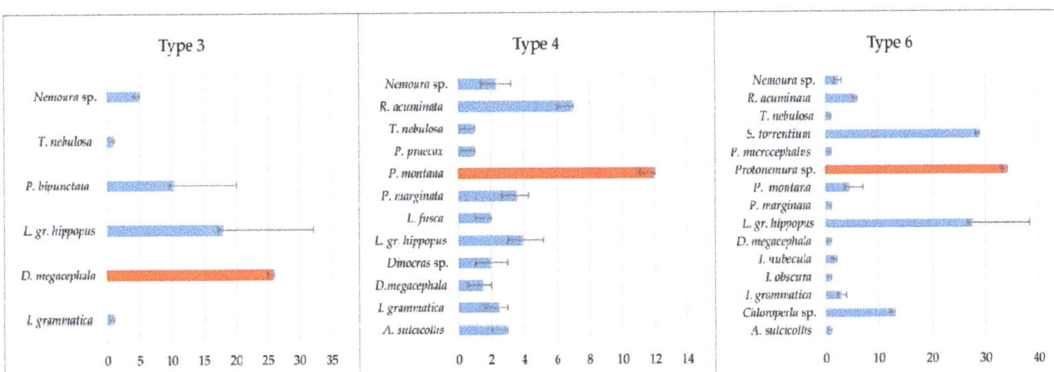

Figure 4. Average abundance of Plecoptera taxa at three types of watercourses (individuals/m^2). The most abundant species are indicated by red bars.

With regard to the participation of the Trichoptera group at the investigated localities, the dominant family was the Limnephilidae with 11 species recorded. Depending on the type of water body, different species were found. *Psychomyia pusilla* (Fabricius, 1781) in type 3, *Micrasema setiferum* (Pictet, 1834) in type 4 and *Chaetopteryx villosa* (Fabricius, 1798) in type 6 were the most abundant species (Figure 5).

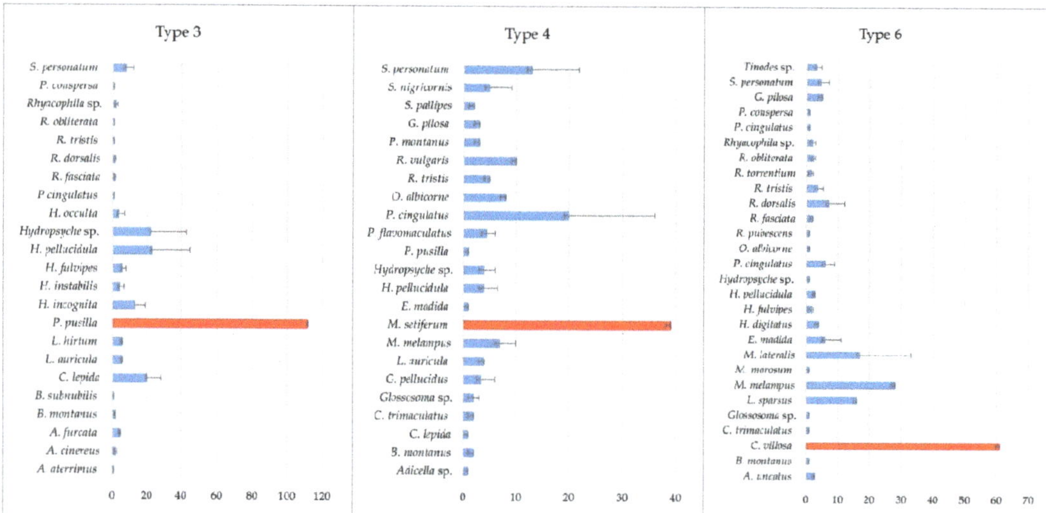

Figure 5. Average abundance of Trichoptera taxa at three types of watercourses (individuals/m^2). The most abundant species are indicated by red bars.

Indicative status assessment was carried out according to the procedure based on the class boundaries and in accordance with current legislation [22]. Analyzed localities were classified in the ecological quality classes from I (high ecological status) to V (bad ecological status).

In the type 3 watercourses, most sites have a poor or bad ecological status, mainly due to the low values of EPT index. Good ecological status was achieved on five sites, while moderate ecological status was recorded on six sites (Table 3).

Table 3. Values of biological indices and assessed ecological quality class for each type 3 locality according to the national legislative of the Republic of Serbia [22].

Metrics/Localities	1	2	3	4	5	6	7	8	9	10	11	12
No. of taxa	28 I	35 I	26 I	12 III	17 II	19 II	40 I	16 II	29 I	17 II	11 III	6 IV
SI	2.389 III	1.769 II	1.99 II	2.196 II	2.589 III	1.919 II	1.759 II	2.433 III	1.818 II	2.256 III	1.947 II	2.179 II
BMWP Score	93 I	88 II	76 II	49 IV	30 IV	75 II	136 I	41 IV	113 I	42 IV	50 III	20 V
ASPT	5.812 II	5.867 II	6.333 II	6.125 II	4.286 III	6.25 II	6.182 II	4.1 III	6.278 II	5.25 II	5.556 II	5 II
H′	2.211 I	1.819 II	1.791 II	1.441 III	2.369 I	1.829 II	2.002 II	1.046 IV	1.059 IV	2.428 I	2.159 II	0.703 IV
EPT index	10 III	11 III	10 III	3 V	2 V	9 III	17 I	1 V	12 II	4 IV	5 IV	2 V
No. of Families	19 I	20 I	15 I	11 II	11 II	13 I	32 I	13 I	21 I	9 III	9 III	4 IV
Tubificinae %	2.78 *	0.00 *	0.34 *	2.13 *	0.00 *	0.00 *	0.14 *	1.65 *	0.00 *	4.27 *	0.00 *	0.00 *
EQC	III	III	III	IV	V	III	II	V	IV	IV	IV	V
Metrics/Localities	13	14	15	16	17	18	19	20	21	22	23	24

Table 3. *Cont.*

Metrics/Localities	1	2	3	4	5	6	7	8	9	10	11	12
No. of taxa	15 II	16 II	51 I	19 II	32 I	30 I	28 I	19 II	26 I	5 IV	9 IV	20 I
SI	1.949 II	1.97 II	2.198 II	2.078 II	1.096 I	1.939 II	1.98 II	2.213 III	1.956 II	2.98 IV	2.705 III	2.493 III
BMWP Score	32 IV	36 IV	128 I	33 IV	121 I	138 I	87 II	82 II	122 I	21 V	11 V	31 IV
ASPT	4 III	6 II	5.333 II	4.714 III	6.722 II	6.9 II	6.214 II	6.308 II	7.625 I	5.25 II	2.75 V	3.875 IV
H′	2.46 I	2.473 I	3.103 I	2.328 I	2.627 I	2.943 I	2.506 I	1.74 II	2.42 I	1.427 III	1.665 II	2.634 I
EPT index	3 V	3 V	11 III	6 IV	20 I	13 II	14 II	8 III	15 II	2 V	1 V	3 V
No. of Families	9 III	7 III	29 I	13 I	22 I	22 I	19 I	17 I	21 I	5 III	6 III	11 II
Tubificinae %	2.13 *	0.00 *	1.47 *	5.11 /	0.00 *	0.00 *	3.35 *	0.41 *	0.00 *	44.44 /	23.91 /	5.88 /
EQC	V	V	III	IV	II	II	II	III	II	V	V	V

Notes: * Good status; / Good was achieved.

In type 4 watercourses, good ecological status was achieved at three analyzed sites. One site was characterized by poor ecological status, while moderate ecological status was recorded at the other two sites (Table 4).

Table 4. Values of biological indices and assessed ecological quality class for each type 4 locality according to the national legislative of the Republic of Serbia [22].

Metrics/Localities	1	2	3	4	5	6
No. of Taxa	39 I	22 I	45 I	44 I	26 I	26 I
SI	1.409 I	1.701 II	2.177 III	1.413 I	1.396 I	2.019 II
BMWP Score	151 I	106 I	130 I	184 I	118 I	70 II
ASPT	7.55 I	7.067 I	6.842 II	6.815 II	6.556 II	7.778 I
H′	2.127 II	1.402 III	2.715 I	2.697 I	1.861 II	2.585 I
No. of sensitive taxa	13 I	8 I	13 I	13 I	6 I	3 III
EPT index	23 I	17 II	19 I	27 I	15 II	9 IV
Tubificinae %	0.00 *	0.00 *	2.59 *	0.00 *	0.00 *	0.00 *
EQC	II	III	III	II	II	IV

Note: * Good status.

Based on the values obtained, it can be concluded that, overall, good ecological status was achieved in the studied type 6 watercourses, based on most of the parameters analyzed. A moderate ecological status was found at four localities (Table 5).

Results of the Spearman correlation test ($p < 0.05$) showed a correlation of EQC with all biological indices (Table 6). It showed negative correlation with BMWP score, H′, No. of taxa, EPT index and No. of sensitive taxa, while positive correlation was observed for the values of SI and Tubificinae %. Results showed a strong negative correlation of EQC with

EPT index, BMWP score and No. of sensitive taxa, while other obtained correlations are of medium strength.

Table 5. Values of biological indices and assessed ecological quality class for each type 6 locality according to the national legislative of the Republic of Serbia [22].

Metrics/Localities	1	2	3	4	5	6	7	8	9	10	11	12	13	14
No. of Taxa	34 I	26 I	19 I	28 I	27 I	23 I	26 I	38 I	48 I	20 I	7 I	4 II	16 I	32 I
SI	1.391 I	1.784 II	2.084 III	1.898 II	1.717 II	2.05 III	1.686 II	1.444 I	1.657 II	1.257 I	1.967 II	1.686 II	2.116 III	1.828 II
EPT index	16 *	13 *	8 *	12 *	12 *	7 *	11 *	23 *	21 *	9 *	5 *	2 *	4 *	13 *
No. of sensitive taxa	11 *	5 *	3 *	7 *	9 *	2 *	8 *	8 *	10 *	6 *	3 *	1 /	2 *	4 *
Tubificidnae %	0.00 *	0.00 *	0.00 *	0.00 *	0.00 *	0.17 *	2.27 *	0.00 *	0.00 *	0.00 *	0.00 *	0.00 *	4.57 *	0.00 *
EQC	II	II	III	II	II	III	II	II	II	II	II	III	III	II

Notes: * Good status; / Good status was achieved.

Table 6. Statistically significant values of Spearman correlation coefficient among values of EQC and SI, BMWP score, H′, No. of taxa, Tubificinae %, EPT index, and No. of sensitive taxa in analyzed watercourses.

Metrics	EQC
SI	0.6818
BMWP score	−0.7356
H′	−0.5239
No. of taxa Tubificinae %	−0.6879 0.4379
EPT Index	−0.8165
No. sensitive taxa	−0.7625

Two principal components were extracted from the biplot and accounted for 77.14% of the total variation in the dataset (Figure 6). The first principle component (PC1) accounted for 60.51% of the variability, with EPT index, BMWP score and No. of sensitive taxa as the parameters that contributed most to the separation (Figure 6, Table 7). The second PC (PC2) explained 16.63% of the total variance, with SI, No. of taxa and Tubificinae % as the parameters that contributed most to the separation (Figure 6, Table 7).

Table 7. Loadings of the variables on the principal components (PC). The parameters that contributed most to the separation are marked in bold.

Metrics	PC 1	PC 2	PC 3	PC 4	PC 5	PC 6	PC 7
SI	0.12873	**0.22732**	0.00423	0.11183	0.49766	0.01228	0.01795
BMWP score	**0.21024**	0.00821	0.00007	0.06461	0.12799	0.14680	0.44208
H′	0.09873	0.07702	0.53114	0.19859	0.09095	0.00101	0.00257
No. of taxa	0.09158	**0.13495**	0.46399	0.23881	0.06641	0.00108	0.00317
Tubificinae %	0.05256	**0.52954**	0.00047	0.29940	0.10380	0.00847	0.00575
EPT Index	**0.21460**	0.01177	0.00003	0.07411	0.00810	0.16843	0.52297
No. of sensitive taxa	**0.20356**	0.01118	0.00008	0.01264	0.10510	0.66194	0.00551

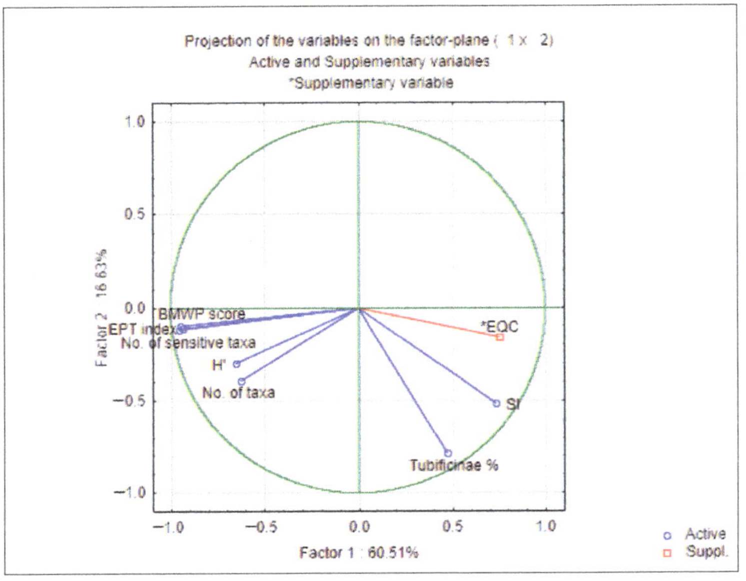

Figure 6. Principal Component Analysis (PCA)—projection of the relative contribution of the biological indices used in the assessment of ecological status and obtained EQC.

4. Discussion

The ecological characteristics of the recorded community at the analyzed sites correspond to the communities usually found in the hilly and mountainous rivers [21,22,35]. The EPT larvae are generally prevalent in the upper reaches of rivers and the assessment of water quality based on these three insect orders is sufficiently accurate [36,37]. According to Pastuchova [38], the composition and distribution of the three orders is determined by their physiological tolerance to a wide range of environmental variables.

The largest number of taxa was found in the Trichoptera group in all three types of analyzed water bodies. Individual taxa within the Trichoptera group, which are numerous, are typical representatives of a particular water body type, in which they occur in large numbers. Trichoptera are a very important component of aquatic ecosystems. Many Trichoptera species are sensitive to pollution, so their presence and relative abundance are used for biological assessment and monitoring of water quality [39].

They are followed by the Ephemeroptera group, which has slightly fewer taxa at the analyzed sites, but more abundant occurrence. A similar structure of the Ephemeroptera community can be observed in the 4 and 6 watercourse types. It is one of the most abundant groups of aquatic macroinvertebrates in all types of freshwater habitats, but its higher species diversity is characteristic of lotic habitats, especially the upper reaches of fast-flowing streams and rivers [40,41].

It can be seen that the Plecoptera community is more diverse and abundant in sites with a higher altitude; it is a significant component of running water ecosystems. Most members from this order are known to be intolerant to variation in their environmental conditions [42,43]. The stonefly sensitivity to variation in abiotic factors may lead to the extinction of taxa. In this regard, the Plecoptera is one of the most endangered groups of aquatic insects [44].

Species *Taeniopteryx nebulosa* (Linnaeus, 1758), *Baetis (Baetis) pavidus* Grandi, 1949 and *Epeorus (Ironopsis) yougoslavicus* (Samal, 1935), classified as strictly protected according to the national legislation, were important from a conservational point of view [34]. These stenovalent species are considered endangered in the lotic habitats of Serbia due to the small number of populations and the relatively distant and isolated biotopes, especially regarding the parameters of temperature, oxygen and water velocity.

The biological metrics analyzed in our research (No. of taxa, SI, BMWP score, H′ index, No. of sensitive taxa, Tubificinae %) are known to be some of the most important indicators of water quality [33,45,46]. All these indices are a measure of the occurrence of taxa that are considered bioindicators of water quality. They vary considerably with the specific type of stressor, which means that they describe changing environmental conditions very well [47].

Shannon–Wiener H′ index, measuring the diversity of species in a community, which takes into account both abundance and evenness, is used to characterize species diversity. Values may vary directly with water quality, and low diversity may indicate an unstable community [31]. Results showed a medium strength correlation for H′ index with EQC (Table 6).

On the other hand, results showed a strong negative correlation of EQC with BMWP score and EPT index. The BMWP is the sum of individual values of all families present in the sample, multiplied by their relative abundance, which are actually indicators of the sensitivity of the taxa to organic pollution. This study confirms the importance of taking into account the family level in ecological assessments and biomonitoring programs development. Indices based on the genus and species taxonomic levels, such as the EPT index among others, are needed to improve the understanding of responses on the family level and the detection of specific pollution [48]. As the EPT group is included in the calculation of the BMWP score, since they have a low tolerance to pollution, many families from these three insect groups are assigned high scores. Therefore, it was expected that the values of the BMWP score would not deviate much from the values of the EPT index [49].

Furthermore, the results showed a strong negative correlation of EQC with the number of sensitive taxa. The number of sensitive taxa is based on the concept of the presence or absence of indicator taxa at the sampling site. Most indicator taxa belong to the EPT group, as they tend to be very sensitive to different forms of pollution [8]. The list was included as part of the Fauna Aquatica Austriaca in 2004 [50]. Taxa that are considered sensitive have a narrow range of environmental requirements (e.g., stenotopic, stenoeceous) and react intolerantly to environmental disturbances. The inventory contains taxa with a wide range of sensitivity to physical, chemical and hydromorphological degradation.

The PCA also confirms previous results, given that biological indices that contributed the most to the separation on the PC1 axis were the EPT index, BMWP score and No. of sensitive taxa (Figure 6, Table 7).

On the other hand, the values of the metrics SI and the percentage participation of the subfamily Tubificinae (Oligochaeta) in the macroinvertebrate community, whose higher values reflect a higher degree of pollution, were accompanied by a decrease in the values of the EPT index at most sampling sites. The values of the SI and Tubificinae indices showed a statistically positive correlation with the values of the EQC (Table 6). The results of PCA analysis are in accordance with those previously mentioned, given that the parameters that contributed the most to the separation on the PC2 axis were No. of taxa and Tubificinae % (Figure 6, Table 7).

The Saprobic Index (SI) is one of the most traditional biological metrics commonly used to assess water quality. It focuses on the tolerance of species to organic pollution and is measured by a combination of the biological oxidation demand of a water sample and the presence of certain indicator organisms in the habitat. High values in the SI indicate a high level of organic pollution and a moderate to poor ecological status [29].

The oligochaete group comprises a large number of species that cover a broad spectrum of pollution sensitivity. Oligochaetes are usually the most dominant taxa in fine/sandy freshwater sediments. Analyzing the percentage of Tubificinae gives an indication of sediment contamination. The presence of the subfamily Tubificinae (Oligochaeta) indicates poor water quality of these sites, but also the presence of a habitat suitable for these organisms (silt, clay mud and sand) [45].

Pollution by organic matter and nutrients, as well as hydromorphological degradation, as the main factors affecting aquatic ecosystems in Serbia are most pronounced in the lowland regions. Population density, agricultural activities and industry in the country are mainly located in the lowlands [51,52]. As the benthic community changes under the various pressure, the abundance of populations of sensitive taxa decreases and the abundance of tolerant species increases [53]. Percentage participation of subfamily Tubificinae and values of SI describe a change in environmental conditions in this type of watercourses very well [45,54].

However, sites where higher values of BMWP score and parameters based on the number of taxa (EPT index and H′) were expected, indicated poor to bad conditions as well, especially at altitudes under 500 m (Tables 3 and 5). Therefore, the overall status of these sites was categorized as poor or bad, taking into account the poorer value of the individual metrics used to assess the overall water quality class [2]. On most investigated sampling sites, the value of the EPT index determined the overall EQC. Results confirmed a strong negative correlation of EQC with the EPT index.

The macroinvertebrate fauna of aquatic ecosystems has changed as a result of various pressures such as organic and chemical pollution, land use and hydromorphological alterations, as well as biological invasions. There is a growing need for cost and time efficient methods that can provide rapid results and assess a wide range of water quality statuses [55,56]. This has led to the use of rapid bioassay protocols [9] and the selection of appropriate biological indices that effectively and adequately reflect the state of the aquatic ecosystem.

The EPT metrics are also included in multimeric indices that have a broader applications for assessing the condition of streams, such as the Benthic Macroinvertebrate Index of Biotic Integrity (B-IBI) [57] and the Invertebrate Community Index (ICI) [58]. Richness metrics, including that of EPT have a direct connection to biodiversity studies conducted in a wide variety of flowing waters, which is why this index is suitable for water assessment outside the country of Serbia as well.

This study has shown that water quality assessments based on the EPT index are more reliable in less polluted watercourses, especially in pristine environments. The EPT index is a relatively accurate and effective tool compared to other indices we have used to detect water disturbance and classify water quality.

5. Conclusions

The representatives of the orders Ephemeroptera, Plecoptera and Trichoptera were recorded at all sampling sites, making more than half of the macroinvertebrate community at investigated hilly and mountainous ecosystems in Serbian waters. The results confirmed the sensitivity of the EPT index, represented by the sum of the taxa richness of these three orders, to changes in the macroinvertebrate communities in these types of water ecosystems. Higher values of the EPT index are found in water bodies with no or low pollution, while lower values of this index indicate an increase in pollution. In addition, these aquatic insects respond to a wide range of potential pollutants and respond to both short-term and long-term conditions that affect water quality.

This study showed a strong negative correlation between the EQC and EPT index, BMWP score and a number of sensitive taxa. Also, results of the PCA analysis are in accordance with the obtained results. As the EPT group is included in the calculation of these two metrics, results showed that the values of the BMWP score and number of sensitive taxa would not deviate much from the values of the EPT index. It can be concluded that the EPT index values effectively and adequately confirm changes in water quality in hilly and mountainous small- to medium-sized streams with predominantly hard bottom substrates. This index reflects the state of the aquatic ecosystem and provides an accurate overall picture of water quality. It proved to be self-sufficient and reliable for water quality assessment in these types of water bodies. In addition, it proved to be an important tool for the prioritization of measures and the revision of the Water Management

Plan for the territory of the Republic of Serbia. Furthermore, this index can also serve as a useful tool for early detection of pollution.

The EPT index should not be used to assess watercourses known to have low EPT taxa richness, such as lowland rivers at altitudes under 500 m, especially large rivers with fine substrate (silt, clay mud and sand) or slow flowing/stagnant water bodies (artificial channels and reservoirs—heavily modified water bodies) where the pollution-tolerant groups are more significant. Moreover, this index is not suitable for the assessment of urban watercourses which are under higher anthropogenic pressure.

Having the above in mind, the overall status assessment is a complex matter. There is still a need for more intensive studies and further testing of the effectiveness of various indices used to assess water quality of different types of watercourses.

Author Contributions: Conceptualization, B.T. and M.P.; methodology, B.T. and M.P.; validation, M.P.; formal analysis, B.T., S.A., K.J., K.Z., J.Č.A. and B.V.; investigation, B.T., K.J., K.Z., B.V., J.Č.A. and S.A.; data curation, B.T. and M.P.; writing—original draft preparation, B.T., M.P. and S.A.; writing—review and editing, B.T. and M.P.; supervision, M.P. All authors have read and agreed to the published version of the manuscript.

Funding: This research was funded by the Ministry of Science, Technological Development and Innovation of the Republic of Serbia; contract No. 451-03-66/2024-03/200007.

Data Availability Statement: Data are contained within the article.

Conflicts of Interest: The authors declare no conflicts of interest.

References

1. Vitecek, S.; Johnson, R.K.; Poikane, S. Assessing the Ecological Status of European Rivers and Lakes Using Benthic Invertebrate Communities: A Practical Catalogue of Metrics and Methods. *Water* **2021**, *13*, 346. [CrossRef]
2. Directive 2000/60/EC of the European Parliament and Council of 23 October 2000 Establishing a Framework for Community Action in the Field of Water Policy. *Off. J. Eur. Commun.* **2000**, *L327/1*, 1–72.
3. Allan, J.D.; Castillo, M.M.; Capps, K.A. *Stream Ecology*; Springer International Publishing: Cham, Switzerland, 2021; ISBN 978-3-030-61285-6.
4. Birk, S.; van Kouwen, L.; Willby, N. Harmonising the Bioassessment of Large Rivers in the Absence of Near-Natural Reference Conditions—A Case Study of the Danube River. *Freshw. Biol.* **2012**, *57*, 1716–1732. [CrossRef]
5. Birk, S.; Bonne, W.; Borja, A.; Brucet, S.; Courrat, A.; Poikane, S.; Solimini, A.; Van De Bund, W.; Zampoukas, N.; Hering, D. Three Hundred Ways to Assess Europe's Surface Waters: An Almost Complete Overview of Biological Methods to Implement the Water Framework Directive. *Ecol. Indic.* **2012**, *18*, 31–41. [CrossRef]
6. Ter Braak, C.J.F.; Prentice, I.C. A Theory of Gradient Analysis. In *Advances in Ecological Research*; Begon, M., Fitter, A.H., Macfadyen, A., Eds.; Academic Press: Cambridge, MA, USA, 1988; Volume 18, pp. 271–317. ISBN 0065-2504.
7. Dahl, J.; Johnson, R.K.; Sandin, L. Detection of Organic Pollution of Streams in Southern Sweden Using Benthic Macroinvertebrates. In *Integrated Assessment of Running Waters in Europe*; Springer: Berlin/Heidelberg, Germany, 2004; pp. 161–172.
8. Rosenberg, D.M.; Resh, V.H. *Freshwater Biomonitoring and Benthic Macroinvertebrates*; Rosenberg, D.M., Resh, V.H., Eds.; Chapman & Hall: New York, NY, USA, 1993; ISBN 0412022516.
9. Barbour, M.T.; Gerritsen, J.; Snyder, B.D.; Stribling, J.B. *Rapid Bioassessment Protocols for Use in Streams and Wadeable Rivers: Periphyton, Benthic Macroinvertebrates and Fish*; California State Water Resources Control Board: Sacramento, CA, USA, 1999.
10. Barbour, M.T.; Graves, C.G.; Plafkin, J.L.; Wisseman, R.W.; Bradley, B.P. Evaluation of EPA's Rapid Bioassessment Benthic Metrics: Metric Redundancy and Variability among Reference Stream Sites. *Environ. Toxicol. Chem.* **1992**, *11*, 437–449. [CrossRef]
11. Couceiro, S.R.M.; Hamada, N.; Forsberg, B.R.; Pimentel, T.P.; Luz, S.L.B. A Macroinvertebrate Multimetric Index to Evaluate the Biological Condition of Streams in the Central Amazon Region of Brazil. *Ecol. Indic.* **2012**, *18*, 118–125. [CrossRef]
12. Edegbene, A.O.; Elakhame, L.A.; Arimoro, F.O.; Osimen, E.C.; Odume, O.N. Development of Macroinvertebrate Multimetric Index for Ecological Evaluation of a River in North Central Nigeria. *Environ. Monit. Assess.* **2019**, *191*, 1–18. [CrossRef] [PubMed]
13. Lenat, D.R. Water Quality Assessment of Streams Using a Qualitative Collection Method for Benthic Macroinvertebrates. *J. N. Am. Benthol. Soc.* **1988**, *7*, 222–233. [CrossRef]
14. Gaufin, A.R.; Tarzwell, C.M. Aquatic Invertebrates as Indicators of Stream Pollution. *Public Health Rep.* **1952**, *67*, 57–64. [CrossRef]
15. Wallace, J.B.; Grubaugh, J.W.; Whiles, M.R. Biotic Indices and Stream Ecosystem Processes: Results from an Experimental Study. *Ecol. Appl.* **1996**, *6*, 140–151. [CrossRef]
16. Herman, M.R.; Nejadhashemi, A.P. A Review of Macroinvertebrate-and Fish-Based Stream Health Indices. *Ecohydrol. Hydrobiol.* **2015**, *15*, 53–67. [CrossRef]

17. Illies, J.; Andrassy, I.; Berzins, B.V.A.; Botosaneanu, L.; Brinck, P.; Brinkhurst, R.O.; Buchholz, K.F.; Dahl, R.; Dahm, A.G.; Fittkau, E.-J. *Limnofauna Europea. Eine Zusammenstellung Aller Die Europäischen Binnengewässer Bewohnenden Mehrzelligen Tierarten Mit Angaben Über Ihre Verbreitung Und Ökologie*; G. Fischer: Stuttgart, Germany, 1967.
18. Petrović, A.; Milošević, D.; Paunović, M.; Simić, S.; Đorđević, N.; Stojković, M.; Simić, V. New Data on the Distribution and Ecology of the Mayfly Larvae (Insecta: Ephemeroptera) of Serbia (Central Part of the Balkan Peninsula). *Turk. J. Zool.* **2015**, *39*, 195–209. [CrossRef]
19. Paunović, M.; Tubić, B.; Kračun, M.; Marković, V.; Simić, V.; Zorić, K.; Atanacković, A. Ecoregions Delineation for the Territory of Serbia. *Water Res. Manag.* **2012**, *2*, 65–74.
20. Hering, D.; Moog, O.; Sandin, L.; Verdonschot, P.F.M. Overview and Application of the AQEM Assessment System. *Hydrobiologia* **2004**, *516*, 1–20. [CrossRef]
21. Regulation on Establishment of Surface and Groundwater Bodies. Official Gazette of the Republic of Serbia. No. 96/2010. 2010. Available online: https://www.pravno-informacioni-sistem.rs/SlGlasnikPortal/eli/rep/sgrs/ministarstva/pravilnik/2010/96/2 (accessed on 11 March 2024).
22. Regulative on Parameters of Ecological and Chemical Status of Surface Waters and Parameters of Chemical and Quantitative Status of Ground Waters. Official Gazette of the Republic of Serbia. No. 67/2011. 2011. Available online: https://www.pravno-informacioni-sistem.rs/SlGlasnikPortal/eli/rep/sgrs/ministarstva/pravilnik/2011/74/4/reg (accessed on 11 March 2023).
23. *EN 27828:1994*; Water Quality—Methods of Biological Sampling—Guidance on Handnet Sampling of Aquatic Benthic Macro-Invertebrates (ISO 7828:1985). European Committee for Standardisation: Brussels, Belgium, 1994.
24. Tubić, B.P.; Popović, N.Z.; Raković, M.J.; Petrović, A.S.; Simić, V.M.; Paunović, M.M. Comparison of the Effectiveness of Kick and Sweep Hand Net and Surber Net Sampling Techniques Used for Collecting Aquatic Macroinvertebrate Samples. *Arch. Biol. Sci.* **2017**, *69*, 233–238. [CrossRef]
25. Lechthaler, W.; Stockinger, W. *Trichoptera: Key to Larvae from Central Europe*; Eutaxa: Vienna, Austria, 2005; ISBN 3950183914.
26. Zwick, P. Key to the West Palaearctic Genera of Stoneflies (Plecoptera) in the Larval Stage. *Limnologica* **2004**, *34*, 315–348. [CrossRef]
27. Bauernfeind, E.; Lechthaler, W. *Ephemeroptera–Key to Larvae from Central Europe*; Eutaxa: Vienna, Austria, 2014; Volume 1.
28. Sivec, I.; Stark, B.P. The Species OfPerla (Plecoptera: Perlidae): Evidence from Egg Morphology. *Scopolia* **2002**, *49*, 1–33.
29. Zelinka, M. Zur Prazisierung Der Biologischen Klassifikation Der Reinheid Fliessender Gewasser. *Arch. Hydrobiol.* **1961**, *57*, 389–407.
30. Armitage, P.D.; Moss, D.; Wright, J.F.; Furse, M.T. The Performance of a New Biological Water Quality Score System Based on Macroinvertebrates over a Wide Range of Unpolluted Running-Water Sites. *Water Res.* **1983**, *17*, 333–347. [CrossRef]
31. Shannon, C.E. A Mathematical Theory of Communication. *Bell Syst. Tech. J.* **1948**, *27*, 379–423. [CrossRef]
32. AQEM. *Manual for the Application of the AQEM System. A Comprehensive Method to Assess European Streams Using Benthic Macroinvertebrates, Developed for the Purpose of the Water Framework Directive, Version 1*; The AQEM Consortium: Duisburg-Essen, Germany, 2002.
33. Čanak Atlagić, J.; Anđus, S.; Đuknić, J.; Milivojević, A.; Kulić, K.; Pokrajac, J.; Tubić, B.; Paunović, M. Assessment of the Impact of Copper Mining and Related Industrial Activities on Aquatic Macroinvertebrate Communities of the Pek River (Serbia). *Arch. Biol. Sci.* **2021**, *73*, 291–301. [CrossRef]
34. The Regulation on Proclamation of Strictly Protected and Protected Wild Species of Plants, Animals and Fungi; Official Gazette of the Republic of Serbia. No. 98/2016. 2016. Available online: https://www.pravno-informacioni-sistem.rs/SlGlasnikPortal/eli/rep/sgrs/ministarstva/pravilnik/2010/5/3/reg (accessed on 11 March 2023).
35. Bylak, A.; Kukuła, K. Impact of Fine-Grained Sediment on Mountain Stream Macroinvertebrate Communities: Forestry Activities and Beaver-Induced Sediment Management. *Sci. Total Environ.* **2022**, *832*, 155079. [CrossRef] [PubMed]
36. Abdul Hamid, S.; Md Rawi, C.S. Influence of Substrate Embeddedness and Canopy Cover on the Distribution of Ephemeroptera, Plecoptera and Trichoptera (EPT) in Tropical Rivers. *Aquat. Insects* **2011**, *33*, 281–292. [CrossRef]
37. Ab Hamid, S.; Fakhrullah Salafusalam Mohd Affendi, M.; Aida Hashim, N.; Author, C. Role of Ephemeroptera, Plecoptera and Trichoptera (Insecta) Functional Feeding Groups in Leaf Decomposition in Tropical River. *J. Biodivers. Environ. Sci.* **2016**, *9*, 204–213.
38. Pastuchová, Z. Macroinvertebrate Assemblages in Conditions of Low-Discharge Streams of the Cerová Vrchovina Highland in Slovakia. *Limnologica* **2006**, *36*, 241–250. [CrossRef]
39. Holzenthal, R.W.; Blahnik, R.J.; Prather, A.L.; Kjer, K.M. Order Trichoptera Kirby, 1813 (Insecta), Caddisflies. *Zootaxa* **2007**, *1668*, 639–698. [CrossRef]
40. Salles, F.F.; Da-Silva, E.R.; Hubbard, M.D.; Serrão, J.E. As Espécies de Ephemeroptera (Insecta) Registradas Para o Brasil. *Biota Neotrop.* **2004**, *4*, 1–34. [CrossRef]
41. Williams, D.D. Applied Aspects of Mayfly Biology. In *Advances in Ephemeroptera Biology*; Flannagan, J.F., Marshall, K.E., Eds.; Springer: Boston, MA, USA; Plenum Press: New York, NY, USA, 1980; pp. 1–17.
42. Elliott, J.M.; Hellawell, J.M. Biological Indicators of Freshwater Pollution and Environmental Management. *J. Anim. Ecol.* **1987**, *56*, 1084. [CrossRef]
43. Lazaridou-Dimitriadou, M. Seasonal Variation of the Water Quality of Rivers and Streams of Eastern Mediterranean. *Web Ecol.* **2002**, *3*, 20–32. [CrossRef]

44. Fochetti, R.; De Figueroa, J.M.T. Notes on Diversity and Conservation of the European Fauna of Plecoptera (Insecta). *J. Nat. Hist.* **2006**, *40*, 2361–2369. [CrossRef]
45. Atanacković, A.D.; Zorić, K.S.; Tomović, J.M.; Vasiljević, B.; Žica, M.; Paunović, M.M. Distributional Patterns of Aquatic Oligochaeta Communities (Annelida: Clitellata) in Running Waters in Serbia. *Arch. Biol. Sci.* **2020**, *72*, 359–372. [CrossRef]
46. Ofenböck, T.; Moog, O.; Gerritsen, J.; Barbour, M. A Stressor Specific Multimetric Approach for Monitoring Running Waters in Austria Using Benthic Macro-Invertebrates. In *Integrated Assessment of Running Waters in Europe*; Springer: Berlin, Germany, 2004; pp. 251–268.
47. Spyra, A.; Kubicka, J.; Strzelec, M. The Use of Biological Indices for the Assessment of the River Quality (Ruda River, Poland). *Ecol. Chem. Eng. S* **2017**, *24*, 285–298. [CrossRef]
48. Tampo, L.; Kaboré, I.; Alhassan, E.H.; Ouéda, A.; Bawa, L.M.; Djaneye-Boundjou, G. Benthic Macroinvertebrates as Ecological Indicators: Their Sensitivity to the Water Quality and Human Disturbances in a Tropical River. *Front. Water* **2021**, *3*, 662765. [CrossRef]
49. Bylak, A.; Kukuła, K.; Ortyl, B.; Hałoń, E.; Demczyk, A.; Janora-Hołyszko, K.; Maternia, J.; Szczurowski, Ł.; Ziobro, J. Small Stream Catchments in a Developing City Context: The Importance of Land Cover Changes on the Ecological Status of Streams and the Possibilities for Providing Ecosystem Services. *Sci. Total Environ.* **2022**, *815*, 151974. [CrossRef] [PubMed]
50. Moog, O.; Graf, W.; Janecek, B.F.U.; Ofenböck, T. Inventory of Sensitive Taxa of Austrian Rivers and Streams. In *Fauna Aquatica Austriaca, Katalog zur Autökologischen Einstufung Aquatischer Organismen Österreichs*; Teil, V., Moog, O., Eds.; Wasserwirtschaftskataster, Bundesministerium für Land- und Forstwirtschaft, Umwelt & Wasserwirtschaft: Wien, Germany, 2003; pp. 1–15.
51. Liška, I.; Wagner, F.; Sengl, M.; Deutsch, K.; Slobodník, J. *Joint Danube Survey 3: A Comprehensive Analysis of Danube Water Quality*; Final Scientific Report; International Commission for the Protection of the Danube River: Vienna, Austria, 2015; Volume 223.
52. Liška, I.; Wagner, F.; Sengl, M.; Deutsch, K.; Slobodník, J.; Paunovic, M. *Joint Danube Survey 4 Scientific Report: A Shared Analysis of the Danube River*; International Commission for the Protection of the Danube River: Vienna, Austria, 2021.
53. Kupilas, B.; Friberg, N.; McKie, B.G.; Jochmann, M.A.; Lorenz, A.W.; Hering, D. River Restoration and the Trophic Structure of Benthic Invertebrate Communities across 16 European Restoration Projects. *Hydrobiologia* **2016**, *769*, 105–120. [CrossRef]
54. Atanacković, A.; Popović, N.; Marinković, N.; Tomović, J.; Đuknić, J.; Stanković, J.; Paunović, M. Effects of Environmental Factors on the Distribution and Diversity of Aquatic Oligochaetes. *Water* **2023**, *15*, 3873. [CrossRef]
55. Tockner, K.; Uehlinger, U.; Robinson, C.T. *Rivers of Europe*; Academic Press: Cambridge, MA, USA, 2009; ISBN 0080919081.
56. Rahmani, H.; Shokri, M.; Janikhalili, K.; Abdoli, A.; Cozzoli, F.; Basset, A. Relationships among Biotic, Abiotic Parameters and Ecological Status in Shahid Rajaee Reservoir (Iran). *Biologia* **2022**, *77*, 3159–3172. [CrossRef]
57. Kerans, B.L.; Karr, J.R. A Benthic Index of Biotic Integrity (B-IBI) for Rivers of the Tennessee Valley. *Ecol. Appl.* **1994**, *4*, 768–785. [CrossRef]
58. DeShon, J.E. Development and Application of the Invertebrate Community Index (ICI). In *Biological Assessment and Criteria—Tools for Water Resource Planning and Decision Making*; Davis, W.S., Simon, T.P., Eds.; Lewis Publication: Boca Raton, FL, USA, 1995; pp. 217–243.

water

MDPI

Article

Microplastics in the Danube River and Its Main Tributaries—Ingestion by Freshwater Macroinvertebrates

Jelena Stanković [1,*], Djuradj Milošević [2], Momir Paunović [1], Boris Jovanović [3], Nataša Popović [1], Jelena Tomović [1], Ana Atanacković [1], Katarina Radulović [4], Davor Lončarević [4] and Maja Raković [1]

1 Department of Hydroecology and Water Protection, Institute for Biological Research "Siniša Stanković", National Institute of Republic of Serbia, University of Belgrade, 11108 Belgrade, Serbia; mpaunovi@ibiss.bg.ac.rs (M.P.); natasa.popovic@ibiss.bg.ac.rs (N.P.); jelena.tomovic@ibiss.bg.ac.rs (J.T.); adjordjevic@ibiss.bg.ac.rs (A.A.); rakovic.maja@ibiss.bg.ac.rs (M.R.)

2 Department of Biology and Ecology, Faculty of Sciences and Mathematics, University of Niš, 18000 Niš, Serbia; djuradj.milosevic@pmf.edu.rs

3 Department of Ecology, Evolution and Organismal Biology, Iowa State University, Ames, IA 50011-2103, USA; prcko@iaestate.edu

4 Institute of Chemistry, Technology and Metallurgy, National Institute of Republic of Serbia, University of Belgrade, 11000 Belgrade, Serbia; kacar@nanosys.ihtm.bg.ac.rs (K.R.); dloncarevic@nanosys.ihtm.bg.ac.rs (D.L.)

* Correspondence: jelena.stankovic@ibiss.bg.ac.rs

Abstract: This study was carried out at the Danube River and its tributaries during the Joint Danube Survey 4 (JDS4) expedition. Three freshwater benthic species were used to estimate the quantity of microplastics (MPs): *Corbicula* spp., *Limnodrilus hoffmeisteri* (Claparede, 1862), and *Polypedilum nubeculosum* (Meigen, 1804). Following the kick and sweep technique, individuals were sampled using a hand net or dredge. In order to estimate the number of MP particles/individual particles/g wet body mass, the body mass and total length of all specimens were measured. Alkaline (*Corbicula* spp. and *L. hoffmaisteri*) and enzymatic (*P. nubeculosum*) protocols were performed for tissue degradation. All samples were filtered through glass microfiber filters (mesh size 0.5 μm). The particles were photographed, measured, and counted. A total of 1904, 169, and 204 MPs were isolated from *Corbicula* spp., *L. hoffmaisteri*, and *P. nubeculosum*, respectively. To confirm the chemical composition of isolated MPs, a subsample of 46 particles of the fragmented particles from 14 sampling sites was analysed via μ-ATR-FTIR spectroscopy analysis. The particles were characterised as polycarbonate (PC), polyethylene terephthalate (PET), polypropylene–polyethylene copolymer (PP-PE), nylon (polyamide-PA) and cellophane, with the domination of PET.

Keywords: microplastic; pollution; Danube River; macroinvertebrates; μFTIR spectroscopy

1. Introduction

Plastic is a synthetic organic polymer which originates from natural derivatives obtained mainly from crude oil, natural gas, or coal. Particles within the size range from 1 μm to 5 mm are referred to as microplastics (MPs) [1]. MPs are created by mechanical erosion, solar radiation, and the biodegradation of larger plastics [2] or are manufactured in extremely small sizes for special products (toothpaste, creams, and cleaning products) that can end up in freshwaters [3]. The global mishandling of synthetic organic polymer waste and low recycling rates have led to a significant increase in plastic pollution. Its ubiquitous presence and non-degradable characteristics are directly related to its persistence in the environment, on a scale range of tens to hundreds of years [4]. A rough estimate predicts that 80% of plastic litter in marine ecosystems is land-based, with rivers serving as its primary pathways [5]. On a global scale, the transport of plastic debris from rivers to the sea is estimated at 1.2 to 2.4 million tons every year [6]. Given the substantial amount of plastics transferring through estuarine systems and the relatively limited reports on microplastic pollution in this ecosystem compared to the marine environment, priority research on MPs abundance, distribution, characteristics, and ecological impacts on estuaries is warranted.

Since the 1950s, 5 billion metric tons (MTs) of plastics have accumulated in the environment [7]. It is estimated that by the end of the century, between 2.5×10^7 and 1.3×10^8 metric tons will be in the ocean [8]. Although the number of ecotoxicological microplastic studies has increased significantly in the last decade [9], 77% of the publications reported results on marine organisms, while only 23% have been based on freshwater research [10]. The research focus of MPs has been directed toward effects on the freshwater ecosystems. The presence and occurrence of MPs were documented in urban rivers [11–13] and lakes [14–17]. Klein et al. [18] analyzed the shoreline sediments of the Rhine-Main area in Germany and Zbyszewski and Corcoran [19] in Lake Huron, Canada, to name just a few. In addition to its ubiquitous presence, MPs have a vector role, as a transport medium for invasive species [20], harmful algal bloom (HAB) [21], or opportunistic pathogens [22,23].

In the present study, the occurrence of MPs in benthic macroinvertebrates was investigated in the Danube River and its tributaries. The Danube River Basin (DRB) flows across 19 countries, occupying an area of 801,463 km^2 with a population of over 80 million inhabitants in its proximity. The DRB covers nine ecoregions, and it is classified as a special case study from the aspects of conservation and management issues [24]. As it passes through different countries, the Danube accumulates MPs, and the number of MPs in the Danube vastly outnumbers fish larvae [12]. MPs in freshwater tend to sink at a much faster rate than in marine environments since freshwater has less density. Biofouling is further increasing the mass of MPs and is aiding in their deposition and accumulation in the sediment. The filtering activities of filter feeders likely have a very significant role in MP circulation by removing MPs from the water column and depositing them in sediments through incorporation into feces and pseudofeces. Previous studies on MPs and biota within freshwater ecosystems have shown that MPs can be ingested by organisms from sediments and that the ability of macroinvertebrates to ingest MP depends on their feeding habits [25]. MPs accumulated in aquatic organisms are transferred via the food chain, and they directly affect the entire aquatic ecosystem [26]. Macroinvertebrates as widespread organisms could be suitable as bioindicators for assessing MP pollution within the different ecological niches, such as the water column and/or freshwater sediment. The reports of MPs in freshwater environments have mainly focused on the biota at higher levels in the food chain, such as fish [27,28]. Several field data have highlighted MP ingestion by freshwater macroinvertebrates [25,29–32]. More recently, Bertoli et al. [33] focused on the influence of feeding guilds and habits of macroinvertebrates on MP ingestion. Few studies have focused on the ingestion of MPs with respect to freshwater insects [34,35] and the role of freshwater benthic macroinvertebrates in MP transfers from aquatic to terrestrial ecosystems [36].

The Asian clam (*Corbicula* spp.) inhabits a wide range of freshwater habitats across the world, including the Danube River [37]. As benthic filter feeders with intensive activities, bivalves accumulate a considerable amount of MP particles from the environment, which is why they have been extensively used in MP studies lately [38]. Bivalves have a longer life cycle compared to other macroinvertebrates, with low mobility, which is suitable for MP studies, especially because they indicate the state in the microhabitat. Bearing in mind that their habitat is the entire Danube in high abundance [39] and other rivers, the Asian clam was selected as the target organism for this MP study. On the other hand, some aquatic oligochaetes are considered suitable for bioaccumulation studies and are included in the standard guidelines because they are easy to culture, have a high biomass yield, are tolerant to various physico–chemical properties of sediments, and are exposed to pollutants via pore water and ingested sediment [40]. *L. hoffmeisteri* (Naididae: Tubificinae) is tolerant to organic pollution and is the dominant species in almost all oligochaete assemblages along the Danube [41], together with *P. nubeculosum*, which are considered suitable bioindicators for assessing the effects of different pollutants on freshwater biota [42]. As non-specific feeders, chironomids can ingest MPs instead of food particles [25,43].

The main aim of the present study was to quantify MPs ingested by *Corbicula* spp., *Limnodrilus hoffmeisteri*, and *Polypedilum nubeculosum* from the Danube and its tributaries. The study investigated the ingestion of MPs under real environmental conditions in different freshwater macroinvertebrates, taking into account particle size, shape and polymer type, and whether it is ingested/detected at different sites along the main course of the Danube and its tributaries.

2. Materials and Methods

2.1. Sampling Sites and Procedure

The study was conducted on the Danube River and its main tributaries. The macroinvertebrates from 23 sampling sites were analyzed, of which 15 were on the Danube, and 8 were on the tributaries (Hron, Tisza, Sava, Velika Morava, Iskar, and Jantra) (Figure 1; Table S1 in the Supplementary Materials).

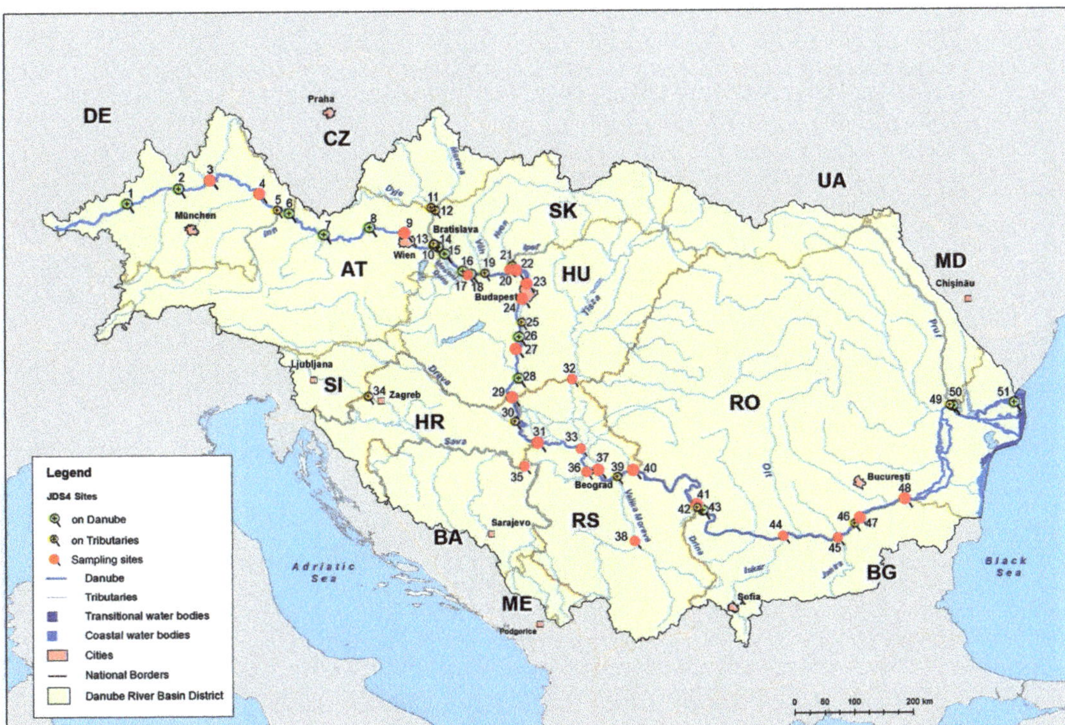

Figure 1. Area of investigation with sampling sites along the Danube River and its tributaries (adapted from https://www.danubesurvey.org/jds4/ (accessed on 2 February 2024).

Samples were collected in summer 2019 by the national Joint Danube Survey 4 (JDS4) teams. Following the multi-habitat procedure [44], individuals were sampled by the kick and sweep (K&S) sampling technique according to European Standards [45] using a hand net (ap. 25 cm × 25 cm, mesh size 500 μm). The iron-forked mouth of the triangle-shaped dredge with a collection net (mesh size 500 μm) was used for the deep water area. The dredge was pulled five times per sampling site. Each transect was considered as a separate sample. The sampling methodology is described in [46].

The samples of freshwater macroinvertebrates were counted, separated, and identified in the laboratory relative to the lowest possible taxonomic level using the following identification keys: Moller Pillot [47,48], Vallenduuk and Moller Pillot [49], Pfleger [50], and Timm [51].

2.2. Preparation of the Samples for MP Isolation

To eliminate MP contamination during the work, deionized water was filtered through 0.5 μm pore size, 47 mm GF/B glass microfibres (Whatman, Kent, United Kingdom). In order to assess potential post-sampling airborne contamination with MPs during the experimental procedure, the digestion processes without tissues ("blank") in 3 replicates per species have been checked for MP contamination.

2.3. Isolation of MPs

In total, 216 specimens of *Corbicula* spp., 130 specimens of *L. hoffmeisteri*, and 79 specimens of *P. nubeculosum* were used for MP isolation. The body mass and total length were measured for all specimens, with the addition of the total mass and total width of the shell of *Corbicula* spp. The tissue of each specimen was rinsed with pre-filtered deionized water and placed into glass beakers. Although numerous approaches have been developed for the extraction of MPs, the alkaline protocol has mostly been used for soft tissues (e.g., *Corbicula* spp. and *L. hoffmaisteri*), while only enzymatic protocols are effective for the digestion of chitinous organisms (e.g., *P. nubeculosum*). All organisms were digested in a pool of 10 specimens or less, depending on the number of individuals collected at the sampling sites.

The samples were processed using the alkaline protocol [52]—treatment with 10% (*w*/*w*) potassium hydroxide (KOH)—and incubated for 12 h at 65 °C in the water bath with a rotation speed of 80 rpm. For the enzymatic protocol, proteinase K was used as per Cole et al. [53].

The solution was filtered through 0.5 μm mesh-size glass microfiber filters. The samples were stored in a rinsed sterile glass Petri dish. The filtrated material was treated with 30% hydrogen peroxide, if needed, in order to remove the remaining organic matter. The minimum size of examined particles was 16 μm in length.

The occurrence of MPs in the organisms is expressed as item/organism and item/g wet weight (ww), where the items are divided based on fibrils and fragments as separate categories.

2.4. Particle Analysis

MPs of a size range from 16 μm to 5 mm were processed. The particles were classified into two main categories: fragments and fibres. All MPs were counted visually and photographed using a Leica MZ16A stereomicroscope (Leica microsystems, Wetzlar, Germany)(10×/21 B ocular and from 20× to 50× objective magnification) with a Leica DFC320 Digital Camera system (Leica microsystems, Wetzlar, Germany), and MPs were measured using calibrated scales in the program ImageJ (version 1.54) [54].

2.5. Micro-Fourier Transform Infrared Spectroscopy (μFTIR)

The fragment particles identified to be the most common in the samples along the Danube River and its tributaries were further chemically characterized by μFTIR. Out of 23 localities, 14 were selected for infrared measurements. Fibres were excluded from the μFTIR analysis due to technical challenges. Due to the high cost of μFTIR analyses on a large sample size, at least the 3 largest fragment particles per sample were randomly selected from the most diverse samples. In total, 46 MP fragments were analysed (see Table S2 in the Supplementary Materials). Infrared measurements were performed with individual manual readings of the particles using a Nicolet iN10 Fourier transform infrared microscope with a micro-attenuated total reflection (ATR) accessory and liquid nitrogen-cooled MCT detector in the ATR mode and carrying out 128 scans at a resolution of 4 cm^{-1}.

The μFTIR method for the identification of MPs provides confident chemical composition information [55].

The OMNIC Pictra Software (version 8.1.0.19) was used to identify the samples, comparing their spectra with the spectra from the Hummel-Polymer Sample Library.

2.6. Data Analysis

The morphometric parameters of the individuals and particles per organism and per g^{-1} body mass were statistically described with an average value and standard deviation (SD). The relation between morphometry-based metrics (length, height, total mass, and body mass) and the number of isolated MPs was obtained using a non-parametric Spearman's rank correlation test.

3. Results

MPs were detected in all samples of *Corbicula* spp. (1904 particles), *L. hoffmeisteri* (169 particles) and *P. nubeculosum* (204 particles). On average, the following were detected per sampling site: from 2.7 to 19.5 fibres/individual and 1.2 to 9.2 fragments/individual in *Corbicula* spp.; from 0.4 to 1.6 fibres/individual and 0.2 to 1.5 fragments/individual in *L. hoffmeisteri*; and from 0.5 to 2.4 fibres/individual and 0.5 to 2.2 fragments/individual in *P. nubeculosum*. In "blank" samples, on average, six fibres were identified, indicating airborne contamination. Therefore, fibres were excluded from μFTIR analyses. Fragments, films, or hard MPs were never present in blank samples. Selected organisms differed according to morphometric parameters (Table 1). The correlation between morphometric parameters and a number of isolated MPs was not significant (Spearman's rank correlation test; $p > 0.05$).

Table 1. Average values of morphometric parameters.

	Corbicula spp.	*L. hoffmeisteri*	*P. nubeculosum*
TL ± SD	14.23 ± 3.78	8.52 ± 6.17	5.56 ± 2.07
BW ± SD	340 ± 0.21	0.76 ± 0.88	0.45 ± 0.74

Note: TL—total length (mm); BW—body weight (mg); SD—standard deviation.

The ingested particles were within the size ranges from 0.016 to 4.67 mm (Table 2). Fibres were the dominant category within *Corbicula* spp. (56.9%) and *L. hoffmeisteri* (58%), while the dominant category in *P. nubeculosum* (50.2%) comprised fragments. In *Corbicula* spp., blue-coloured fibres were dominant among fibres within all species, while transparent fragmented MPs were found to be the most abundant in *Corbicula* spp., and black-coloured particles were observed in *L. hoffmeisteri* and *P. nubeculosum*.

Table 2. Minimum and maximum length of fibres and fragments.

	Corbicula spp.		*L. hoffmeisteri*		*P. nubeculosum*	
	Min	**Max**	**Min**	**Max**	**Min**	**Max**
Fibres	0.08	4.67	0.049	4.61	0.031	4.13
Fragments	0.02	3.22	0.018	0.288	0.016	0.0241

Note: All measures are in mm.

In order to estimate the accumulation of MP particles for each species (216 specimens of *Corbicula* spp., 130 specimens of *L. hoffmeisteri*, and 79 specimens of *P. nubeculosum*), the number of isolated fibres and fragments was calculated per organism and per g^{-1} ww (Figure 2).

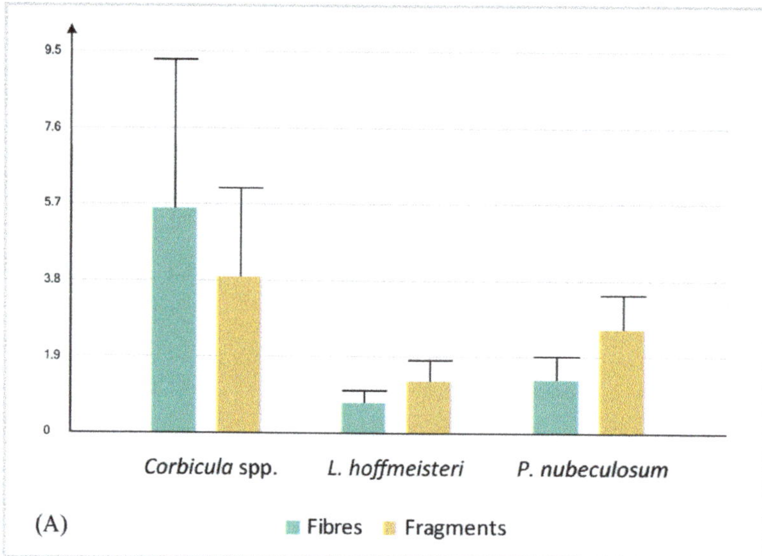

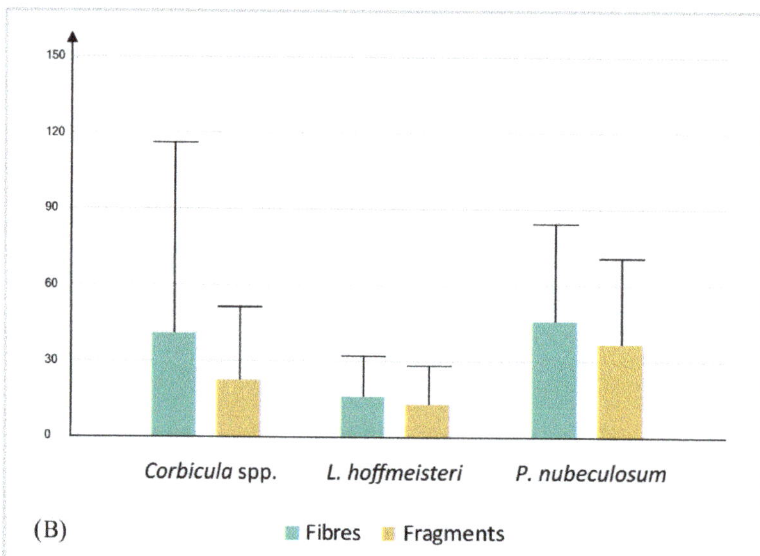

Figure 2. An average number of fibres and fragments (bars) per (**A**) organism and (**B**) g^{-1} ww for each species. Lines represent the variance in the number of particles.

The data show a higher abundance of MPs at sampling sites JDS4–3, JDS4–23, JDS4–24, JDS4–40, and JDS4–41 in *Corbicula* spp.; JDS4–37 in *L. hoffmeisteri*; and JDS4–31 in *P. nubeculosum* along the Danube. The abundance of MPs was higher at sampling sites JDS4–20, JDS4–35, JDS4–36, and JDS4–38 in *Corbicula* spp.; JDS4–38 in *L. hoffmeisteri*; and JDS4–33 and JDS4–35 in *P. nubeculosum* on tributaries (Figure 3A).

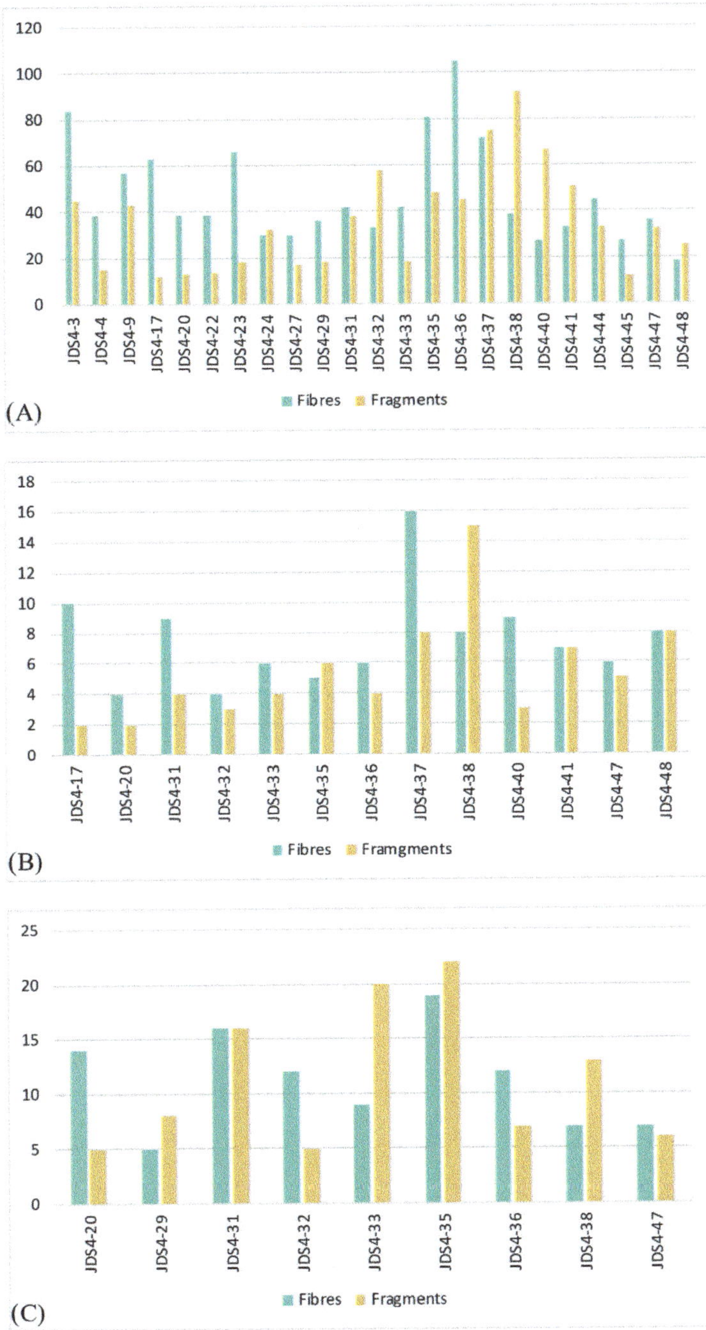

Figure 3. The quantities of fibres and fragments in (**A**) *Corbicula* spp., (**B**) *L. hoffmeisteri*, and (**C**) *P. nubeculosum* at sampling sites along the Danube and its tributaries.

In order to confirm that extracted particles were indeed MPs and to determine the polymer type of the isolated MPs, *Corbicula* spp. samples were selected for the μ–ATR–

FTIR analysis. From the total of 46 particles, 40 were confirmed as plastic polymers via μ-ATR–FTIR. Analyzed MPs were identified as polycarbonate (12/40), polypropylene–polyethylene co-polymer (3/40), nylon (Polyamide) (1/40), cellophane (2/40), and PET, which was the most dominant with 21 particles out of 40 (Figure 4; see Tables S2 and S3 in the Supplementary Materials).

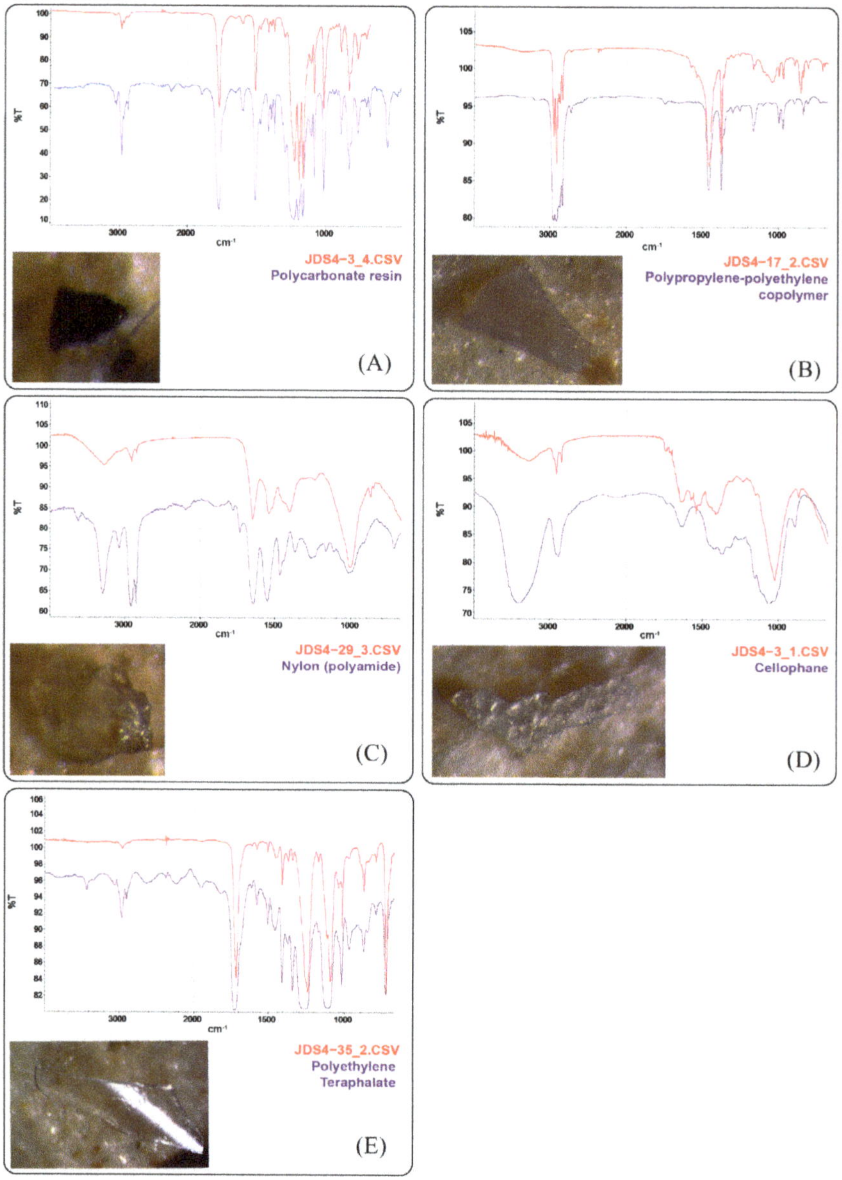

Figure 4. Results of μ–ATR–FTIR analyses of the MP sample (red line); the chemical substance standard database (blue/purple line): (**A**) polycarbonate—JDS4-3_4; (**B**) polypropylene–polyethylene copolymer—JDS4-17_2; (**C**) nylon (polyamide)—JDS4-29_3; (**D**) cellophane—JDS4-3_1; (**E**) polyethylene terephthalate—JDS4-35_2.

4. Discussion

The JDS4 MP study with respect to biota provided comparable information on the MPs in the biota along 2040 rkm of the Danube, contributing to the general knowledge of their distribution in biological systems. To the best of our knowledge, besides the study of Su et al. [56], this is the second study on MPs in benthic macroinvertebrates in large-spatial-scale rivers.

Our results confirmed that there is no correlation between the morphometric parameters of the organisms and the quantity of ingested MPs. Similar findings are documented for fish [57] and shellfish [58]. In the case of selected species, the size of the organism does not affect the quantity of ingested MPs.

Hohenblum et al. [59] reported MPs in the Danube's Austrian stretch in water samples with a concentration range of 0.039–0.205 mg/m^3 and 0.029–0.516 mg/m^3, specifically in the entry and exit points, respectively. Accordingly, the annual average range of transport of MPs is estimated from 6 to 66 kg per day in the Austrian Danube River. Isolated particles in their study are categorized as fragments (over 50%), pellets (4–10%), and green lenticular flakes (2.1–2.8%), while in this study, the category of fibres was the most dominant.

Scherer et al. [25] demonstrated the ingestion of polystyrene among different freshwater invertebrates: *Physella acuta* (Draparnaud, 1805) (Mollusca), *Lumbriculus variegatus* (Müller, 1774) (Oligochaeta), and *Chironomus riparius* (Meigen, 1804) (Diptera: Chironomidae). Hurley et al. [60] isolated particles from the tissue of *Tubifex tubifex* (Müller, 1774) from the Salford Quays basin (Manchester City, UK), where the majority are fibres (87%), while the rest of the particles are fragments. This is similar to our findings where fibres were dominant in the samples of *Corbicula* spp. (56.9%) and *L. hoffmeisteri* (58%). Lin et al. [61] detected microgranules (0–28%), microfilms (0–16%), microfragments (3–47%), and microfibres (40–64%) within the midge larvae (Diptera: Chironomidae) at five sampling sites in the Wu River basin, Taiwan, while in this study, fragments were the dominant category of particles in *P. nubeculosum* (50.2%). As the dominant category in most studies, the origin of fibres has become questionable. Some authors suggested that the majority of fibres have natural origins, as they are cotton fibres [62]. On the other hand, the extent of plastic use in the textile industry is dominant, as it is estimated to be the third largest industry for plastics worldwide [7]. Synthetic textile fibres dominate, with a share of over 60% of total global textile fibre production in 2019, with 52% being polyester fibres (58 million tons), followed by polyamides (nylon) with 5% (5.6 million tons) [63].

The negative effects of MPs on freshwater biota have been documented previously. Exposure to polystyrene MPs caused a decrease in the weight of *L. hoffmeisteri* and induced inflammatory responses and sediment-avoidance behaviour [64]. When exposed to MPs, Asian clams not only showed statistically significant fitness reduction and an increase in lipid oxidative damage, but neurotoxicity was also detected [65]. In association with polychlorinated biphenyls (PCBs), MP exposure leads to tubular dilation [66]. No data have been published for MP ingestion by *Polypedilum nubeculosum* species, but various studies have demonstrated negative effects on chironomids [43,67,68].

Su et al. [56] provided the results of MP analyses in *C. fluminea* in the Middle and Lower Yangtze River Basin. Their results show that Asian clams are a good bioindicator for describing MP pollution, especially for sediments, as in 61 out of 63 samples of Asian clams, MPs are detected. Su et al. [56] reported abundance ranges from 0.3 to 4.9 items/g wet body mass and from 0.4 to 5.0 items/individual per site. MP pollution in *C. fluminea* from Taihu Lake (China) resulted within the range of 0.2–12.5 items/g wet mass [69]. Baldwin et al. [70] have isolated 18 to 105 MP particles per individual, with a mean value of 51.7 items/organism. Our results show a higher abundance of MP particles in the *Corbicula* spp. of DRB in comparison to the mentioned studies in China, indicating the pressure caused by plastic pollution in the Danube Basin. The size range of all MPs in this study was from 0.016 to 4.67 mm, which is a very similar size range of particles ingested by Asian Clams from the middle–lower Yangtze River Basin, from 0.021 to

4.02 mm [56], and the Taihu Lake, from 0.05 to 5 mm [69]. In addition, all previous studies [56,69,70] documented the dominance of fibres ingested by *C. fluminea*.

Cellophane was dominant in *C. fluminea* from the Taihu Lake, followed by PET, polyester, and polypropylene [69]. Our study showed that PET particles (used for the production of plastic bottles) were dominant in the *Corbicula* spp. samples from the Danube, while cellophane was present with lower abundances. It should be noted that cellophane particles can be difficult to identify via infrared spectroscopy in the presence of cellulose because of their similar molecular structure [71]. For the infrared identification of cellophane, we selected larger irregular-shaped particles that are not characteristic of natural regularly ordered structures like fibres and compared them with spectra from the library. These particles achieve a spectral matching score of 60% for cellophane and 37% for cellulose. The domination of PET fragments in the samples, besides potentially high quantities in the Danube, could be due to the clam's preference for ingestion. Li et al. [52] demonstrated a higher intake of PET fibres (4.1 items/g) than five other polymer types (1 item/g or less). In addition, natural colour particles (brown and white) are identified as calcium carbonate. Thus, these subcategories should be analysed with much more scrutiny in future analysis in order to avoid the misidentification of MPs with inorganic materials.

The results are reasonable because PET is used in manufacturing plastic bottles, which could be found floating in rivers. Over time, mechanical forces, sunlight, and biological processes altogether break down macroscopic plastic bottles into smaller pieces, which became available to smaller organisms. The specific density of PET is 1.38 g cm^{-3}, which is already dense by itself and sufficient for sinking to the sediment. In addition, after the microbial or algal inhabitancy of the plastic surface, MPs became even denser and can sink into the sediments easier. Their increased presence in the downstream region of the Danube could not only be explained by local pollution but also by the floating plastic debris from upper stream regions, which are more developed. The data from this study could indicate that tributaries greatly contribute to MP loads in the Danube. Furthermore, the higher presence of MP debris on sites JDS4-37, JDS4-40 and JDS4-41 could indicate the influence of Belgrade, as well as the tributaries Sava and Velika Morava.

Lechner et al. [12] estimated an average of 7.5 g of plastic litter per 1000 m^3 s^{-1} with respect to the mean flow (4.2 t per day or 1533 t per year), transported via the Danube to the Black Sea.

Some of the bivalves are widely used in the human diet. At the time of consumption, commercially grown bivalves *Mytilus edulis* and *Crassostrea gigas* contain on average 0.36 ± 0.07 and 0.47 ± 0.16 particles g^{-1} ww (wet weight), respectively [72]. The same study concluded that 250 g ww of mussels or 100 g ww of oysters results in the ingestion of 90 or 50 particles of MPs, respectively. When estimated per year, humans ingest 11,000 MP particles just through a diet of bivalves [72]. The presence of marine MPs in seafood could potentially be a threat to food safety because of the additives in plastics, mainly endocrine disruptors phthalates and bisphenol A [73] or the adsorption of POPs, PCBs, PAHs, organo-halogenated pesticides, nonylphenol, and dioxins [74–76] on the MPs' surface.

Freshwater Asian clams are invasive species [77] and are useful bioindicators of emerging contaminants [78,79], including MPs [69]. Due to increasing synthetic pollution in aquatic environment, there is a need to include MPs in the standard procedures of water analysis in order to gather more data on this problem.

5. Conclusions

The present study is the first field study to investigate 2040 km of the Danube River, and to the best of our knowledge, the second study on MPs in macroinvertebrates in the river. MPs isolated from *Corbicula* spp. showed the presence and bioavailability of five types of polymers—cellophane, polyamide, polypropylene–polyethylene copolymer, polycarbonate, and PET, which was the most dominant polymer (58%).

The analyzed parameters (number of MPs per site, mean number of MPs per individual per site, and mean number of MPs per body mass—g/wet weight) indicated a higher MP load for tributaries, as well as an important influence of tributaries and settlements on the presence of MP debris in the Danube.

The results of the JDS4 MPs study confirmed that bivalves are suitable test organisms for the assessment of MP loads in the aquatic environment.

As benthic macroinvertebrates are an important component of food chains, as well as the basis of many services and functions of freshwater ecosystems, such as nutrient cycling and water quality, assessing the ecological risk of MPs in freshwater ecosystems is crucial for the successful implementation of strategies to ensure clean water supplies and to prevent the loss of biodiversity in freshwater. Therefore, further standardised studies providing comparable data on MPs in biota within the Danube River Basin using Asian clams are not only needed but other test organisms are also needed in order to assess the MP load and possible consequences more accurately. In addition, the uptake of particles, pathways, and quantities and their relation to particular size need to be further studied.

Supplementary Materials: The following supporting information can be downloaded at https://www.mdpi.com/article/10.3390/w16070962/s1; Table S1: List of sampling sites with codes and information for each site; Table S2. Number of *C. fluminea* specimens per sample and analysed particles using micro-ATR-FTIR spectroscopy; Table S3. Description of the analyzed particles.

Author Contributions: Conceptualization, M.P.; methodology, M.R., D.M. and J.S.; formal analysis, J.S., B.J., N.P., K.R. and D.L.; investigation, J.S., N.P., J.T. and A.A.; resources, M.P.; data curation, J.S.; writing—original draft preparation, J.S., D.M. and M.R.; writing—review and editing, D.M., M.P., B.J., N.P., J.T., A.A. and M.R.; visualization, J.S., B.J., N.P., K.R. and D.L.; supervision, D.M., M.P. and M.R.; project administration, M.P.; funding acquisition, M.P. All authors have read and agreed to the published version of the manuscript.

Funding: This research was funded by the Ministry of Science, Technological Development and Innovation of the Republic of Serbia (Contract No. 451-03-47/2023-01/200007).

Data Availability Statement: Data is contained within the article (and Supplementary Materials).

Acknowledgments: We are grateful to the ICPDR for familiarizing us with the physiognomy of the River Danube during several Joint Danube Surveys. Special thanks should be given to Igor Liška (ICPDR Secretariat), Jaroslav Slobodník (logistic officer of the four JDS missions), and Franz Wagner (chairman of the Monitoring and Assessment Working Group of ICPDR), who all provided essential help for the successful actions of the Core Team during all Danubian activities of our group.

Conflicts of Interest: The authors declare no conflicts of interest.

References

1. Hale, R.C.; Seeley, M.E.; La Guardia, M.J.; Mai, L.; Zeng, E.Y. A Global Perspective on Microplastics. *J. Geophys. Res. Ocean.* **2020**, *125*, e2018JC014719. [CrossRef]
2. Andrady, A.L. Microplastics in the Marine Environment. *Mar. Pollut. Bull.* **2011**, *62*, 1596–1605. [CrossRef] [PubMed]
3. Lambert, S.; Wagner, M. Microplastics Are Contaminants of Emerging Concern in Freshwater Environments: An Overview. In *Freshwater Microplastics: Emerging Environmental Contaminants*; Wagner, M., Lambert, S., Eds.; Springer International Publishing: Cham, Switzerland, 2018; pp. 1–23, ISBN 978-3-319-61615-5.
4. Gregory, M.R.; Andrady, A.L. Plastics in the Marine Environment. In *Plastics and the Environment*; John Wiley & Sons: Hoboken, NJ, USA, 2003; pp. 379–401, ISBN 9780471721550.
5. Bowmer, T.; Kershaw, P.J. (Eds.) GESAMP. In Proceedings of the GESAMP International Workshop on Plastic Particles as a Vector in Transporting Persistent, Bio-Accumulating and Toxic Substances in the Oceans, Paris, France, 28–30 June 2010; GESAMP Reports & Studies No. 82. pp. 1–68.
6. Lebreton, L.C.M.; van der Zwet, J.; Damsteeg, J.-W.; Slat, B.; Andrady, A.; Reisser, J. River Plastic Emissions to the World's Oceans. *Nat. Commun.* **2017**, *8*, 15611. [CrossRef] [PubMed]
7. Geyer, R.; Jambeck, J.R.; Law, K.L. Production, Use, and Fate of All Plastics Ever Made. *Sci. Adv.* **2017**, *3*, e1700782. [CrossRef] [PubMed]
8. Everaert, G.; Van Cauwenberghe, L.; De Rijcke, M.; Koelmans, A.A.; Mees, J.; Vandegehuchte, M.; Janssen, C.R. Risk Assessment of Microplastics in the Ocean: Modelling Approach and First Conclusions. *Environ. Pollut.* **2018**, *242*, 1930–1938. [CrossRef] [PubMed]
9. Sorensen, R.M.; Jovanović, B. From Nanoplastic to Microplastic: A Bibliometric Analysis on the Presence of Plastic Particles in the Environment. *Mar. Pollut. Bull.* **2021**, *163*, 111926. [CrossRef] [PubMed]

10. Katare, Y.; Singh, P.; Sankhla, M.S.; Singhal, M.; Jadhav, B.E.; Parihar, K.; Nikalje, B.T.; Trpathi, A.; Bhardwaj, L. Microplastics in Aquatic Environments: Sources, Ecotoxicity, Detection & Remediation. *Biointerface Res. Appl. Chem.* **2022**, *12*, 3407–3428.
11. Dris, R.; Gasperi, J.; Rocher, V.; Saad, M.; Renault, N.; Tassin, B. Microplastic Contamination in an Urban Area: A Case Study in Greater Paris. *Environ. Chem.* **2015**, *12*, 592–599. [CrossRef]
12. Lechner, A.; Keckeis, H.; Lumesberger-Loisl, F.; Zens, B.; Krusch, R.; Tritthart, M.; Glas, M.; Schludermann, E. The Danube so Colourful: A Potpourri of Plastic Litter Outnumbers Fish Larvae in Europe's Second Largest River. *Environ. Pollut.* **2014**, *188*, 177–181. [CrossRef]
13. Morritt, D.; Stefanoudis, P.V.; Pearce, D.; Crimmen, O.A.; Clark, P.F. Plastic in the Thames: A River Runs through It. *Mar. Pollut. Bull.* **2014**, *78*, 196–200. [CrossRef]
14. Eriksen, M.; Mason, S.; Wilson, S.; Box, C.; Zellers, A.; Edwards, W.; Farley, H.; Amato, S. Microplastic Pollution in the Surface Waters of the Laurentian Great Lakes. *Mar. Pollut. Bull.* **2013**, *77*, 177–182. [CrossRef] [PubMed]
15. Corcoran, P.L.; Norris, T.; Ceccanese, T.; Walzak, M.J.; Helm, P.A.; Marvin, C.H. Hidden Plastics of Lake Ontario, Canada and Their Potential Preservation in the Sediment Record. *Environ. Pollut.* **2015**, *204*, 17–25. [CrossRef] [PubMed]
16. Biginagwa, F.J.; Mayoma, B.S.; Shashoua, Y.; Syberg, K.; Khan, F.R. First Evidence of Microplastics in the African Great Lakes: Recovery from Lake Victoria Nile Perch and Nile Tilapia. *J. Great Lakes Res.* **2016**, *42*, 146–149. [CrossRef]
17. Imhof, H.K.; Ivleva, N.P.; Schmid, J.; Niessner, R.; Laforsch, C. Contamination of Beach Sediments of a Subalpine Lake with Microplastic Particles. *Curr. Biol.* **2013**, *23*, R867–R868. [CrossRef] [PubMed]
18. Klein, S.; Worch, E.; Knepper, T.P. Occurrence and Spatial Distribution of Microplastics in River Shore Sediments of the Rhine-Main Area in Germany. *Environ. Sci. Technol.* **2015**, *49*, 6070–6076. [CrossRef]
19. Zbyszewski, M.; Corcoran, P.L. Distribution and Degradation of Fresh Water Plastic Particles Along the Beaches of Lake Huron, Canada. *Water Air Soil Pollut.* **2011**, *220*, 365–372. [CrossRef]
20. Gregory, M.R. Environmental Implications of Plastic Debris in Marine Settings—Entanglement, Ingestion, Smothering, Hangers-on, Hitch-Hiking and Alien Invasions. *Philos. Trans. R. Soc. London. Ser. B Biol. Sci.* **2009**, *364*, 2013–2025. [CrossRef] [PubMed]
21. Masó, M.; Garcés, E.; Pagès, F.; Camp, J. Drifting Plastic Debris as a Potential Vector for Dispersing Harmful Algal Bloom (HAB) Species. *Sci. Mar.* **2003**, *67*, 107–111. [CrossRef]
22. McCormick, A.; Hoellein, T.J.; Mason, S.A.; Schluep, J.; Kelly, J.J. Microplastic Is an Abundant and Distinct Microbial Habitat in an Urban River. *Environ. Sci. Technol.* **2014**, *48*, 11863–11871. [CrossRef]
23. Zettler, E.R.; Mincer, T.J.; Amaral-Zettler, L.A. Life in the "Plastisphere": Microbial Communities on Plastic Marine Debris. *Environ. Sci. Technol.* **2013**, *47*, 7137–7146. [CrossRef]
24. Sommerwerk, N.; Hein, T.; Schneider-Jacoby, M.; Baumgartner, C.; Ostojić, A.; Siber, R.; Bloesch, J.; Paunović, M.; Tockner, K. Chapter 3—The Danube River Basin. In *Rivers of Europe*; Tockner, K., Uehlinger, U., Robinson, C.T., Eds.; Academic Press: London, UK, 2009; pp. 59–112, ISBN 978-0-12-369449-2.
25. Scherer, C.; Brennholt, N.; Reifferscheid, G.; Wagner, M. Feeding Type and Development Drive the Ingestion of Microplastics by Freshwater Invertebrates. *Sci. Rep.* **2017**, *7*, 17006. [CrossRef] [PubMed]
26. Li, X.; Bao, L.; Wei, Y.; Zhao, W.; Wang, F.; Liu, X.; Su, H.; Zhang, R. Occurrence, Bioaccumulation, and Risk Assessment of Microplastics in the Aquatic Environment: A Review. *Water* **2023**, *15*, 1768. [CrossRef]
27. Foekema, E.M.; De Gruijter, C.; Mergia, M.T.; van Franeker, J.A.; Murk, A.J.; Koelmans, A.A. Plastic in North Sea Fish. *Environ. Sci. Technol.* **2013**, *47*, 8818–8824. [CrossRef]
28. Horton, A.A.; Jürgens, M.D.; Lahive, E.; van Bodegom, P.M.; Vijver, M.G. The Influence of Exposure and Physiology on Microplastic Ingestion by the Freshwater Fish Rutilus Rutilus (Roach) in the River Thames, UK. *Environ. Pollut.* **2018**, *236*, 188–194. [CrossRef]
29. Setälä, O.; Norkko, J.; Lehtiniemi, M. Feeding Type Affects Microplastic Ingestion in a Coastal Invertebrate Community. *Mar. Pollut. Bull.* **2016**, *102*, 95–101. [CrossRef]
30. Bour, A.; Haarr, A.; Keiter, S.; Hylland, K. Environmentally Relevant Microplastic Exposure Affects Sediment-Dwelling Bivalves. *Environ. Pollut.* **2018**, *236*, 652–660. [CrossRef] [PubMed]
31. Sfriso, A.A.; Tomio, Y.; Rosso, B.; Gambaro, A.; Sfriso, A.; Corami, F.; Rastelli, E.; Corinaldesi, C.; Mistri, M.; Munari, C. Microplastic Accumulation in Benthic Invertebrates in Terra Nova Bay (Ross Sea, Antarctica). *Environ. Int.* **2020**, *137*, 105587. [CrossRef] [PubMed]
32. Pan, Z.; Liu, Q.; Jiang, R.; Li, W.; Sun, X.; Lin, H.; Jiang, S.; Huang, H. Microplastic Pollution and Ecological Risk Assessment in an Estuarine Environment: The Dongshan Bay of China. *Chemosphere* **2021**, *262*, 127876. [CrossRef]
33. Bertoli, M.; Pastorino, P.; Lesa, D.; Renzi, M.; Anselmi, S.; Prearo, M.; Pizzul, E. Microplastics Accumulation in Functional Feeding Guilds and Functional Habit Groups of Freshwater Macrobenthic Invertebrates: Novel Insights in a Riverine Ecosystem. *Sci. Total Environ.* **2022**, *804*, 150207. [CrossRef]
34. Nel, H.A.; Froneman, P.W. Presence of Microplastics in the Tube Structure of the Reef-Building Polychaete Gunnarea Gaimardi (Quatrefages 1848). *Afr. J. Mar. Sci.* **2018**, *40*, 87–89. [CrossRef]
35. Windsor, F.M.; Durance, I.; Horton, A.A.; Thompson, R.C.; Tyler, C.R.; Ormerod, S.J. A Catchment-Scale Perspective of Plastic Pollution. *Glob. Change Biol.* **2019**, *25*, 1207–1221. [CrossRef] [PubMed]

36. Yıldız, D.; Yalçın, G.; Jovanović, B.; Boukal, D.S.; Vebrová, L.; Riha, D.; Stanković, J.; Savić-Zdraković, D.; Metin, M.; Akyürek, Y.N.; et al. Effects of a Microplastic Mixture Differ across Trophic Levels and Taxa in a Freshwater Food Web: In Situ Mesocosm Experiment. *Sci. Total Environ.* **2022**, *836*, 155407. [CrossRef] [PubMed]
37. Paunović, M.; Csányi, B.; Simonović, P.; Zorić, K. Invasive Alien Species in the Danube. In *The Danube River Basin*; Liska, I., Ed.; Springer: Berlin/Heidelberg, Germany, 2015; pp. 389–409, ISBN 978-3-662-47739-7.
38. Li, J.; Green, C.; Reynolds, A.; Shi, H.; Rotchell, J.M. Microplastics in Mussels Sampled from Coastal Waters and Supermarkets in the United Kingdom. *Environ. Pollut.* **2018**, *241*, 35–44. [CrossRef] [PubMed]
39. Paunović, M.; Csányi, B.; Knežević, S.; Simić, V.; Nenadić, D.; Jakovčev-Todorović, D.; Stojanović, B.; Cakić, P. Distribution of Asian Clams Corbicula Fluminea (Müller, 1774) and C. Fluminalis (Müller, 1774) in Serbia. *Aquat. Invasions* **2007**, *2*, 99–106. [CrossRef]
40. Rodriguez, P.; Reynoldson, T.B. *The Pollution Biology of Aquatic Oligochaetes*, 1st ed.; Springer: Dordrecht, The Netherlands, 2011; pp. 225–261, ISBN 978-94-007-1717-6.
41. Atanacković, A.D.; Šporka, F.; Csányi, B.; Vasiljević, B.M.; Tomović, J.M.; Paunović, M.M. Oligochaeta of the Danube River—A Faunistical Review. *Biologia* **2013**, *68*, 269–277. [CrossRef]
42. Pinder, L.C. V Biology of Freshwater Chironomidae. *Annu. Rev. Entomol.* **1986**, *31*, 1–23. [CrossRef]
43. Stanković, J.; Milošević, D.; Savić-Zdraković, D.; Yalçın, G.; Yildiz, D.; Beklioğlu, M.; Jovanović, B. Exposure to a Microplastic Mixture Is Altering the Life Traits and Is Causing Deformities in the Non-Biting Midge Chironomus Riparius Meigen (1804). *Environ. Pollut.* **2020**, *262*, 114248. [CrossRef] [PubMed]
44. AQEM. Consortium Manual for the Application of the AQUEM System. *A Compr. Method Assess Eur. Streams Using Benthic Macroinvertebrates Dev. Purp. Water Framew. Dir.* **2002**, *1*, 202.
45. *EN 27828:1994*; Water Quality—Methods of Biological Sampling—Guidance on Handnet Sampling of Aquatic Benthic MacroInvertebrates (ISO 7828:1985). European Committee for Standardisation: Brussels, Belgium, 1994.
46. Liška, I.; Wagner, F.; Sengl, M.; Deutsch, K.; Slobodník, J. *Joint Danube Survey 3: A Comprehensive Analysis of Danube Water Quality*; International Commission for the Protection of the Danube River (ICPDR): Vienna, Austria, 2015; pp. 1–369.
47. Moller Pillot, H.K.M. De Larven Der Nederlandse Chironomiae (Diptera). (Inleiding, Tanypodinae En Chironominitle). *Ned. Faun. Meded.* **1984**, *1A*, 1–277.
48. Moller Pillot, H.K.M. De Larven Der Nederlandse Chironomiae (Diptera) (Orthocladiinae Sensu Lato). *Ned. Faun. Meded.* **1984**, *1B*, 1–175.
49. Vallenduuk, H.J.; Moller Pillot, H.K.M. *Chironomidae Larvae of the Netherlands and Adjacent Lowlands: General Ecology and Tanypodinae*; Chironomidae Larvae of the Netherlands and Adjacent Lowlands; KNNV: Zeist, The Netherlands, 2007; ISBN 9789050112598.
50. Pfleger, V. *Molluscs*; Field guide in colour to; Silverdale Books: Devon, UK, 2000; ISBN 9781856054492.
51. Timm, T. A Guide to the Freshwater Oligochaeta and Polychaeta of Northern and Central Europe. *Lauterbornia* **2009**, *66*, 1–235. Available online: https://www.researchgate.net/publication/285323193_A_guide_to_the_freshwater_Oligochaeta_and_Polychaeta_of_Northern_and_Central_Europe (accessed on 4 March 2024).
52. Li, L.; Su, L.; Cai, H.; Rochman, C.M.; Li, Q.; Kolandhasamy, P.; Peng, J.; Shi, H. The Uptake of Microfibers by Freshwater Asian Clams (Corbicula Fluminea) Varies Based upon Physicochemical Properties. *Chemosphere* **2019**, *221*, 107–114. [CrossRef] [PubMed]
53. Cole, M.; Webb, H.; Lindeque, P.K.; Fileman, E.S.; Halsband, C.; Galloway, T.S. Isolation of Microplastics in Biota-Rich Seawater Samples and Marine Organisms. *Sci. Rep.* **2014**, *4*, 4528. [CrossRef] [PubMed]
54. Schindelin, J.; Arganda-Carreras, I.; Frise, E.; Kaynig, V.; Longair, M.; Pietzsch, T.; Preibisch, S.; Rueden, C.; Saalfeld, S.; Schmid, B.; et al. Fiji: An Open-Source Platform for Biological-Image Analysis. *Nat. Methods* **2012**, *9*, 676–682. [CrossRef]
55. Hidalgo-Ruz, V.; Gutow, L.; Thompson, R.C.; Thiel, M. Microplastics in the Marine Environment: A Review of the Methods Used for Identification and Quantification. *Environ. Sci. Technol.* **2012**, *46*, 3060–3075. [CrossRef] [PubMed]
56. Su, L.; Cai, H.; Kolandhasamy, P.; Wu, C.; Rochman, C.M.; Shi, H. Using the Asian Clam as an Indicator of Microplastic Pollution in Freshwater Ecosystems. *Environ. Pollut.* **2018**, *234*, 347–355. [CrossRef] [PubMed]
57. Güven, O.; Gökdağ, K.; Jovanović, B.; Kıdeyş, A.E. Microplastic Litter Composition of the Turkish Territorial Waters of the Mediterranean Sea, and Its Occurrence in the Gastrointestinal Tract of Fish. *Environ. Pollut.* **2017**, *223*, 286–294. [CrossRef] [PubMed]
58. Naji, A.; Nuri, M.; Vethaak, A.D. Microplastics Contamination in Molluscs from the Northern Part of the Persian Gulf. *Environ. Pollut.* **2018**, *235*, 113–120. [CrossRef]
59. Hohenblum, P.; Liebmann, B.; Liedermann, M. *Plastic and Microplastic in the Environmen*; Report REP-0551; Umweltbundesamt GmbH: Vienna, Austria, 2015; ISBN 978-3-99004-363-9. Available online: https://www.umweltbundesamt.at/fileadmin/site/publikationen/rep0551.pdf (accessed on 4 March 2024).
60. Hurley, R.R.; Woodward, J.C.; Rothwell, J.J. Ingestion of Microplastics by Freshwater Tubifex Worms. *Environ. Sci. Technol.* **2017**, *51*, 12844–12851. [CrossRef]
61. Lin, C.-T.; Chiu, M.-C.; Kuo, M.-H. Effects of Anthropogenic Activities on Microplastics in Deposit-Feeders (Diptera: Chironomidae) in an Urban River of Taiwan. *Sci. Rep.* **2021**, *11*, 400. [CrossRef]
62. Stanton, T.; Johnson, M.; Nathanail, P.; MacNaughtan, W.; Gomes, R.L. Freshwater and Airborne Textile Fibre Populations Are Dominated by 'Natural', Not Microplastic, Fibres. *Sci. Total Environ.* **2019**, *666*, 377–389. [CrossRef] [PubMed]
63. TextileExchange. 2020 Annual Report. Texas, USA, 2020. pp. 1–7. Available online: http://textileexchange.org/ (accessed on 4 March 2024).

64. Lu, G.; Xue, Q.; Ling, X.; Zheng, X. Toxic Interactions between Microplastics and the Antifungal Agent Ketoconazole in Sediments on Limnodrilus Hoffmeistteri. *Process Saf. Environ. Prot.* **2023**, *172*, 250–261. [CrossRef]
65. Oliveira, P.; Barboza, L.G.A.; Branco, V.; Figueiredo, N.; Carvalho, C.; Guilhermino, L. Effects of Microplastics and Mercury in the Freshwater Bivalve Corbicula Fluminea (Müller, 1774): Filtration Rate, Biochemical Biomarkers and Mercury Bioconcentration. *Ecotoxicol. Environ. Saf.* **2018**, *164*, 155–163. [CrossRef]
66. Rochman, C.M.; Parnis, J.M.; Browne, M.A.; Serrato, S.; Reiner, E.J.; Robson, M.; Young, T.; Diamond, M.L.; Teh, S.J. Direct and Indirect Effects of Different Types of Microplastics on Freshwater Prey (Corbicula Fluminea) and Their Predator (Acipenser Transmontanus). *PLoS ONE* **2017**, *12*, e0187664. [CrossRef]
67. Silva, C.J.M.; Silva, A.L.P.; Gravato, C.; Pestana, J.L.T. Ingestion of Small-Sized and Irregularly Shaped Polyethylene Microplastics Affect Chironomus Riparius Life-History Traits. *Sci. Total Environ.* **2019**, *672*, 862–868. [CrossRef] [PubMed]
68. Silva, C.J.M.; Beleza, S.; Campos, D.; Soares, A.M.V.M.; Patrício Silva, A.L.; Pestana, J.L.T.; Gravato, C. Immune Response Triggered by the Ingestion of Polyethylene Microplastics in the Dipteran Larvae Chironomus Riparius. *J. Hazard. Mater.* **2021**, *414*, 125401. [CrossRef] [PubMed]
69. Su, L.; Xue, Y.; Li, L.; Yang, D.; Kolandhasamy, P.; Li, D.; Shi, H. Microplastics in Taihu Lake, China. *Environ. Pollut.* **2016**, *216*, 711–719. [CrossRef] [PubMed]
70. Baldwin, A.K.; Spanjer, A.R.; Rosen, M.R.; Thom, T. Microplastics in Lake Mead National Recreation Area, USA: Occurrence and Biological Uptake. *PLoS ONE* **2020**, *15*, e0228896. [CrossRef] [PubMed]
71. Cai, H.; Du, F.; Li, L.; Li, B.; Li, J.; Shi, H. A Practical Approach Based on FT-IR Spectroscopy for Identification of Semi-Synthetic and Natural Celluloses in Microplastic Investigation. *Sci. Total Environ.* **2019**, *669*, 692–701. [CrossRef]
72. Van Cauwenberghe, L.; Janssen, C.R. Microplastics in Bivalves Cultured for Human Consumption. *Environ. Pollut.* **2014**, *193*, 65–70. [CrossRef]
73. Oehlmann, J.; Schulte-Oehlmann, U.; Kloas, W.; Jagnytsch, O.; Lutz, I.; Kusk, K.O.; Wollenberger, L.; Santos, E.M.; Paull, G.C.; Van Look, K.J.W.; et al. A Critical Analysis of the Biological Impacts of Plasticizers on Wildlife. *Philos. Trans. R. Soc. B Biol. Sci.* **2009**, *364*, 2047–2062. [CrossRef] [PubMed]
74. Endo, S.; Takizawa, R.; Okuda, K.; Takada, H.; Chiba, K.; Kanehiro, H.; Ogi, H.; Yamashita, R.; Date, T. Concentration of Polychlorinated Biphenyls (PCBs) in Beached Resin Pellets: Variability among Individual Particles and Regional Differences. *Mar. Pollut. Bull.* **2005**, *50*, 1103–1114. [CrossRef] [PubMed]
75. Ogata, Y.; Takada, H.; Mizukawa, K.; Hirai, H.; Iwasa, S.; Endo, S.; Mato, Y.; Saha, M.; Okuda, K.; Nakashima, A.; et al. International Pellet Watch: Global Monitoring of Persistent Organic Pollutants (POPs) in Coastal Waters. 1. Initial Phase Data on PCBs, DDTs, and HCHs. *Mar. Pollut. Bull.* **2009**, *58*, 1437–1446. [CrossRef]
76. Hirai, H.; Takada, H.; Ogata, Y.; Yamashita, R.; Mizukawa, K.; Saha, M.; Kwan, C.; Moore, C.; Gray, H.; Laursen, D.; et al. Organic Micropollutants in Marine Plastics Debris from the Open Ocean and Remote and Urban Beaches. *Mar. Pollut. Bull.* **2011**, *62*, 1683–1692. [CrossRef] [PubMed]
77. Sousa, R.; Antunes, C.; Guilhermino, L. Ecology of the Invasive Asian Clam Corbicula Fluminea (Müller, 1774) in Aquatic Ecosystems: An Overview. *Ann. Limnol. Int. J. Lim.* **2008**, *44*, 85–94. [CrossRef]
78. dos Santos, K.C.; Martinez, C.B.R. Genotoxic and Biochemical Effects of Atrazine and Roundup®, Alone and in Combination, on the Asian Clam Corbicula Fluminea. *Ecotoxicol. Environ. Saf.* **2014**, *100*, 7–14. [CrossRef]
79. Cid, A.; Picado, A.; Correia, J.B.; Chaves, R.; Silva, H.; Caldeira, J.; de Matos, A.P.A.; Diniz, M.S. Oxidative Stress and Histological Changes Following Exposure to Diamond Nanoparticles in the Freshwater Asian Clam Corbicula Fluminea (Müller, 1774). *J. Hazard. Mater.* **2015**, *284*, 27–34. [CrossRef]

water

Article

Tackling the Phylogeny of Lampreys—Insight from the Croatia's Danube Basin

Sara Pleše * and Ivana Buj

Faculty of Science, University of Zagreb, Rooseveltov trg 6, 10000 Zagreb, Croatia; ivana.buj@biol.pmf.hr
* Correspondence: sara.plese@biol.pmf.hr

Abstract: This research addresses the pressing issue of protecting endangered lamprey species in Croatia, a crucial element in preserving biodiversity, particularly in the face of increasing human-induced impacts on natural ecosystems due to global warming. Lampreys, a group of vertebrates with an ancient lineage, are not fully understood taxonomically, posing a challenge to conservation efforts. In the Danube and Adriatic basins of Croatia, where lampreys are found, the lack of modern molecular methods and analyses has hindered an accurate determination of species numbers. This study aimed to bridge this knowledge gap by assessing the genetic diversity and structure of identified lamprey species and lineages in Croatia using the gene for cytochrome *b*. The research revealed four distinct lineages within the species *Eudontomyzon vladykovi* Oliva and Zanandrea, 1959 and confirmed the presence of the species *Eudontomyzon danfordi* Regan, 1911 in Croatia. Genetic diversity and differentiation tests, coupled with molecular diagnostic analyses, indicated moderate to high levels of genetic diversity within and between the identified species and lineages, emphasizing the deep structuring within *Eudontomyzon vladykovi* species. These results highlight the significance of understanding lamprey taxonomy and genetic diversity for effective conservation. The study provides important insights into the intricate relationships and conservation needs of lampreys, and provides a basis for future discussions involving additional genetic markers. By gaining a comprehensive understanding of the taxonomy, ecology, and genetic diversity of lampreys, we can ensure their conservation and that of associated ecosystems.

Keywords: lamprey diversity; phylogenetic insights; taxonomy; evolution; freshwater conservation; *Eudontomyzon* genus

1. Introduction

The Croatian Danube basin is home to a remarkable diversity of fish species, a result of the interplay of historical events and geological processes. The Danube River and its tributaries, which cover a large part of the country, are home to a wide variety of species as a result of biogeographical separation and evolutionary processes. Geological forces, including the uplift of the Dinarides, tectonic activity, and the formation of the Central Paratethys Basin, have significantly influenced the area. These events, coupled with the effects of glacial and interglacial periods, have facilitated the creation of harbors for fish populations during glacial periods and their subsequent dispersal to rivers in central and northern Europe after the last glaciation [1]. With its distinctive geomorphological and hydrological features, the Danube basin and its tributaries are emerging as a hotspot of European ichthyofaunal diversity, housing an impressive 81 species of freshwater fish species (64 native and 17 invasive species) [2–5].

Among this great diversity of fish in the Croatian Danube basin, a fascinating group of animals, the lampreys, can be found, either buried in the substrate or swimming freely in the water. Lampreys are elongated and scaleless organisms widely distributed in temperate waters, with the exception of tropical and polar regions [6]. Their fascinating life cycle has attracted the interest of scientists [7–12] and varies among species and is influenced by diet and ecological conditions. During the larval stage, known as ammocoetes, lampreys burrow in the sand for several years before metamorphosing and growing, making them ecosystem engineers [13]. Some species migrate to the sea to feed or reproduce, while others remain in freshwater. During the trophic phase, lampreys feed as parasites, and in some cases, this phase is absent. These peculiar life cycles and behaviors explain why they are rarely caught in their natural habitats and why they are so little studied. Lampreys belong to the Agnatha class, which means that they lack jaws and have a permanently open mouth [6,14,15]. Despite belonging to a distinct superclass (Agnatha) that has not undergone significant anatomical change in the past 360 million years [16], lampreys are often lumped together with fishes (superclass: Gnathostomata) in the literature.

The reported number of lamprey species in Croatia varies in different publications from the last decade and earlier [3,4,17]. Globally, the Petromyzontidae family comprises 10 genera and 40 described species [18], and 13 species are found in Europe, with one species now extinct [17]. Earlier field studies in Croatia identified five species, including *Lethenteron zanandreai* (Vladykov, 1955), *Petromyzon marinus* Linnaeus, 1758, *Eudontomyzon danfordi* Regan, 1911, *Eudontomyzon mariae* (Berg, 1931), and *Lampetra planeri* (Bloch, 1784) [4]. However, a recent list of Croatian freshwater ichthyofauna only includes three species: *Eudontomyzon vladykovi* Oliva and Zanandrea, 1959, *Lampetra soljani* Tutman, Freyhof, Duli, Glamuzina, and Geiger, 2017, and *Petromyzon marinus* Linnaeus, 1758 [3]. The difference in the number of species is explained by incorrect determination, the frequent use of synonyms in the literature, and the lack of molecular genetic analysis [3]. The identification of lamprey is usually based on morphological observations, such as the position of the dorsal fins, the relative position of the cloaca and dorsal fins, the position of the eyes, the structure of the oral funnel, the teeth and laminae, and the number of myomers [17–19]. As morphlogical determinations are not always accurate and lampreys lack countable structures such as fin rays, shells, or ossified structures [18,20], it is necessary to use molecular and genetic methods in order to resolve their taxonomy. To solve the problem, modern genetic techniques, molecular analyses, and phylogenetic reconstructions are necessary but have not yet been carried out for lampreys in Croatia.

The purpose of this research is to provide important information on the genetic diversity and distribution of lamprey species in the Danube basin in Croatia, which will help in their conservation and management. Furthermore, the results of this study will contribute to a better understanding of the evolutionary history and biogeography of lampreys in Europe and the Black Sea basin. Owing to the combination of methods applied to the Petromyzontidae family, the results of the analyses conducted will finally reveal some of the secrets hidden in the genetic material of these secretive organisms and provide accurate data on the phylogeny of lampreys in the study area.

2. Materials and Methods

2.1. Sampling

Individuals of adult lampreys and ammocoetes were sampled from 20 streams (a total of 39 different localities along 13 different river catchments): Bednja, Mura, Drava, Voćinska river, Krapina, Sutla, Mrežnica, Dobra, Kupa, Kupčina, Korana, Radonja, Una, Glina, Ilova, Česma, Sava, Danube, Orljava, and Bosna (Figure 1). A total of 103 individuals were caught using the electrofishing method or randomly by hand (Table 1). A small part of the dorsal fin was cut off with scissors, and the individuals were released. The collected fin samples were stored in tubes filled with 96% ethanol, together with information on the location of the catch. The samples were stored in a freezer at −20 °C (to prevent degradation of the DNA sample) until further analysis.

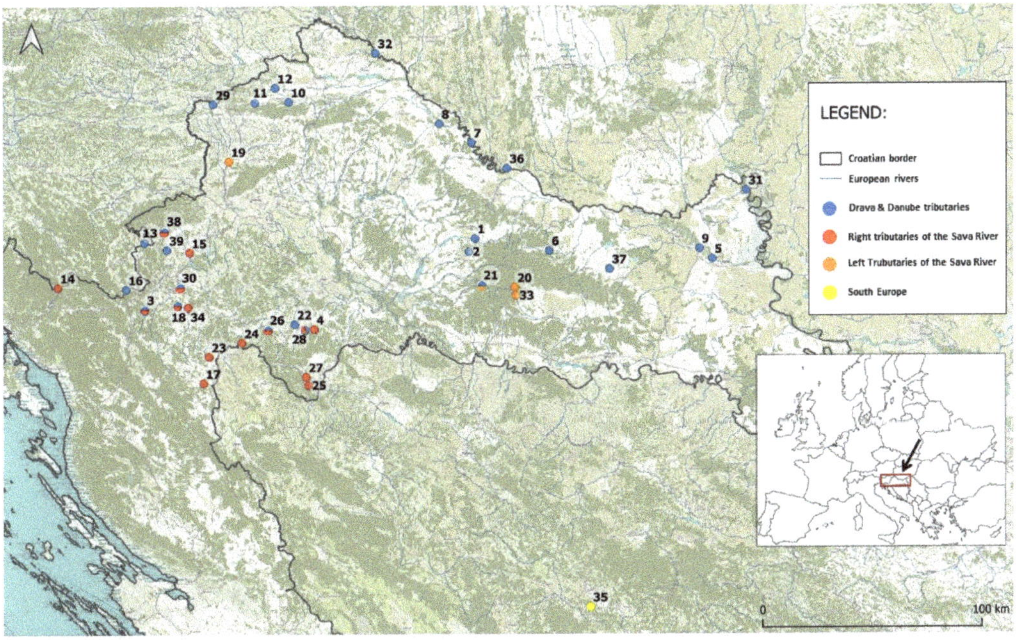

Figure 1. Map of the research area with sampling localities marked with numbers from 1 to 39. Location numbers correspond with numbers in Table 1, where more information on localities is presented.

Table 1. Distribution (loacalities) and number of samples of cytochrome *b* haplotypes of lampreys in the rivers of the Danube basin.

Locality (Name & Number)	Drainage System	Number of Samples (*n*)	cyt *b* Haplotype	GenBank Accession Numbers
Maslenjača (1)	Ilova	1	DUN2	PP661384
Toplica, Daruvar (2)	Ilova	2	DUN2	PP661384
Lešće (3)	Dobra	7	DOB1, DUN2	PP661393, PP661384
Sunja (4)	Sava	8	KUP6, SAV5	PP661399, PP661390
Osijek (5)	Drava	1	DRA1	PP661378
Voćin (6)	Drava	7	DRA2, DRA3, DRA5	PP661379, PP661380, PP661382
Štorgač (7)	Drava	1	DUN1	PP661383
Repaški most (8)	Drava	1	DRA4	PP661381
Petrijevci (9)	Drava	1	DRA1	PP661378
Ivanečka Željeznica (10)	Bednja	1	BED1	PP661376
Bednja (11)	Bednja	2	BED1	PP661376
Voća (12)	Bednja	4	BED1, BED2	PP661376, PP661377
Bubnjarci (13)	Kupa	1	DUN 2	PP661384

Table 1. *Cont.*

Locality (Name & Number)	Drainage System	Number of Samples (n)	cyt *b* Haplotype	GenBank Accession Numbers
Brod na Kupi (14)	Kupa	2	KUP1, KUP2	PP661394, PP661395
Lazina (15)	Kupa	1	KUP3	PP661396
Pribanjci (16)	Kupa	1	DUN2	PP661384
Furjašnica (17)	Korana	2	KUP4	PP661397
Brezova Glava (18)	Korana	2	KUP5, DUN2	PP661398, PP661384
Stubička Slatina (19)	Sava	1	KRA1	PP661385
Brzaja (20)	Sava	5	SAV4	PP661389
Pakra (21)	Sava	7	PAK1, PAK2, PAK3, DUN2	PP661402, PP661403, PP661404, PP661384
Bručina (22)	Kupa	3	DUN2	PP661384
Ruševnica (23)	Kupa	1	KUP6	PP661399
Glinica (24)	Kupa	1	KUP6	PP661399
Čemernica (25)	Kupa	1	UNA1	PP661400
Buzeta, Šibine (26)	Kupa	4	KUP6, DUN2	PP661399, PP661384
Žirovnica (27)	Una	2	UNA1, UNA2	PP661400, PP661401
Petrinjčica (28)	Kupa	5	SAV3, SAV4, DUN2	PP661388, PP661389, PP661384
Lupinjak (29)	Sava (Sutla)	3	SAV1, SAV2	PP661386 PP661387
Mostanje (30)	Mrežnica	2	MRE1, DUN2	PP661392, PP661384
Batina (31)	Dunav	1	DUN1	PP661383
Goričan (32)	Mura	1	DUN1	PP661383
Orljava (33)	Orljava	4	SAV4, SAV6	PP661389, PP661391
Živković kosa (34)	Korana	1	RAD1	PP661407
Ušće Lašve (35)	Bosna	2	BOS1, BOS2	PP661405, PP661406
Terezino polje (36)	Drava	5	DRA3	PP661380
Bukvik (37)	Drava	5	DRA3	PP661380
Jaševnica (38)	Kupa	2	KUP3, DUN2	PP661396, PP661384
Ozalj (39)	Kupa	2	DUN2	PP661384

Note(s): GenBank accession numbers will be provided in the proof, however, alignments can be sent to the editor, if needed.

2.2. Laboratory Protocols

Laboratory procedures included DNA isolation, gene amplification by polymerase chain reaction, and verification of amplification by agarose gel electrophoresis. DNA isolation was conducted using a DNeasy Blood & Tissue chemical kit according to the protocol of the kit manufacturer, QIAGEN (Amsterdam, Netherlands). The protocol for gene amplification by polymerase chain reaction was optimized to obtain the best products, and PCR conditions and primers are listed in Table 2. PCR products obtained and verified by electrophoresis were sent to the Macrogen Service Centre for sequencing using the same primers that were used for PCR amplification.

Table 2. Protocol for PCR reaction and used primers.

Genetic Marker	**cyt *b***
PCR conditions	35 cycles of 45 s at 92 °C, following 90 s at 48 °C, following 105 s at 72 °C
PCR primers	ProK (5′TTATTTAATGTTAAGATRCTAGCTTTGG3′) Pak-Glu F (5′CACCGTTGTAGAATTCAA CTATAAG3′)

The molecular marker chosen for analysis in this study was the cytochrome *b* (cyt *b*) gene. As part of mitochondrial DNA and also a coding gene, its mutation rate is usually lower than other gene markers, making it suitable for phylogenetic analyses at species and supraspecific levels. It was chosen because it has shown the best performance and yielded the most reliable results in previous studies of phylogenetic structure, relationships, and taxonomic research [2,21], particularly in recovering phylogenetic relationships among closely related freshwater fish taxa [22,23].

2.3. Data Analyses

2.3.1. Sequence Alignment

Sequences were aligned using the BioEdit Sequence Alignment Editor version 7.2.5 [24] and compared to previously published sequences. Visual checks were performed on chromatograms and alignments to ensure accuracy.

2.3.2. Neutrality Test and Analyses of Population Genetic Diversity and Polymorphism

Genetic diversity within the identified lineages and species was assessed by estimating measures of DNA sequence polymorphisms using the DnaSP 5.10 program [25]. To gain an understanding of gene flow and genetic differentiation of lineages, estimates of genetic differentiation (χ2, Hst, Kst, Kst*, Z, Z*, and Snn) of the lineages and their support were computed in addition to specific measurements of DNA polymorphisms. Sequences obtained from the captured individuals were used as input data for genetic statistical tests of diversity. Neutrality tests Tajima's D, Fu and Li's D, and F tests were performed and showed that the data set is in mutation-drift equilibrium (not statistically significant, $p < 0.05$) also using the DnaSP 5.10 program [25].

2.3.3. Phylogeographic and Evolutionary Analyses

Phylogenetic reconstruction was performed to confirm the phylogenetic position of lampreys from the Croatian Danube basin within the phylogenetic tree of the genus *Eudontomyzon*, to reveal relationships among them, and to test whether the phylogenetic position of each population is consistent with reported taxonomic hypotheses. Two different phylogenetic reconstruction methods were employed: maximum parsimony (MP) and maximum likelihood (ML), implemented in PAUP v.4.0a [26]. For MP analysis, a heuristic search mode with 100 replicates was used, with a randomized input order of taxa and tree bisection-reconnection (TBR) branch swapping, giving equal weight to all codon sites and nucleotide substitution types weighted equally. ML analysis was performed under the heuristic search option using the TBR branch-swapping algorithm. Branch support (BS) was assessed by nonparametric bootstrapping (1000 pseudoreplicates, ten additional sequence replicates). The length of the MP tree was 163, with a consistency index of 0.6994, a homoplasy index of 0.3006, and a retention index of 0.7461. With haplotypes as input sequences for phylogenetic tree and network construction, sequences from the GenBank were also used (Table 3). Sequences of several *Eudontomyzon* species from the GenBank were included in the phylogenetic analyses to provide a clearer insight into the phylogenetic position of the lampreys from the study area in relation to their European relatives, while the sequence of *Lampetra fluviatilis* (accession number: GQ206175.1; Table 3) was used as an outgroup. The median-joining approach (MJ) was employed as an additional method using the computer program Network 4.5.1.6 [27]. In the resulting phylogenetic network, horizontal gene transfers were observed, which can be useful in reconstructing phylogenetic relationships among closely related taxa.

Table 3. Lamprey cyt *b* sequences retrieved from the GenBank and included in the phylogenetic and evolutionary history reconstruction.

Species	bp (Number of Base Pairs)	Locality	GenBank Accession Number	Reference
Eudontomyzon danfordi Regan, 1911	1133	Zdychava River (SVK)	GQ206158.1	[28]
Eudontomyzon mariae (Berg, 1931)	1133	Ivianka River (UKR)	GQ206162.1	[28]
Eudontomyzon stankokaramani (Karaman, 1974)	1133	Zeta River (MNE)	GQ206189.1	[28]
Eudontomyzon vladykovi Oliva and Zanandrea, 1959	1133	Studenec Creek (SVK)	GQ206161.1	[28]
Lampetra fluviatilis (Linnaeus, 1758)	16,159	Garonne Estuary (FRA)	NC001131.1	[29]
Eudontomyzon lanceolata (Kux and Steiner, 1972)	1133	Chakhtsutsyr River (RUS)	GQ206176.1	[28]
Lampetra planeri (Bloch, 1784)	1133	Kalte Moldau River (DEU)	GQ206149.1	[28]
Lethenteron zanandreai (Vladykov, 1955)	1133	Vipava River (SLO)	GQ206184.1	[28]
Eudontomyzon sp. *Dnieper*	1164	Ugra River (RUS)	KP135487.1	[30]
Eudontomyzon sp. *Dnieper*	1164	Rudyanka River (RUS)	KP135483.1	[30]
Eudontomyzon sp. *Dnieper*	1164	Sigosa River (RUS)	KP135485.1	[30]
Eudontomyzon sp. *Dnieper*	1164	Vyazma River (RUS)	KP135482.1	[30]
Eudontomyzon hellenicus (Vladykov, Renaud, Kott and Economidis, 1982)	1133	Strymon River (GRC)	GQ206160.1	[28]
Lampetra fluviatilis (Linnaeus, 1758)	1133	Luga River (RUS)	GQ206175.1	[28]
Eudontomyzon stankokaramani (Karaman, 1974)	1191	unknown river (SLO)	KX787432.1	[31]
Eudontomyzon lanceolata (Kux and Steiner, 1972)	1191	unknown river (TUR)	KX787431.1	[31]

Divergence times between phylogenetic lineages were estimated by a Bayesian MCMC coalescent method using BEAST 2.4.7 software [32]. The analysis was conducted on the cyt *b* data set. Rate homogeneity across phylogenetic lineages was assessed by the log-likelihood ratio test (LRT), comparing the likelihood of phylogenetic trees (reconstructed using the maximum likelihood approach) with and without molecular clock enforcement in PAUP software version 4.0. As the likelihood scores were not the same in both cases for the cyt *b* data set, we applied a relaxed molecular clock. Molecular clock calibration was based on the divergence rate of the cyt *b* gene in Petromyzontidae of 0.12% per lineage per million years, as reported by [33]. The divergence rates were drawn from an uncorrelated lognormal distribution and a Yule speciation prior to a random starting tree. The substitution model used was HKY with a gamma-site heterogeneity model. We used default prior distributions for kappa, frequencies, and alpha, while substitution rate parameters were unlinked across codon positions. The number of MCMC steps (the length of the chain) was ten million.

The ancestral distribution ranges of lineages were reconstructed using Statistical Dispersal-Vicariance Analysis implemented in the S-DIVA software version 1.9 [34]. This method reconstructs the ancestral distribution in a phylogeny by optimizing a three-dimensional cost matrix, in which extinctions and dispersals 'cost' more than vicariance and determines the statistical support for the ancestral range [34]. Reconstruction of ancestral geographic ranges was conducted using the cyt *b* data set. An input set of trees was generated using Bayesian analysis (BAY). A total of six recent geographic ranges for *Eudontomyzon* were denoted (4 of them are marked in Figure 1): A—right tributaries of the Sava River, C—North Europe, D—tributaries of the Drava and Danube Rivers, E—Central Europe, F—South Europe.

3. Results

3.1. Phylogenetic Relationships and Distribution of Lampreys in Continental Croatia

The phylogenetic reconstruction of the Petromyzontidae family in the area of continental Croatia (the Danube basin) is based on the sequences of the gene for cytochrome *b* (cyt *b*) of sampled lampreys and sequences retrieved from the GenBank (Tables 2 and 3). The cyt *b* gene sequences were 1191 base pairs long, with 68 variable and 46 parsimony significant sites (sites with at least two nucleotides occurring at least twice).

A total of 103 newly obtained sequences were combined with sequences retrieved from GenBank (Table 3) and included in the analyses. With the Croatian sample set, 32 haplotypes were identified, and five phylogenetic lineages were recovered in the phylogenetic tree (Figure 2). Four of these lineages, including samples from both the Sava and Drava River basins in Croatia, belong to *E. vladykovi*, according to current taxonomy. However, several samples from the Drava R. basin, particularly from the easternmost part of this river basin in Croatia and from the Bednja River (the largest left tributary of the Drava River), clustered together with sequences of *E. danfordi* from Slovakia in a separate phylogenetic lineage. As this lineage corresponds to the species *E. danfordi*, this study confirms for the first time that the Croatian Drava and Bednja rivers are part of the distribution range of this species.

Another unexpected result is the high amount of cryptic diversity within *E. vladykovi*, with at least four clearly separated phylogenetic units whose geographic distributions are also largely separated. *Eudontomyzon vladykovi* lineage I is so far, and based on our results, only repoted from Croatian waters, where it mainly inhabits the northwest part (area of Hrvatsko zagorje), but also the Mrežnica and Sunja rivers, so that it was found in both the Sava and Drava river basins. The second lineage, although found in a small number of samples, occupies a wide but fragmented distribution area. It was also found in both river basins in continental Croatia. The third lineage within *E. vladykovi* comprises the largest number of specimens from Croatia (it is present only in the Sava River basin, but it was found in a large number of localities), but also from Bosnia and Herzegovina. Lineage IV is also restricted to the Sava River basin, but is much rarer.

The MJ phylogenetic network (Figure 3), based on the cyt *b* gene sequences obtained in this study, corroborates the results visible in the phylogenetic trees. A clear separation of two species—*Eudontomyzon danfordi* and four lineages within the species *Eudontomyzon vladykovi*—is evident. Once again, there is a particularly high level of diversity within the *E. vladykovi* III lineage that needs to be recognized.

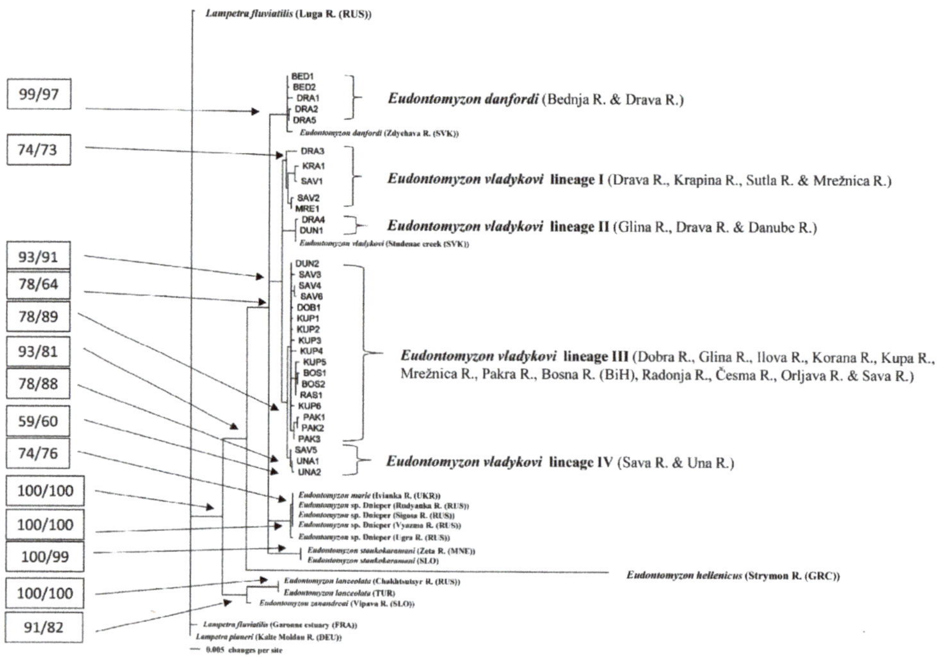

Figure 2. ML phylogram showing the position of *Eudontomyzon* species in the Danube basin, together with sequences retrieved from the GenBank based on the cyt *b* gene. Numbers at nodes represent ML and MP bootstrap values.

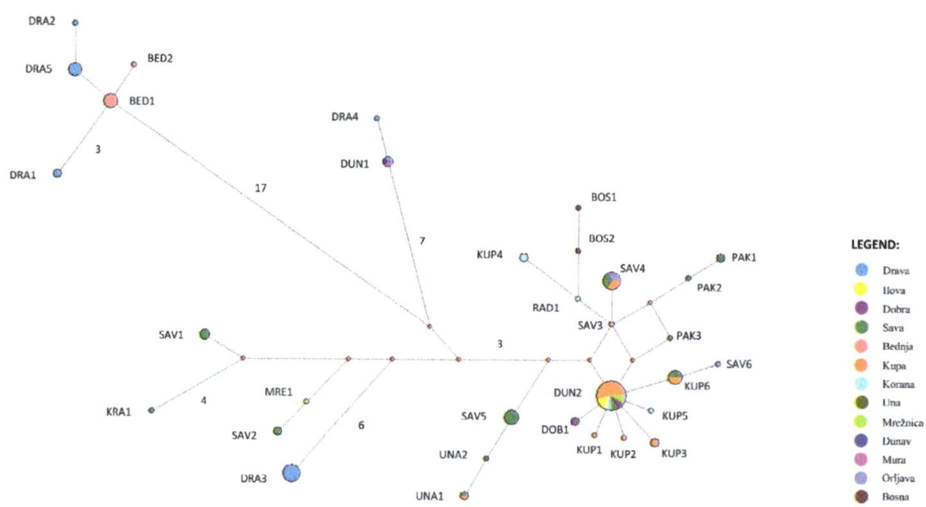

Figure 3. Median-joining network of cyt *b* haplotypes. Red circles represent median vectors. The number of mutations is indicated by branches when there are more than two mutations.

3.2. Inter- and Intraspecific Genetic Diversity and Differentiation

The genetic diversity of both species distributed in continental Croatia, as well as within each of the four *E. vladykovi* lineages, is moderate to high. A large number of haplotypes were determined, as was their high diversity. Furthermore, the average number of nucleotide differences for the *Eutontomyzon vladykovi* species and the vladykovi I lineage is extremely high, indicating a deeper structure and the possibility of the presence of cryptic diversity (Table 4). It is noteworthy that the haplotype diversity of *E. danfordi* is only 15% lower than the haplotype diversity of *E. vladykovi*, although its distribution area in Croatia is much smaller and it is present in a much lower number of individuals (15 vs. 88) caught in this study. However, the average number of nucleotide differences is almost five times higher in *E. vladykovi*, indicating that deep structuring is present, in contrast to the uniform genetic structure of *E. danfordi*. Moreover, the number of shared polymorphisms is very low, and they exist only between the *E. vladykovi* lineage III and all other units, including *E. danfordi* (Table 5).

Table 4. Measures of genetic polymorphism for the *Eudontomyzon vladykovi* lineages and for the *Eudontomyzon danfordi* species. N—number of sequences, h—number of haplotypes, S—number of polymorphic sites, Hd—haplotype diversity, k—average number of nucleotide differences, π—nucleotide diversity, η—total number of mutations.

Species/Lineage	N	h	S	Hd	k	π	η
E. danfordi	15	5	6	0.752	1.524	0.00128	6
E. vladykovi	88	30	53	0.895	7.158	0.00601	54
vladykovi I	16	5	19	0.667	6.983	0.00586	19
vladykovi II	4	2	1	0.5	0.5	0.00042	1
vladykovi III	59	20	23	0.8	2.414	0.00203	23
vladykovi IV	9	3	2	0.556	0.722	0.00061	2

Table 5. Shared polymorphisms on the cyt *b* gene between lineages.

Species/Lineage	Vladykovi I	Vladykovi II	Vladykovi III	Vladykovi IV
E. danfordi	0	0	1	0
vladykovi I	-	0	3	0
vladykovi II	-	-	1	0
vladykovi III	-	-	-	0

The pairwise distances (p-distances) (Table 6) between species and lineages are always higher than the p-distance within *E. danfordi* and each of the *E. vladykovi* lineages. *Eudontomyzon vladykovi* lineages III and IV are the most closely related lineages, which is reflected in the lowest p-distance between them. On the other hand, as expected, p-distances between *E. danfordi* and all *E. vladykovi* lineages are the highest.

Table 6. *p*-values between and inside lineages (expressed as a percentage).

Species/Lineage	*E. danfordi*	Vladykovi I	Vladykovi II	Vladykovi III	Intraspecific/Intralineage
E. danfordi					0.08–0.41 (0.20)
vladykovi I	1.84–2.68 (2.26)				0.08–1.25 (0.70)
vladykovi II	2.01–2.35 (2.18)	1.09–1.51 (1.30)			/ (0.10)
vladykovi III	1.93–2.35 (2.14)	0.83–1.59 (1.21)	1.09–1.42 (1.255)		0.80–0.58 (0.30)
vladykovi IV	1.84–2.09 (1.97)	0.83–1.34 (1.09)	0.92–1.09 (1.01)	0.33–0.75 (0.54)	0.08–0.16 (0.10)

The distinction not only of species but also of lineages of *E. vladykovi* lineages is corroborated by the results of the genetic differentiation tests (Table 7), since there was no gene flow between species and lineages, indicating that there is no reproduction of individuals belonging to different lineages.

Table 7. Estimation of the number of migrants per generation between the lamprey lineages of the Danube basin in Croatia according to [35–37].

Species/Lineage	Nm [35]	Nm [36]	Nm [37]
E. danfordi:E. vladykovi	0.61	0.1	0.11
vladykoviI:vladykoviII	0.67	0.15	0.15
vladykoviI:vladykoviIII	0.53	0.26	0.26
vladykoviI:vladykoviIV	0.51	0.21	0.21
vladykoviII:vladykoviIII	0.74	0.6	0.6
vladykoviII:vladykoviIV	0.06	0.03	0.03
vladykoviIII:vladykoviIV	1.19	0.2	0.2

3.3. Molecular Identification of Species and Lineages

Molecular diagnostic analyses identified sites of fixed differences based on the cyt *b* gene. More fixed differences (Table 8) show that the species/lineages diverged earlier and indicate a large genetic difference between the species/lineages. As expected, the highest number of fixed differences was observed between *Eudontomyzon danfordi* and the phylogenetic lineages of *Eudontomyzon vladykovi*. However, there are also fixed differences present between all phylogenetic lineages of the *Eudontomyzon vladykovi* species. The table of diagnostic sites (Table 9) shows exactly which sites are involved. It is noteworthy that each of the *E. vladykovi* lineages can be unambiguously recognized on the basis of diagnostic sites in cyt *b* gene.

Table 8. Fixed differences based on the cyt *b* gene between lineages.

Species/Lineage	Vladykovi I	Vladykovi II	Vladykovi III	Vladykovi IV
E. danfordi	19	24	21	21
vladykovi I	-	9	6	6
vladykovi II	-	-	12	10
vladykovi III	-	-	-	3

Table 9. Fixed differences on the cyt *b* gene with specified nucleotide sites of fixed differences and bases present in each lineage indicated. Diagnostic sites for a particular lineage are represented by nucleotides in colored boxes.

	Species/Lineage				
Nucleotide Spot	*E. danfordi*	Vladykovi I	Vladykovi II	Vladykovi III	Vladykovi IV
36	T	C	C	C	C
57	C	T	T	T	T
72	C	T	T	T	T
156	G	A	A	A	A
157	C	T	T	T	T
162	T	C	C	C	C
312	A	G	A	A	A
378	A	A	G	A	A
390	T	T	C	T	T
480	A	G	G	G	G

Table 9. *Cont.*

	Species/Lineage				
Nucleotide Spot	*E. danfordi*	Vladykovi I	Vladykovi II	Vladykovi III	Vladykovi IV
483	A	G	G	G	G
531	C	C	C	C	T
588	A	A	G	A	A
630	T	T	C	T	T
639	G	A	A	A	A
675	T	C	C	C	C
726	C	T	T	T	T
771	A	C	C	C	C
807	G	A	G	G	G
813	C	C	C	T	C
909	G	A	A	A	A
1155	G	G	A	G	G

3.4. Evolutionary History of the Family Petromyzontidae in Europe

Evolutionary history analyses based on the cyt *b* gene (Figure 4) revealed an ancient origin and long-term evolutionary history of both *E. vladykovi* and *E. danfordi*. Separation of the clade comprising all *E. vladykovi* lineages from its sister clade comprising *E. danfordi*, *E. stankokaramani* and *E.* sp. Dnieper occurred in the early Miocene, about 22 mya (million years ago), whereas these two clades started to diversify around 17 mya. The origin of the different species and *E. vladykovi* lineages occurred mostly during the Middle Miocene, 11.6–8.5 mya.

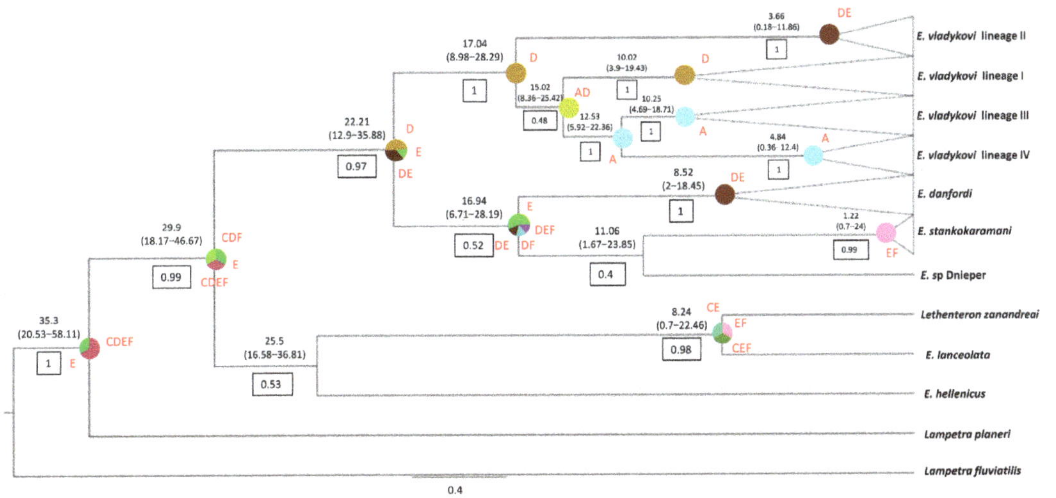

Figure 4. Estimates of divergence time based on cyt *b* sequences of lamprey species. The timing of divergence events is presented as a mean and a 95% credible interval (in millions of years ago). Lower and upper bounds of the highest posterior density (HPD) interval; the HPD is the shortest interval containing 95% of the sampled values. Numbers in squares are bootstrap values. Colored circles represent the highest probability of the ancestral range (A—right tributaries of the Sava River, C—North Europe, D—tributaries of the Drava and Danube Rivers, E—Central Europe, F—South Europe).

4. Discussion

4.1. Diversity of the Petromyzontidae Family in the Danube Basin in Croatia

As this study was the first to investigate the structure of lamprey populations in continental Croatia at the molecular genetic level, it provided new and very important insights. Our results show significantly higher diversity on all levels—species, intraspecific, and genetic diversity. In addition to *E. vladykovi*, which was thought to be the only lamprey species in continental Croatia [3], this study proved that *E. danfordi* is also present and that its distribution range includes localities in the Drava River basin. Since its current distribution range in Croatia comprises the Bednja River (located in the western part of the Drava River basin) and the easternmost localities (Osijek and Petrijevci), with no reports of its presence in the Drava River in between, it is possible that the current localities are remnants of the once wider distribution range of this species, which has been reduced as a consequence of anthropogenic activities, especially habitat modifications. Regardless of its small and fragmented distribution range in Croatia, the genetic diversity of *E. danfordi* is moderately high, which is consistent with its long-term diversification in the Miocene.

Eudontomyzon vladykovi has a wider distribution in Croatia, inhabiting localities in both the Drava and Sava river basins. Its genetic diversity is high, and its intraspecific structure is also very pronounced. Four *E. vladykovi* lineages can be observed in the phylogenetic structure of this species, with no observed gene flow between them and mostly distinct distribution ranges. The origin of *E. vladykovi* can be dated to the early Miocene (22 mya). The intraspecific divergences of this species are also of very ancient origin and started already in the Miocene, 17 mya.

It is noteworthy that although *E. vladykovi* now has a larger distribution area in the Sava River basin, our results suggest that its ancestral distribution area was located in the Drava and Danube River basins, similar to *E. danfordi*, whose ancestral area was in the Drava and Danube River basins in Croatia or even further north, in Central Europe. The time frame of the origin of these two species (early and middle Miocene) corresponds to the development and evolution of the Danube River basin, which is concordant with the results of their ancestral ranges. Namely, the Proto-Danube was formed in the North Alpine foreland basin around 19–18 mya and started its southward migration [38,39]. Interestingly, the period between 17 and 18 mya, which was recovered in this study as the time of divergence of the *E. vladykovi* lineages and the origin and evolution of the species *E. danfordi*, *E. stankokaramani* and *E.* sp. Dnieper, has already been observed as an important period in the evolutionary history of the fish species distributed in the Danube River basin. The first major diversification event within the genus *Barbus* was also associated with this period [5]. Since the Proto-Danube progressed slowly, reaching the Vienna basin at 14–13 mya and the Pannonian basin at 10–5 mya [39], it is possible that the evolutionary history of both *E. danfordi* and *E. vladykovi* was largely shaped by the evolution of the Danube River basin, especially since 10–5 mya is the exact time frame when the origin of *E. vladykovi* lineages and the origin of diversification of the Croatian *E. danfordi* population occurred. Moreover, the ancestral range of the Croatian *E. danfordi* and *E. vladykovi* lineages was reconstructed in the Drava and Danube River basins. On the other hand, the ancestral ranges of *E. vladykovi* lineages I and II are associated with the right tributaries of the Sava River, indicating the colonization of the Sava River basin at the same time and the subsequent evolution and diversification of *E. vladykovi* there. Another important observation is that some *E. vladykovi* lineages inhabit the same rivers and river basins, with no obvious barriers between their localities, that would prevent migration and gene flow. This implies the presence of reproductive isolation between lineages, but it is also consistent with their ecology and behavior, particularly their connection with small patches of suitable habitats. Discovering various aspects of life cycles and diversities in the ecology and behavior of different *E. vladykovi* lineages and populations should be the focus of further investigation.

Given the long-term evolutionary history of both, *E. vladykovi* and *E. danfordi*, their high and moderately high genetic diversity is not surprising and is positive from a conservational perspective. Moreover, the absence of population genetic patterns that would imply bottleneck events indicates that both species have survived glaciations in the waters of continental Croatia and that glacial events did not have a significant effect on them. Namely, there is no abruption in their evolutionary history; the intraspecific structure of *E. vladykovi*, which originated in the Miocene, is well preserved until now, and intraspecific divergences of *E. danfordi* also started in the Miocene and continued until recent times. The waters of the Danube River basin in Croatia have been reported to serve as glacial refugia for other freshwater fish species, e.g., *Cobitis elongatoides* Bacescu & Mayer, 1969 [40], and it is likely that they also served as glacial refugia for lampreys. Nevertheless, lamprey populations with high genetic diversity have also been recognized as candidates for protection at several sites in Italy, Ireland, and Spain [7].

4.2. Taxonomic Implications

Besides revealing a high amount of cryptic diversity within lampreys in the Danube basin in Croatia, the results of all analyses conducted in this investigation indicate the necessity of a taxonomic revision of *E. vladykovi* (also observed by [33] at the European level) and the possibility that a species complex exists under this name, which, on the other hand, also has strong consequences for the practical conservation of lampreys in Croatia. The deep structuring within *E. vladykovi* observed by phylogenetic reconstruction and evolutionary history analyses (at least four phylogenetically distant and evolutionarily independent lineages were observed in this study) is comparable to the structuring within its sister clade, which comprises *E. danfordi*, *E. stankokaramani*, and *E.* sp. Dnieper. The origins of both clades occurred in a similar time frame, and the emergence of lineages included within both clades occurred mostly in the Miocene (11.6–10.25 mya), so it is surprising that lineages within one clade are all considered to belong to a single species (*E. vladykovi*), whereas lineages within the second clade each represent a separate species. Importantly, no gene flow was detected between the different *E. vladykovi* lineages, each of which can be recognized by molecular diagnostic sites in the cyt *b* gene. Although the above results are strong indications that each of the recovered *E. vladykovi* lineages may represent a separate species, a detailed comparison of morphological characteristics is necessary to find morphological diagnostic characters. Furthermore, although cyt *b* recovered cryptic diversity within *E. vladykovi*, it is necessary to investigate more genetic markers, including nuclear genes, in order to reliably resolve the taxonomy of this species or species complex.

4.3. Conservation Recommendations

The channeling of watercourses has been highlighted as the primary cause of endangerment at the sites where individuals were captured [41,42]. To preserve these populations, which harbor such a high level of genetic diversity and contribute to the biodiversity and ecosystem stability of Croatia, it is necessary to conserve, possibly protect, and prevent further endangerment of these sites. There is no reason why protective measures should not be implemented even before taxonomic issues are resolved. Among the sites highlighted in this study as unique and worthy of protection because of their natural richness are the areas inhabited by populations of the *E. danfordi* species, namely the Bednja River area, and the sites where the Vladykovi II lineage is found, specifically the Mura River basin and the confluence of the Mura River and Drava rivers. In addition to their genetic diversity, these sites collectively yielded the fewest individuals captured, suggesting the possible presence of smaller populations that, due to their size (currently indeterminable), may be more vulnerable to threats than some populations found in the tributaries of the Sava River. Furthermore, individuals of the Vladykovi II lineage indicate that populations in these areas may be crucial for unraveling the evolutionary history of the species *E. vladykovi* in Croatia.

Continued conservation efforts are essential to protecting lamprey populations and their habitats. Habitat restoration initiatives should be prioritized to improve the quality and connectivity of aquatic environments, particularly in regions where lamprey populations are threatened by habitat degradation [8]. Furthermore, the establishment of protected areas or reserves in critical lamprey habitats is imperative to ensure their long-term survival and to mitigate human disturbance. Monitoring programs must be developed and used to track population trends, habitat changes, and human impacts over time, providing essential data for adaptive management strategies [43]. In addition, outreach and education initiatives are essential to raise awareness of the ecological significance of lampreys and to build support for conservation efforts among local communities and among stakeholders. Cross-border collaboration is also necessary, working with other countries and international organizations to coordinate conservation efforts across transboundary river basins, recognizing the migratory nature of lamprey species and the interconnectedness of their habitats [44].

Author Contributions: S.P.: Conceptualization, investigation, writing—original draft; I.B.: conceptualization, resources, investigation, writing—original draft. All authors have read and agreed to the published version of the manuscript.

Funding: This research received no external funding.

Data Availability Statement: The original data presented in this study are openly available at https://www.ncbi.nlm.nih.gov/genbank/.

Acknowledgments: The authors are grateful to all colleagues from the Laboratory for Vertebrates and Nature protection, Division of Zoology, Department of Biology, Faculty of Science, University of Zagreb; namely: Davor Zanella, Perica Mustafić, Marko Ćaleta, Zoran Marčić, Sven Horvatić, Lucija Ivić, Lucija Onorato, Roman Karlović and Siniša Vajdić for their strong support of this investigation and their continuous efforts in providing samples for this research over the years.

Conflicts of Interest: The authors declare no conflict of interest.

References

1. Hewitt, G.M. Post-glacial re-colonization of European biota. *Biol. J. Linn. Soc.* **1999**, *68*, 87–112. [CrossRef]
2. Buj, I.; Podnar, M.; Mrakovčić, M.; Choleva, L.; Šlechtova, V.; Tvrtković, N.; Ćaleta, M.; Mustafić, P.; Marčić, Z.; Zanella, D.; et al. Genetic diversity and phylogenetic relationships of spined loaches (genus *Cobitis*) in Croatia based on mtDNA and allozyme analyses. *Folia Zool.* **2008**, *57*, 71–82.
3. Ćaleta, M.; Marčić, Z.; Buj, I.; Zanella, D.; Mustafić, P.; Duplić, A.; Horvatić, S. A review of extant Croatian freshwater fish and lampreys: Annotated list and distribution. *Croat. J. Fish.* **2019**, *77*, 137–234.
4. Mrakovčić, M.; Brigić, A.; Buj, I.; Ćaleta, M.; Mustafić, P.; Zanella, D. *Red Book of Freshwater Fish of Croatia*; Ministry of Culture, State Institute for Nature Protection: Zagreb, Croatia, 2006.
5. Raguž, L.; Buj, I.; Marčić, Z.; Veble, V.; Ivić, L.; Zanella, D.; Horvatić, S.; Mustafić, P.; Ćaleta, M.; Sabolić, M. First look into the evolutionary history, phylogeographic and population genetic structure of the Danube barbel in Croatia. *Knowl. Manag. Aquat. Ecosyst.* **2021**, *422*, 13. [CrossRef]
6. Kottelat, M.; Freyhof, J. *Handbook of European Freshwater Fishes*, 1st ed.; Kottelat: Cornol, Switzerland; Freyhof: Berlin, Germany, 2007; p. 646.
7. De Cahsan, B.; Nagel, R.; Schedina, I.M.; King, J.J.; Bianco, P.G.; Tiedemann, R.; Ketmaier, V. Phylogeography of the European brook lamprey (*Lampetra planeri*) and the European river lamprey (*Lampetra fluviatilis*) species pair based on mitochondrial data. *J. Fish Biol.* **2020**, *96*, 905–912. [CrossRef]
8. Mateus, C.S.; Stange, M.; Berner, D.; Roesti, M.; Quintella, B.R.; Alves, M.J.; Almeida, P.R.; Salzburger, W. Strong genome-wide divergence between sympatric European river and brook lampreys. *Curr. Biol.* **2013**, *23*, R649–R650. [CrossRef] [PubMed]
9. Rüber, L.; Gandolfi, A.; Foresti, D.; Paltrinieri, L.; Splendiani, A.; Seehausen, O. Phylogenetic and biogeographic history of brook lampreys (*Lampetra*: Petromyzontidae) in the river basins of the Adriatic Sea based on DNA barcode data. *Ecol. Evol.* **2023**, *13*, e10496. [CrossRef] [PubMed]
10. Zanandrea, G. Speciation among lampreys. *Nature* **1959**, *184*, 380. [CrossRef]
11. Zanandrea, G.I.S. Recenti ricerche sulle forme appaiate di Lamprede dell'Italia e del Danubio. *Bolletino Zool.* **1959**, *26*, 545–554. [CrossRef]
12. Zanandrea, G. Studies on European lampreys. *Evolution* **1961**, *15*, 523–534. [CrossRef]
13. Boeker, C.; Geist, J. Lampreys as ecosystem engineers: Burrows of *Eudontomyzon* sp. and their impact on physical, chemical, and microbial properties in freshwater substrates. *Hydrobiologia* **2016**, *777*, 171–181. [CrossRef]
14. Pough, F.H.; Janis, C.H.; Heiser, J.B. *Vertebrate Life*, 9th ed.; Prentice Hall: Upper Saddle River, NJ, USA, 2013; pp. 1–69.
15. Young, J.Z. *The Life of Vertebrates*, 3rd ed.; Clarendon Press: Oxford, UK, 1985; pp. 76–109.

16. Gess, R.W.; Coates, M.I.; Rubidge, B.S. A lamprey from the Devonian period of South Africa. *Nature* **2006**, *443*, 981–984. [CrossRef] [PubMed]
17. Tutman, P.; Freyhof, J.; Dulčić, J.; Glamuzina, B.; Geiger, M. *Lampetra soljani*, a new brook lamprey from the southern Adriatic Sea basin (Petromyzontiformes: Petromyzontidae). *Zootaxa* **2017**, *4273*, 531–548. [CrossRef] [PubMed]
18. Vladykov, V.D.; Kott, E. *List of the Northern Hemisphere Lampreys (Petromyzondidae) and Their Distribution*, 3rd ed.; Miscellaneous Special Publications; Department Fisheries and Oceans: Ottawa, Canada, 1979; pp. 1–30.
19. Gill, H.S.; Renaud, C.B.; Chapleau, F.; Mayden, R.L.; Potter, I.C. Phylogeny of living parasitic lampreys (Petromyzontiformes) based on morphological data. *Copeia* **2003**, 687–703. [CrossRef]
20. Krappe, M. Quantitative Analysen Populationsbiologischer Phänomene im Lebenszyklus des Bachneunauges *Lampetra planeri* (Bloch, 1784). Ph.D. Thesis, Fakultät der Universität Rostock, Rostock, Germany, 2005.
21. Perea, S.; Böhme, M.; Zupančič, P.; Freyhof, J.; Šanda, R.; Özuluğ, M.; Abdoli, A.; Doadrio, I. Phylogenetic relationships and biogeographical patterns in Circum-Mediterranean subfamily Leuciscinae (Teleostei, Cyprinidae) inferred from both mitochondrial and nuclear data. *BMC Evol. Biol.* **2010**, *10*, 265. [CrossRef] [PubMed]
22. Castresana, J. Cytochrome *b* phylogeny and the taxonomy of great apes and mammals. *Mol. Biol. Evol.* **2001**, *18*, 465–471. [CrossRef]
23. Patwardhan, A.; Ray, S.; Roy, A. Molecular Markers in Phylogenetic Studies—A Review. *J. Phylogen. Evol. Biol.* **2014**, *2*, 131. [CrossRef]
24. Hall, T.A. BioEdit: A user-friendly biological sequence alignment editor and analysis program for Windows 95/98/NT. *Nucleic Acids Symp. Ser.* **1999**, *41*, 95–98.
25. Librado, P.; Rozas, J. DnaSP v5: A software for comprehensive analysis of DNA polymorphism data. *Bioinformatics* **2009**, *25*, 1451–1452. [CrossRef]
26. Swofford, D.L. *PAUP*: Phylogenetic Analysis Using Parsimony (*and Other Methods), Version 4*; Sinauer Associates: Sunderland, MA, USA, 2002.
27. Nakhleh, L.; Guohua, J.; Fengmei, Z.; Mellor-Crummey, J. Reconstructing Phylogenetic Networks Using Maximum Parsimony. In Proceedings of the IEEE Computational Systems Bioinformatics Conference (CSB'05), Stanford, CA, USA, 8–11 August 2005.
28. Lang, N.J.; Roe, K.J.; Renaud, C.B.; Gill, H.S.; Potter, I.C.; Freyhof, J.; Naseka, A.M.; Cochran, P.A.; Pérez, H.E.; Habit, E.M.; et al. Novel Relationships among Lampreys (Petromyzontiformes) Revealed by a Taxonomically Comprehensive Molecular Data Set. *Am. Fish. Soc. Symp.* **2009**, *72*, 41–55.
29. Delarbre, C.; Escriva, H.; Gallut, C.; Barriel, V.; Kourilsky, P.; Janvier, P.; Laudet, V.; Gachelin, G. The complete nucleotide sequence of the mitochondrial DNA of the agnathan *Lampetra fluviatilis*: Bearings on the phylogeny of cyclostomes. *Mol. Biol. Evol.* **2000**, *17*, 519–529. [CrossRef] [PubMed]
30. Levin, B.; Ermakov, A.; Ermakov, O.; Levina, M.; Sarycheva, O.; Sarychev, V. Ukrainian Brook Lamprey *Eudontomyzon mariae* (Berg): Phylogenetic position, genetic diversity, distribution, and some data on biology. In *Jawless Fishes of the World*, 2nd ed.; Orlov, A., Beamish, R., Eds.; Cambridge Scholar Publishing, Lady Stephenson Library: Newcastle upon Tyne, UK, 2016; Volume 1, pp. 154–196.
31. Fox, A.M. A Molecular Phylogenetic Analysis of the Lamprey Genera Eudontomyzon, Lampetra, and Lethenteron. Ph.D. Thesis, University of Manitoba, Department of Biological Sciences, Winnipeg, MB, Canada, 29 August 2016.
32. Bouckaert, R.; Heled, J.; Kuhnert, D.; Vaughan, T.; Wu, C.-H.; Xie, D.; Suchard, M.A.; Rambaut, A.; Drummond, A.J. BEAST 2: A software platform for Bayesian evolutionary analysis. *PLoS Comput. Biol.* **2014**, *10*, e1003537. [CrossRef]
33. Pereira, A.M.; Levy, A.; Vukić, J.; Šanda, R.; Levin, B.A.; Freyhof, J.; Geiger, M.; Choleva, L.; Francisco, S.M.; Robalo, J.I. Putting European lampreys into perspective: A global-scale multilocus phylogeny with a proposal for a generic structure of the Petromyzontidae. *J. Zool. Syst. Evol. Res.* **2021**, *59*, 1982–1993. [CrossRef]
34. Yu, Y.; Harris, A.J.; He, X. S-DIVA (Statistical Dispersal-Vicariance Analysis): A tool for inferring biogeographic histories. *Mol. Phylogenet. Evol.* **2010**, *56*, 848–850. [CrossRef] [PubMed]
35. Nei, M. Evolution of human races at the gene level. *Prog. Clin. Biol. Res.* **1982**, *103 Pt A*, 167–181.
36. Lynch, M.; Crease, T.J. The analysis of population survey data on DNA sequence variation. *Mol. Biol. Evol.* **1990**, *7*, 377–394. [PubMed]
37. Hudson, R.R.; Slatkin, M.; Maddison, W.P. Estimation of levels of gene flow from DNA sequence data. *Genetics* **1992**, *132*, 583–589. [CrossRef] [PubMed]
38. Krezsek, C.; Olariu, C. Filling of sedimentary basins and the birth of large rivers: The lower Danube network in the Dacian Basin, Romania. *Glob. Planet. Chang.* **2021**, *197*, 103391. [CrossRef]
39. Magyar, I.; Radivojević, D.; Sztano, O.; Synak, R.; Ujszaszi, K.; Pocsik, M. Progradation of the paleo-Danube shelf margin across the Pannonian Basin during the Late Miocene and Early Pliocene. *Glob. Planet. Chang.* **2013**, *103*, 168–173. [CrossRef]
40. Buj, I.; Šanda, R.; Marčić, Z.; Ćaleta, M.; Mrakovčić, M. Combining morphology and genetics in resolving taxonomy—a systematic revision of spined loaches (genus *Cobitis*; Cypriniformes, Actinopterygii) in the Adriatic Watershed. *PLoS ONE* **2014**, *9*, e99833. [CrossRef]
41. Meyer, L.; Brunken, H. Historical occurrence and current distribution of migrating fishes and lampreys (Osteichthyes et Cyclostomata) in the drainage system of the River Aller (Lower Saxony) with an evaluation of future developments of their stocks. *Braunschweiger Naturkundliche Schriften* **1997**, *5*, 281–303. (in German).

42. Ojutkangas, E.; Aronen, K.; Laukkanen, E. Distribution and abundance of river lamprey (*Lampetra fluviatilis*) ammocoetes in the regulated river Perhonjoki. *Regul. Rivers Res. Manag.* **1995**, *10*, 239–245. [CrossRef]
43. Kelly, F.L.; King, J.J. A review of the ecology and distribution of three lamprey species, *Lampetra fluviatilis* (L.), *Lampetra planeri* (Bloch) and *Petromyzon marinus* (L.): A context for conservation and biodiversity considerations in Ireland. In *Biology and Environment: Proceedings of the Royal Irish Academy*; Royal Irish Academy: Dublin, Ireland, 2001; Volume 101, p. 101.
44. Clemens, B.J.; Arakawa, H.; Baker, C.; Coghlan, S.; Kucheryavyy, A.; Lampman, R.; Lança, M.J.; Mateus, C.S.; Miller, A.; Nazari, H.; et al. Management of anadromous lampreys: Common threats, different approaches. *J. Great Lakes Res.* **2021**, *47* (Suppl. S1), S129–S146. [CrossRef]

water

MDPI

Article

Study on the Spatiotemporal Distribution Characteristics of Meiofauna in Baiyangdian Lake and Its Influencing Factors

Yingkun Cao [1,2], Jiandong Mu [3], Zhe Pan [2], Futang Ma [2], Jianxia Liu [2], Haojun Dong [2], Wei Zhang [1,*,†] and Liqing Wang [1,*,†]

1 Key Laboratory of Exploration and Utilization of Aquatic Genetic Resources of the Ministry of Education, Engineering Research Center of Environmental DNA and Ecological Water Health Assessment, Shanghai Ocean University, Shanghai 201306, China; yingkun-cao@163.com

2 Ocean College, Hebei Agricultural University, Qinhuangdao 066000, China; lfpanzhe@163.com (Z.P.); futang_m@126.com (F.M.); ljx001130@126.com (J.L.); dhj1254493244@gmail.com (H.D.)

3 Key Laboratory of Marine Living Resources and Environment in Hebei Province, Hebei Ocean and Fisheries Science Research Institute, Qinhuangdao 066200, China; jesse.mu@163.com

* Correspondence: weizhang@shou.edu.cn (W.Z.); lqwang@shou.edu.cn (L.W.)

† These authors contributed equally to this work.

Abstract: Baiyangdian Lake, the largest freshwater shallow lake on the North China Plain, plays a pivotal role in maintaining the regional ecological balance and biodiversity. Meiofauna are integral components of Baiyangdian Lake; however, their community characteristics and relationship with environmental factors have not yet been studied. The aim of the following study was to evaluate the density, spatiotemporal patterns, and habitat response dynamics of meiofauna in Baiyangdian Lake. A field investigation was conducted at 33 sites spanning various habitats, including aquatic plant-dominant, trench, and pelagic areas, across the spring, summer, and autumn seasons of 2021. The results revealed that the meiofauna in Baiyangdian Lake primarily comprise freshwater nematodes (91.78%), ostracods, and copepods, with a mean abundance of 69.40 ± 35.20 ind. 10 cm^{-2}, peaking in the spring, followed by summer and autumn. The mean biomass was 164.95 ± 99.39 dwt. 10 cm^{-2}, with that of ostracods being the most substantial and that of copepods being the least, with both of them exhibiting seasonal fluctuations. Notably, in the summer, the abundance of meiofauna was positively correlated with the water depth and negatively correlated with ammonia nitrogen levels (R^2 = 0.13 and R^2 = 0.24, respectively; $p < 0.05$ and $p < 0.01$; n = 33). The results of our study indicate that the distribution and abundance of meiofauna are significantly affected by environmental factors, with the water depth and ammonia nitrogen levels being potential key determinants. The results of the present study are conducive to evaluating the health status of the Baiyangdian ecosystem, protecting biodiversity, and studying the impacts of anthropogenic activities and environmental changes on the lake, and can also provide scientific support for its ecological restoration and governance as well as the assessment of ecological service functions.

Keywords: meiofauna; spatiotemporal distribution; environmental factors; Baiyangdian Lake

1. Introduction

Meiofauna are crucial to ecosystems, linking material cycles and energy flows and serving as key indicators of environmental changes [1]. Initially, the nesting and feeding activities of meiofauna can alter the transportation of particles and microorganisms as well as the transfer of organic matter, thereby modifying the physical, chemical, and biological properties of sediments [2]. Meanwhile, their activities significantly influence the service processes of benthic ecosystems, such as sediment stability, biogeochemical cycles, and waste emissions, and also exert a direct or indirect positive or negative impact on the dynamics of food webs [1]. Furthermore, owing to the characteristics of meiofauna, such as its short life cycle (3–5 generations per year), stable feeding types, and sensitivity to disturbances such as pollution, meiofauna have emerged as an important indicator organism for marine environmental quality [3]. In particular, free-living nematodes, owing to their biological and ecological traits, can act as indicators of ecosystem health [4]. At present, significant progress has been made in research on the environmental indication of meiofauna. For instance, scholars in countries such as Poland, Japan, Switzerland, and Spain have utilized meiofaunal assemblages to indicate environmental changes, such as water acidification, anoxic events, water depth variations, and an altered water nutrient status [5–10]. However, studies on freshwater meiofauna in China, which started in the 1970s, remain limited, with only occasional reports on species [11] and descriptions of community characteristics [12–17]. Research on the distribution of meiofauna and their response to environmental changes remains limited.

Meiofauna can function as environmental indicator organisms because meiofaunal assemblages and spatiotemporal distribution are substantially influenced by environmental changes [18]. The distribution of meiofauna is notably influenced by environmental factors, such as transparency, chlorophyll a (Chl a), water depth, water temperature, dissolved oxygen (DO), ammonia nitrogen (NH_3-N), nitrate nitrogen, and redox potential, and thus displays a high degree of spatial and temporal variability [19–26]. Meiofaunal assemblages may constitute a useful bioindicator of a lake's trophic state [6,10,26,27]. For example, when comparing three lakes with different trophic statuses (oligo-, meso-, and eutrophic), Schroeder discovered the highest meiofaunal abundances in the mesotrophic lake [10]. Kurashov compared meio- and macrofaunal productions in lakes with varying trophic states and demonstrated that meiofaunal production was highest in the littoral zone of eutrophic lakes and was twice as high as the production of benthic macroinvertebrates in the shallow zone of a mesotrophic lake [28]. The density of the numerically dominant nematodes declined upon nutrient enrichment, whereas ostracods became more abundant. Other taxa, including copepods, attained a maximum at intermediate nutrient levels or, in the case of oligochaetes, were nearly unaffected by nutrient enrichment [29].

Baiyangdian Lake is the largest freshwater lake in the North China Plain [30]. Its ecosystem health and biodiversity are crucial for the regional ecological environment and the maintenance of ecological balance [30]. Over the past 12 years, the habitat of Baiyangdian Lake has undergone continuous degradation, leading to a decrease in macrofauna diversity, with pollution-tolerant species becoming dominant [31–33]. However, research on meiofauna, which play a significant role in ecosystem dynamics and serve as important environmental indicators, remains largely explored [34].

We hypothesized that differences in the spatiotemporal distribution of meiofauna are caused by different environmental factors and that, therefore, differences in the distribution of meiofauna can indicate changes in environmental factors in the study waters. Based on the above scientific hypotheses, we analyzed the physical and chemical factors of Baiyangdian Lake water, the species composition, quantity distribution and assemblage of meiofauna, and their relationships. Our study results will be beneficial for assessing the health status of the Baiyangdian Lake ecosystem, determining how to protect its biodiversity, studying the impact of human activities and environmental changes, and providing scientific support for its ecological restoration and governance, as well as the evaluation of ecological service functions.

2. Materials and Methods

2.1. Study Sites

In the following study, 33 sampling sites (Figure 1, Table 1) were set up across Baiyangdian Lake according to different habitat types, such as aquatic plant-dominant areas (abbreviated as S), trench areas (abbreviated as H), and pelagic areas (abbreviated as K), for data collection across the spring (April), summer (July), and autumn (October) in 2021 to investigate and measure the meiofauna and physicochemical factors of the water body. Sample collection and processing were carried out with reference to the method of Ristau et al. (2012) [29].

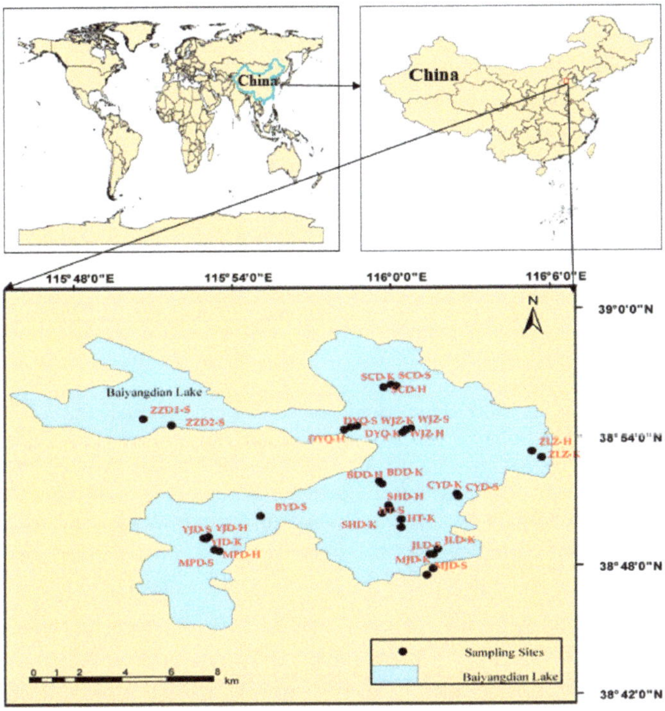

Figure 1. Sampling sites and study area (note: 33 sampling sites were set up across Baiyangdian Lake according to different habitat types, such as aquatic plant-dominant areas (abbreviated as S), trench areas (abbreviated as H), and pelagic areas (abbreviated as K)).

Table 1. List of site names, locations, habitat types, and abbreviations of the 33 sampling sites.

Sampling Site Name	Longitude (°)	Latitude (°)	Habitat Type	Abbreviations
Mapengdian	115.886	38.814	Aquatic plant-dominant areas	MPD-S
Mapengdian	115.883	38.815	Trench areas	MPD-H
Yangjiaodian	115.876	38.824	Pelagic areas	YJD-K
Yangjiaodian	115.878	38.824	Aquatic plant-dominant areas	YJD-S
Yangjiaodian	115.879	38.825	Trench areas	YJD-H
Baiyangdian	115.912	38.841	Aquatic plant-dominant areas	BYD-S
Houtang	116.001	38.838	Aquatic plant-dominant areas	HT-S
Houtang	116.001	38.832	Pelagic areas	HT-K
Julongdian	116.019	38.811	Trench areas	JLD-H
Julongdian	116.021	38.811	Aquatic plant-dominant areas	JLD-S
Julongdian	116.024	38.815	Pelagic areas	JLD-K
Mengjiadian	116.021	38.8	Pelagic areas	MJD-K
Mengjiadian	116.017	38.795	Aquatic plant-dominant areas	MJD-S
Badadian	115.989	38.866	Trench areas	BDD-H
Badadian	115.987	38.868	Pelagic areas	BDD-K
Shihoudian	115.989	38.843	Pelagic areas	SHD-K
Shihoudian	115.994	38.846	Aquatic plant-dominant areas	SHD-S
Shihoudian	115.993	38.849	Trench areas	SHD-H
Dayaquan	115.969	38.91	Trench areas	DYQ-H
Dayaquan	115.973	38.911	Pelagic areas	DYQ-K
Dayaquan	115.965	38.908	Aquatic plant-dominant areas	DYQ-S
Zaolinzhuang	116.089	38.886	Pelagic areas	ZLZ-K
Zaolinzhuang	116.083	38.891	Trench areas	ZLZ-H

Table 1. *Cont.*

Sampling Site Name	Longitude (°)	Latitude (°)	Habitat Type	Abbreviations
Chiyudian	116.036	38.858	Pelagic areas	CYD-K
Chiyudian	116.037	38.856	Aquatic plant-dominant areas	CYD-L
Shaochedian	115.99	38.941	Trench areas	SCD-H
Shaochedian	115.998	38.942	Pelagic areas	SCD-K
Shaochedian	115.995	38.943	Aquatic plant-dominant areas	SCD-S
Wangjiazhai	116.002	38.906	Trench areas	WJZ-H
Wangjiazhai	116.004	38.908	Pelagic areas	WJZ-K
Wangjiazhai	116.007	38.909	Aquatic plant-dominant areas	WJZ-S
Zaozhadian1	115.838	38.917	Aquatic plant-dominant areas	ZZD1-S
Zaozhadian2	115.856	38.912	Aquatic plant-dominant areas	ZZD2-S

2.2. Environmental Parameters

Field measurements of the physicochemical factors of water quality, namely the depth (m), temperature (°C), pH, and DO (mg L^{-1}) of the water at each station, were taken using a portable multi-parameter water quality meter (YSI ProPlus, YSIInc., Yellow Springs, OH, USA). Turbidity FTU (FNU) was measured using a portable suspended solids monitor (Hach TSS Portable 2100Q, HACH, Loveland, CO, USA). Petroleum hydrocarbon TPHs (μg L^{-1}) were measured using a portable oil meter (Oil Tech121A, Environ., Boston, MA, USA). The water body Chl a was measured using a hand-held chlorophyll meter (Chloro Tech 121A, Environ., Boston, MA, USA). Transparency (represented by Secchi disk depth, SDD) (m) was measured using a Secchi disk (Shanghai Biaozhuo Scientific Instrument Co., Ltd., Shanghai, China).

After the collection of the water samples, the pH, dissolved oxygen (DO), Chl a, ammonia nitrogen (NH_3-N), TN, TP, and COD values were analyzed in the laboratory based on "Surface Water Environmental Quality Standard GB 3838-2002" [35].

The calculation formula of the comprehensive trophic level index (TLI) of the water body is as follows [36,37]:

$$TLI = \sum_{i=1}^{m} \omega i \times TLI(i)$$

$$\omega i = \frac{rij^2}{\sum_{i=1}^{m} rij^2}$$

In the formula, TLI(i) represents the trophic level index of the i-th indicator; W_i represents the relevant weight of the trophic level index of the i-th indicator; r_{ij} represents the correlation coefficient between the i-th indicator and the reference parameter Chl a; and m represents the number of evaluation indicators [37].

We evaluated the study area regarding the comprehensive trophic level index (Table 2).

Table 2. The corresponding relationship between the TLI and nutrition category [37].

Category	Oligotrophic	Mesotrophic	Lightly Eutrophic	Moderately Eutrophic	Severely Eutrophic	Extremely Eutrophic
TLI	<30	$30 \leq TLI \leq 50$	$50 < TLI \leq 60$	$60 < TLI \leq 70$	$70 < TLI \leq 80$	TLI > 80

Note: within the same trophic state, the higher the index value, the greater the degree of eutrophication.

2.3. Meiofauna Analyses

In April, July, and October 2021, sediment from each station was collected using a Peterson mud sampler (with a sampling area of 1/16 m^2) and carefully placed on a tray. A modified sampling tube (15 cm in length and 2.6 cm in inner diameter) was used to collect core samples of 0–5 cm [38]. Three parallel samples were collected at each station, and the parallel samples were mixed and placed in a sampling bottle, which was immediately fixed with 75% ethanol solution and returned to the laboratory.

Sample staining: During this stage, 5 mL of Rose Bengal staining solution (0.1 g of Rose Bengal dye dissolved in 100 mL of distilled water) was added to the sample, shaken well, and allowed to stand for 24 h [39].

Sample rinsing: The stained sample was placed on a set of sieves composed of an upper layer with a 500 μm aperture and a lower layer with a 42 μm aperture and was slowly rinsed with filtered running water. The sieve was gently tapped and oscillated to remove ethanol and fine sediments until the filtrate was clear [40].

Sample sorting and counting: The sample was transferred to a Petri dish with equally wide parallel lines drawn on it and then sorted and counted by taxon under a dissecting microscope (Nikon SMZ800, Melville, NY, USA). The sorted samples were added to 75% ethanol-fixing solution for preservation and labeled appropriatey.

2.4. Data Analysis

The abundance calculation formula is as follows:

$$X = 10a\pi \times (d/2)^2$$

The abundance of meiofauna is expressed in units of "ind. 10 cm^{-2}", which represents the number of meiofauna per 10 cm^2. It was used to calculate the abundance of a certain type of meiofauna in a sample. "X" represents abundance, "d" represents the inner diameter of the sampler in "d" cm, and "a" represents the number of organisms obtained in the experiment.

The estimation of biomass "B" utilizes a method of multiplying the empirical value of the average dry weight of individuals in each group by the abundance of each group [41–43]. The empirical values of the average dry weight for the different groups of meiofauna in the present study were derived from previous studies [42,44–46], measured in units of μg dwt. 10cm^{-2} (Table 3).

Table 3. Empirical values of the individual average dry weights for different meiofauna groups.

Group	Individual Dry Weight (μg)
Nematodes	0.40
Ostracods	26.00
Copepods	1.86

Linear regression analysis and plotting of the correlations between the measured environmental factors and benthic organisms were performed using Origin 2021. The environmental factors were analyzed using the R studio 4.3.0 FactoMineR and Factoextra packages, and PCA plots were generated using ggplot2 and corrplot.

3. Results

3.1. Physical and Chemical Parameters of the Water Body

The various environmental factors of the water in the three different habitats (pelagic areas abbreviated as K, aquatic plant-dominant areas abbreviated as S, and trench area abbreviated as H) across the different seasons are presented in Figure 2.

The PCA results showed that in the spring (Figure 3a), Dim1 explained 26.9% of the environmental variability, whereas Dim2 explained 17.9%, with them jointly explaining 44.8% of the environmental variability. There was a certain clustering phenomenon of the meiofaunal groups in the different habitats. Although there was a more significant overlap when the habitats in the trench, pelagic, and aquatic plant-dominant areas clustered, the trench area showed a separation phenomenon compared to the pelagic and aquatic plant-dominant areas, with higher habitat overlap. This finding suggests that in the spring, the habitat difference between the pelagic and aquatic plant-dominant areas is relatively small; in comparison, there were significant differences between the trench, pelagic, and aquatic plant-dominant areas. In addition, DO contributed the most to Dim1, showing a negative correlation. Total nitrogen, water depth, and Chl a contributed significantly to Dim1 and were negatively correlated, except for Chl a. Turbidity contributed the most to Dim2, showing a negative correlation. Total phosphorus, water depth, water temperature, and petroleum hydrocarbons contributed significantly to Dim2, among which total phosphorus was negatively correlated, and the other factors were positively correlated.

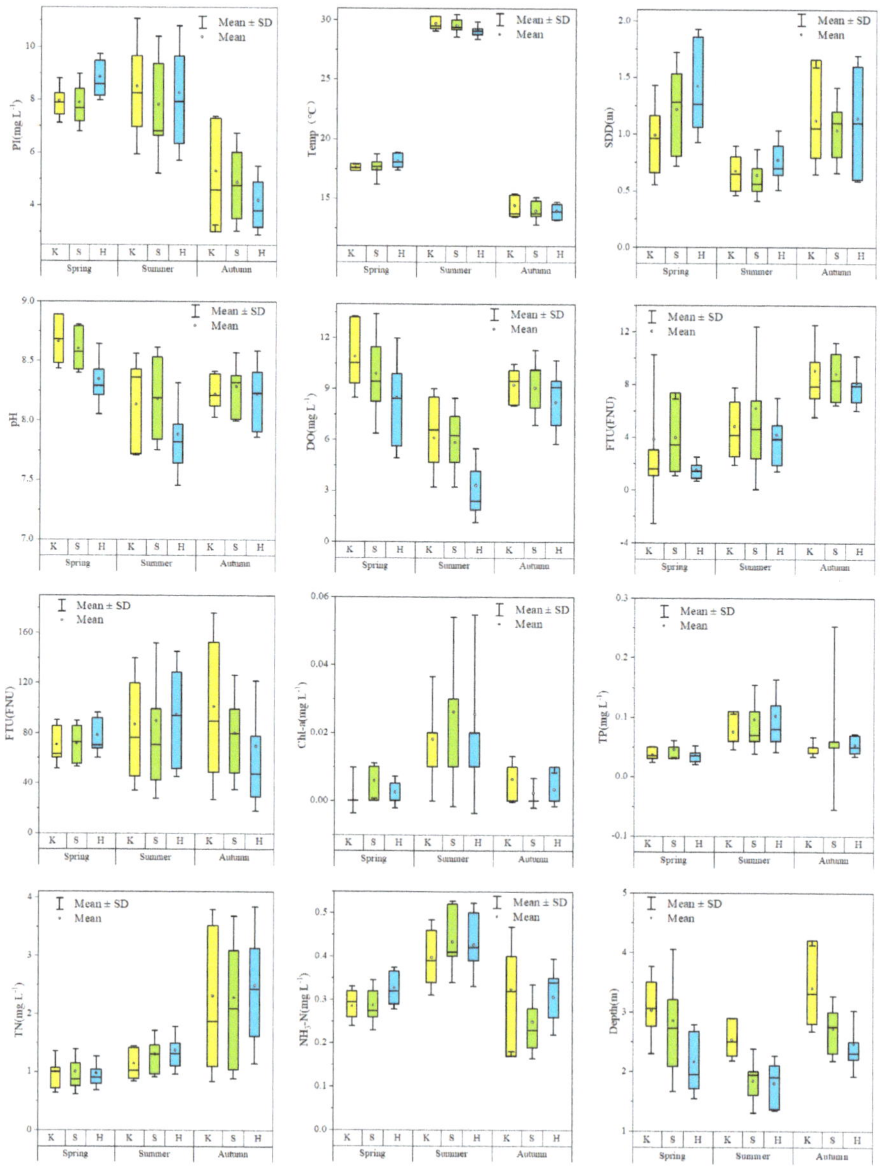

Figure 2. Boxplots of the water-related environmental factors in the different seasons and habitats of Baiyangdian Lake.

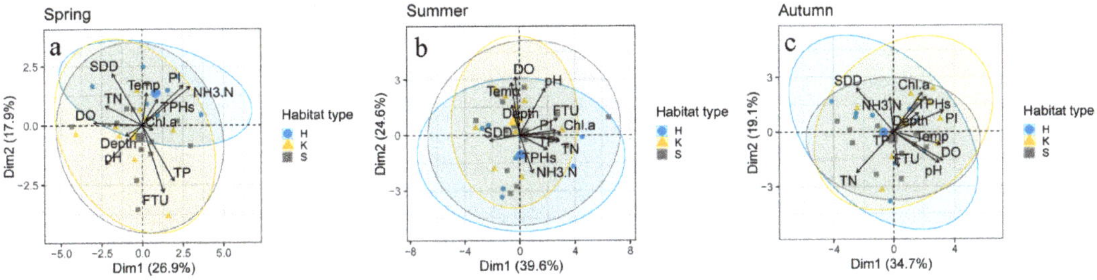

Figure 3. Evolution of the complement system: (**a**) spring, (**b**) summer, and (**c**) autumn.

In the summer (Figure 3b), Dim1 explained 39.6% of the environmental variability, while Dim2 explained 24.6%, with them jointly explaining 64.2% of the environmental variability. After clustering the habitats in the trench, pelagic, and aquatic plant-dominant areas at a confidence interval of 95%, there was a large overlap, indicating a relatively small habitat difference. Among them, Chl a contributed the most to Dim1 and was positively correlated. In addition, transparency contributed significantly to Dim1 and was negatively correlated. DO was found to make the greatest contribution to Dim2 and was positively correlated; in comparison, ammonia nitrogen contributed significantly to Dim2 and was negatively correlated.

In the autumn (Figure 3c), Dim1 explained 34.7% of the environmental variability, while Dim2 explained 19.1%, with them jointly explaining 53.8% of the environmental variability. Similar to the observations made in the summer, there was a large overlap between the habitats and a small difference. The permanganate index contributed significantly to Dim1 and was positively correlated, while transparency and total nitrogen contributed significantly to Dim1 and were negatively correlated. Transparency contributed significantly to Dim2 and was positively correlated; in comparison, turbidity contributed significantly to Dim2 and was negatively correlated.

In the spring, data from a total of 30 sampling sites were collected, with only one station being slightly eutrophic, 2 sites being poor in nutrients, and the other 27 sites reaching a medium nutrient level. In the summer, data from a total of 33 sites were collected, with 6 sites being moderately eutrophic, 19 sites being slightly eutrophic, and 8 sites reaching a medium nutrient level. In the autumn, data from a total of 33 sites were collected, with 5 sites being slightly eutrophic and the other 28 sites all reaching a medium nutrient level. Throughout the year (excluding the winter), Baiyangdian Lake had poor to medium nutrient levels, accounting for 67.71%. Among the seasons, the spring showed higher levels compared to the autumn, which, in turn, showed higher levels than the summer.

3.2. Distribution of the Meiofaunal Assemblages

Surveys of meiofauna at various sites in Baiyangdian Lake were conducted in the spring, summer, and autumn. The main groups identified were freshwater nematodes (hereinafter referred to as nematodes), ostracods, and copepods. In terms of abundance, nematodes emerged as the most dominant group in all seasons followed by ostracods, with copepods exhibiting the lowest numbers (Table 4).

The abundance of meiofauna showed seasonal variations, with spring > summer > autumn. The mean abundance of meiofauna was 69.40 ± 35.20 ind. 10 cm^{-2}, and there were also differences across the sites (Figure 4). In the spring, the mean abundance of nematodes at each station was 84.95 ± 39.19 ind. 10 cm^{-2}, with the highest abundance in the Mengjiadian aquatic plant-dominant area (water inlet) at 973.63 ind. 10 cm^{-2}, followed by 231.79 ind. 10 cm^{-2} in the Badadian trench area, 174.00 ind. 10 cm^{-2} in the Wangjiazhai pelagic area, 168.31 ind. 10 cm^{-2} in the Julongdian aquatic plant-dominant area, 101.13 ind. 10 cm^{-2} in the Badadian pelagic area, and no meiofauna in the Chiyudi reed or

Julongdian trench areas. In the summer, the mean abundance of nematodes at each station was 74.01 ± 20.44 ind. 10 cm^{-2}, with the highest abundance in the Shiyihdian aquatic plant-dominant area at 479.28 ind. 10 cm^{-2}, followed by 410.81 ind. 10 cm^{-2} in the Shaochedian trench area, 290.83 ind. 10 cm^{-2} in the Mengjiadian pelagic area, 170.23 ind. 10 cm^{-2} in the Shiyihdian trench area, and 168.97 ind. 10 cm^{-2} in the Shaochedian aquatic plant-dominant area, whereas no meiofauna were found in the Yangjiaodian aquatic plant-dominant area. In the autumn, the mean abundance of nematodes at each station was 27.25 ± 5.81 ind. 10 cm^{-2}, with the highest abundance in the Badadian pelagic area at 108.04 ind. 10 cm^{-2}, followed by 103.02 ind. 10 cm^{-2} in the Mengjiadian pelagic area, 87.31 ind. 10 cm^{-2} in the Shiyihdian trench area, 54.02 ind. 10 cm^{-2} in the Dayaquan aquatic plant-dominant area, and 101.13 ind. 10 cm^{-2} in the Badadian pelagic areas, whereas no meiofauna were found in the Yangjiaodian aquatic plant-dominant or Baiyangdian Lake lotus areas.

Table 4. Various groups of meiofauna by abundance across the different seasons.

Group	Item	Spring	Summer	Autumn
Nematodes	Abundance (ind. 10 cm^{-2})	84.95 ± 39.19	74.01 ± 20.44	27.25 ± 5.81
	Relative proportion of abundance (%)	93.86	88.56	92.91
	Biomass (dwt. 10 cm^{-2})	33.98 ± 15.68	29.60 ± 8.18	0.11 ± 2.32
	Relative proportion of biomass (%)	19.65	11.38	0.21
Ostracods	Abundance (ind. 10 cm^{-2})	5.32 ± 1.90	8.81 ± 2.01	1.95 ± 1.20
	Relative proportion of abundance (%)	5.89	10.55	6.65
	Biomass (dwt. 10 cm^{-2})	138.49 ± 49.30	229.14 ± 52.23	50.69 ± 31.27
	Relative proportion of biomass (%)	80.10	88.09	99.78
Copepods	Abundance (ind. 10 cm^{-2})	0.23 ± 0.12	0.74 ± 0.18	0.13 ± 0.09
	Relative proportion of abundance (%)	0.25	0.89	0.44
	Biomass (dwt. 10 cm^{-2})	0.42 ± 0.22	1.38 ± 0.34	0.24 ± 0.17
	Relative proportion of biomass (%)	0.24	0.53	0.00

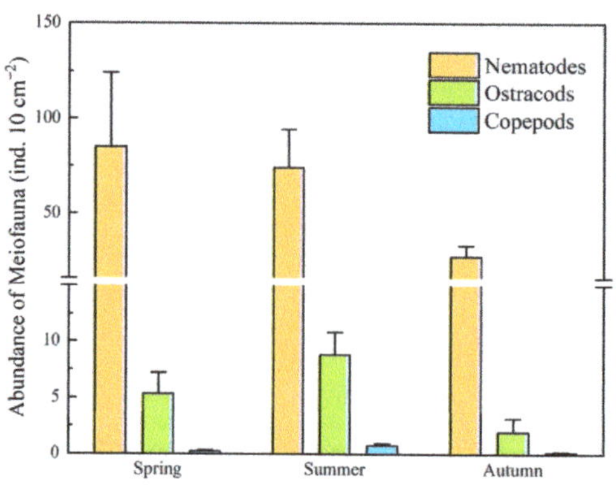

Figure 4. Seasonal variation in the abundance of meiofauna in Baiyangdian Lake.

The mean biomass of meiofauna was 164.95 ± 99.39 dwt. 10 cm^{-2}, following the order of ostracods > nematodes > copepods. The biomass level across different seasons followed the order of summer > spring > autumn (Figure 5). In the spring, the biomass levels of nematodes and ostracods in the aquatic plant-dominant area were the highest at 76.71 and 1.00 µg dwt. 10 cm^{-2}, respectively. The biomass levels of nematodes and copepods in the trench area were the lowest at 16.19 and 0 µg dwt. 10 cm^{-2}, respectively. The biomass levels of ostracods in the aquatic plant-dominant area were the lowest at 81.66 µg dwt. 10 cm^{-2}. In the summer, the biomass levels of ostracods were the highest in the pelagic area at 313.28 µg dwt. 10 cm^{-2} and the lowest in the trench area at 139.73 µg dwt. 10 cm^{-2}. In comparison, the biomass levels of nematodes and copepods were similar in the trench, pelagic, and aquatic plant-dominant areas. In the autumn, the biomass levels of ostracods and copepods in the trench area were the lowest at 14.29 and 0 μg dwt. 10 cm^{-2}, respectively. Moreover, the biomass levels of nematodes and copepods in the pelagic area were the highest at 15.48 and 0.73 μg dwt. 10 cm^{-2}, respectively. The biomass levels of ostracods in the aquatic plant-dominant area were the highest at 109.42 μg dwt. 10 cm^{-2}; in comparison, the biomass levels of nematodes were the lowest at 9.4 μg dwt. 10 cm^{-2}. According to the above analysis, the season has a certain influence on the biomass of meiofauna, with the highest biomass levels of ostracods and copepods occurring in the summer and the lowest levels occurring in the autumn. The biomass levels of nematodes followed the order of spring > summer > autumn.

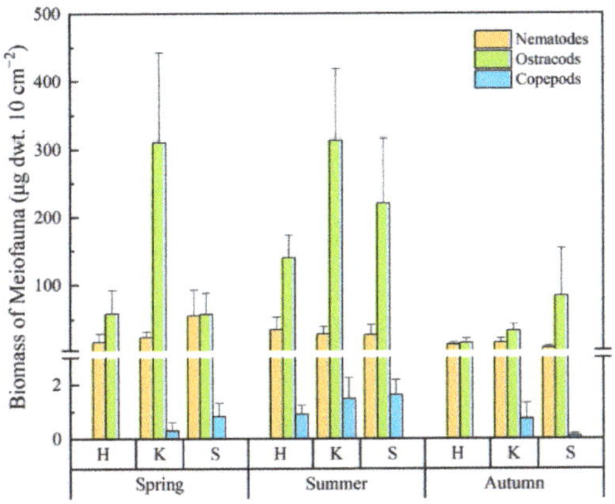

Figure 5. Seasonal variation in the biomass of meiofauna in Baiyangdian Lake.

3.3. Relationships between Meiofauna and Physicochemical Parameters

Correlation analysis was conducted to study the abundance of meiofauna across different seasons and 12 environmental factors (Table 5, Figure 6). In the spring, the abundances of meiofauna, nematodes, and ostracods did not show significant correlations with the various environmental factors. The abundance of copepods was significantly positively correlated with total nitrogen ($p < 0.01$) and significantly negatively correlated with water temperature ($p < 0.01$). In the summer, the total abundance of meiofauna was significantly positively correlated with water depth ($R^2 = 0.13$, $p < 0.05$, n = 33) and significantly negatively correlated with ammonia nitrogen ($R^2 = 0.24$, $p < 0.01$, n = 33). The abundance of ostracods was significantly positively correlated with dissolved oxygen (DO) ($p < 0.05$). In the autumn, the total abundances of meiofauna, nematodes, and ostracods was not significantly correlated with the various environmental factors. The abundance of copepods was significantly positively correlated with water depth and turbidity ($p < 0.05$, $p < 0.01$). The abundances of meiofauna and nematodes in each season were not significantly correlated with the comprehensive trophic status index (TLI) ($p > 0.05$).

Table 5. Correlation coefficient between the abundance of meiofauna and environmental factors of Baiyangdian Lake in the spring, summer, and autumn.

Environmental Factor	Season											
	Spring				Summer				Autumn			
	Nematodes	Ostracods	Copepods	Total	Nematodes	Ostracods	Copepods	Total	Nematodes	Ostracods	Copepods	Total
Depth	0.371	0.243	0.033	0.384	0.349 *	0.302	−0.293	0.365 *	0.231	−0.113	0.387 *	0.215
Temp	0.140	−0.006	−0.519 **	0.107	0.099	0.284	−0.053	0.123	0.217	0.127	−0.015	0.245
SDD	−0.351	0.230	0.175	−0.348	0.173	0.083	−0.009	0.176	−0.118	−0.282	0.264	−0.174
pH	0.236	0.096	0.305	0.248	0.114	0.215	−0.225	0.129	0.057	0.075	−0.019	0.072

Table 5. *Cont.*

Environmental Factor	Season											
	Spring				Summer				Autumn			
	Nematodes	Ostracods	Copepods	Total	Nematodes	Ostracods	Copepods	Total	Nematodes	Ostracods	Copepods	Total
DO	0.218	0.104	0.063	0.224	0.288	0.367 *	−0.257	0.312	0.140	0.107	−0.049	0.163
FNU	0.184	−0.232	−0.050	0.181	−0.012	−0.142	−0.045	−0.025	0.087	0.272	0.682 **	0.155
TPHs	0.040	−0.151	0.060	0.054	−0.309	−0.283	−0.294	−0.329	0.088	−0.112	−0.046	0.064
Chl-a	−0.071	−0.159	0.118	−0.072	−0.177	−0.074	−0.221	−0.181	0.261	−0.076	−0.164	0.245
TP	0.160	−0.299	0.280	0.160	−0.162	−0.320	−0.182	−0.189	0.134	0.005	−0.014	0.136
TN	−0.254	0.268	0.622 **	−0.219	−0.067	−0.182	−0.220	−0.085	−0.233	0.136	0.083	−0.205
NH_3-N	−0.206	−0.131	−0.274	−0.206	−0.482 **	−0.282	0.126	−0.494 **	0.248	−0.155	0.051	0.218
COD_{Mn}	−0.287	−0.174	−0.395	−0.289	−0.216	0.072	−0.160	−0.204	0.147	−0.105	−0.041	0.126

Notes: Depth: water depth; Temp: water temperature; SDD: Secchi disk depth (transparency); pH: pH value (acidity and alkalinity value); DO: dissolved oxygen; FTU: turbidity; TPHs: total petroleum hydrocarbons; Chl-a: chlorophyll-a; TP: total phosphorus; TN: total nitrogen; NH_3-N: ammonia nitrogen; COD_{Mn}: permanganate index. The symbol * indicates significant correlation at the 0.05 level (two sides), and ** indicates significant correlation at the 0.01 level (two sides).

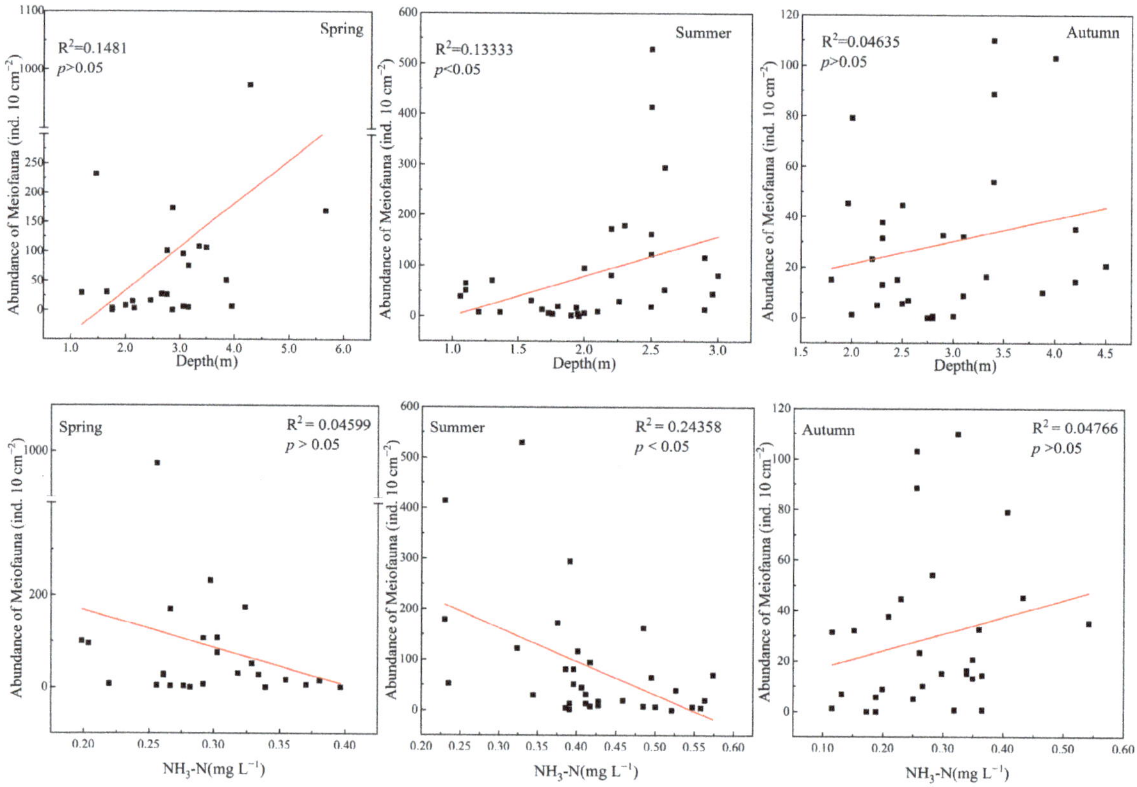

Figure 6. Diagrams showing the relationship between the abundance of meiofauna, ammonia nitrogen, and the water depth of Baiyangdian Lake across different seasons.

4. Discussion

4.1. Spatiotemporal Distribution of Meiofauna in Baiyangdian Lake

In the present study, our results revealed that in Baiyangdian Lake, meiofaunal assemblages only include nematodes, ostracods, and copepods, and nematodes are also the predominant group. Nematodes outnumber other multicellular animals on the ocean floor and in inland waters and soils, making them essential components of nearly all ecosystems on Earth [1,47,48]. Nematodes exhibit significant ecological dominance, comprising the majority of the meiofaunal assemblage. The majority of the current knowledge on freshwater nematode ecology is based on research in European freshwater bodies only. In recent years, the relative importance of nematodes in freshwater habitats has attracted growing attention; however, the study of nematode ecology in freshwater lakes remains particularly limited [1,19,49–53]. In China, Jihua Wu first reported the annual qualitative and quantitative research results for nematodes in East Lake, Wuhan City, indicating that nematodes are a distinct dominant group and have a close relationship with the total phosphorus in the sediment [54]. In addition, among the investigated meiofaunal groups in Chinese water bodies such as Biandan Pond, Honghu Lake, East Lake, Dongchang Lake, Weishan Lake, and Dongtang Sun Lake, free-living nematodes are the most dominant group in terms of abundance [12–15,17,55].

Further analysis revealed that the abundance of meiofauna, particularly the dominant group of nematodes, exhibited a seasonal pattern, with the highest abundance in the spring, followed by summer, and the lowest in the autumn. Across different habitat types, the pelagic areas had the highest abundance followed by the trench areas, with the lowest abundance observed in the aquatic plant-dominant areas. This distribution is consistent with the findings of Traunspurger et al. (2020) [56], which suggest that habitats rich in periphyton support a higher meiofaunal diversity and abundance, whereas oligotrophic lakes have a lower biomass. Additionally, the results of the study by Peters [57] also indicate that in the littoral zone communities of lakes, meiofauna are abundant, with nematodes being the primary component.

In terms of biomass, in the present study, a mean biomass of 164.95 ± 99.39 μg dwt. 10 cm^{-2} was recorded for meiofauna, with the highest values observed in the summer, followed by spring, and the lowest values observed in the autumn. Ostracods exhibited the greatest biomass, whereas copepods showed the lowest. Similar findings were reported in a study on Lake Obersee [24], where oligochaetes contributed most to the biomass, and nematodes contributed most to benthic abundance.

Notably, in this study, nematodes in Baiyangdian Lake accounted for an impressive 91.55% of the total abundance of meiofauna, with their abundance peaking in the spring at 93.86%, followed by autumn at 92.91%, and the lowest abundance in the summer at 88.56%. This phenomenon suggests that the abundance of nematodes directly influences the overall abundance of meiofauna. Furthermore, in the summer, the total abundance of meiofauna was positively correlated with the water depth, which may be attributed to the better tolerance of meiofauna to low-oxygen environments. Under hypoxic conditions, the predation pressure may be reduced, favoring reproduction in deeper habitats [58,59]. Additionally, in comparison to the summer meiofauna survey results from Weishan Lake in China [15], we found that in Baiyangdian Lake, ostracods, rather than nematodes, had the highest biomass across all three seasons, which differs from the findings of the above study conducted in Weishan Lake. Concurrently, sampling research on meiofaunal abundance in Dongchang Lake during the summer and autumn by Huang Yong and colleagues [14] showed significant differences compared to the results of the present study. In their study, Wu (1999) [54], for the first time, reported the annual qualitative and quantitative research results of nematodes in East Lake, indicating that nematodes are a distinctly dominant group and have a close relationship with the total phosphorus content in sediment.

In summary, the findings of the present study demonstrate that nematodes dominate the meiofaunal assemblages in Baiyangdian Lake in terms of abundance, and their diversity is closely associated with seasonal and environmental factors. In mesotrophic or slightly eutrophic lakes, the distribution and abundance of meiofauna are influenced not only by the level of nutrient enrichment but also by other environmental factors, such as aquatic plant-dominant growth and dissolved oxygen (DO) levels in the water [24,56,57,60].

4.2. The Characteristics of the Meiofauna Community and Their Relationships with Environmental Factors

The results of the present study indicated that in the summer, the total abundance of meiofauna showed a significant positive correlation with the water depth ($p < 0.05$) and a significant negative correlation with the ammonia nitrogen concentration ($p < 0.01$). These findings suggest that water depth may be a key environmental factor affecting the distribution and abundance of meiofauna in the summer, while increased ammonia nitrogen levels may exert pressure on their survival. The results of the present study are similar to those of Ardeshir (2014) and Bianca et al. (2015) on the relationship between meiofaunal community and water depth; that is, in shallow lakes, the abundance of nematodes increased with increasing trophic levels and water depth [6,61].

For the specific meiofaunal groups of ostracods and copepods, we observed differences in their responses to environmental factors across seasons. Particularly in the summer, the abundance of ostracods was positively correlated with dissolved oxygen (DO) levels ($p < 0.05$); in comparison, copepods showed a significant positive correlation with total nitrogen levels in the spring ($p < 0.01$) and a significant negative correlation with water temperature ($p < 0.01$). The above results indicate that different groups of meiofauna may have distinct adaptive strategies to specific environmental factors.

A correlation analysis was conducted between the biomass of meiofauna and the various groups across different seasons and the total lake index (TLI), a comprehensive measure of trophic status. No significant correlations were found ($p > 0.05$). This finding contrasts with some results in the existing literature [62,63], which suggest that oligotrophic lakes often support greater biomass in lake habitats, while the opposite is true for eutrophic lakes. In these studies, the biomass of nematodes is considered a reflection of abundance, and biomass is reported to vary with increasing lake depths, indicating that depth may be an important factor influencing biomass. Furthermore, the environmental factors that affect the abundance of nematodes, the abundance of meiofauna, and the biomass of meiofauna also include the pH value of bottom water, the silt–clay content of sediments, and the organic matter content [64–66]. The total abundance of meiofauna in Baiyangdian Lake showed no significant correlation with water-related environmental factors in the spring and autumn. However, the cluster analysis results showed evident seasonal differences, which might be caused by factors such as a sedimentary environment.

Additionally, we observed significant differences in the water-related environmental factors across the various habitat types, which may be associated with the physical characteristics and biological processes of the habitats. For instance, the environmental factor values in the trench areas were lower, whereas ammonia nitrogen levels were higher in the aquatic plant-dominant areas. These differences could potentially lead to variations in the biomass of meiofauna across different habitats.

The results of the present study reveal the assemblages of meiofauna and their correlation with water-related environmental factors and show differences in different seasons and habitats. In the future, in-depth studies can be conducted on the interaction relationships of these differences with environmental factors (such as meteorology, hydrology, and sediment environment), food resources, and other organisms in the ecosystem, as well as the specific impact mechanisms on them, to provide more information for an in-depth understanding of the structure and function of shallow lake ecosystems and offer references for assessing the health and biodiversity of lake ecosystems.

In the design of our study, we adopted conventional methods for this type of research. These methods can reflect the correlation between environmental factors and the distribution of meiofauna to a certain extent; however, using such methods may lead to the neglect of the influence of sediments on them. Moreover, in recent studies, we found that some scholars have begun to pay attention to this issue. In future studies, we will continuously improve our research methods.

5. Conclusions

The results of the present study suggest that the meiofaunal community exhibits distinct seasonal variations in Baiyangdian Lake. The primary groups of meiofauna were nematodes, ostracods, and copepods, with their abundances characterized by the highest levels in the spring and the lowest levels in the autumn. Water depth and ammonia nitrogen are potential key factors affecting meiofaunal distribution and abundance. Our research results provide important information on the ecology of meiofauna in large, shallow lakes in China.

Author Contributions: L.W. and W.Z.: Writing—review and editing and Funding acquisition. Y.C.: Writing—original draft. J.M.: Monitoring environmental data and Funding acquisition. Z.P.: Formal analysis. F.M.: Data curation. J.L.: Visualization and Investigation. H.D.: Software and Validation. All authors have read and agreed to the published version of the manuscript.

Funding: This research was funded by the Project of Shanghai Science and Technology Commission (19DZ1204504), the Key R&D Projects in Hebei Province (20373603D), the Fundamental Research Funds for Hebei Province Universities (KY2021011), and Hebei Provincial Natural Science Foundation Outstanding Youth Fund Project (C2023204161).

Institutional Review Board Statement: The study presented in this article involves the investigation of meiofauna, specifically small benthic invertebrates, in Baiyangdian Lake. Given the nature of the study and the organisms involved, which are non-vertebrate and do not possess a centralized nervous system capable of experiencing pain or distress, the research does not fall under the purview of Institutional Review Board (IRB) regulations that govern the ethical treatment of animals in research.

Informed Consent Statement: Not applicable.

Data Availability Statement: Data are contained within the article.

Acknowledgments: We would like to express our gratitude to Yao Li, Xiaolong Hou from the Hebei Ocean and Fisheries Science Research Institute, and Jiale Han, Xinchao Ping, and Leyu Fan from Ocean College, Hebei Agricultural University, for their support and assistance with this article. We also would like to express our sincere gratitude to the anonymous reviewers for their valuable insights and constructive feedback, which significantly contributed to the improvement and refinement of this manuscript. Their expertise and meticulous attention to detail have been instrumental in enhancing the quality and clarity of our work.

Conflicts of Interest: The authors declare no conflicts of interest.

References

1. Rundle, S.D.; Robertson, A.L.; Schmid-Araya, J.M. *Freshwater Meiofauna: Biology and Ecology*; Backhuys Publishers: Leiden, The Netherlands, 2002.
2. Moens, T.; Beninger, P.G. Meiofauna: An inconspicuous but important player in mudflat ecology. In *Mudflat Ecology*; Springer: Cham, Switzerland, 2018; pp. 91–147.
3. Kennedy, A.D.; Jacoby, C.A. Biological indicators of marine environmental health: Meiofauna—A neglected benthic component? *Environ. Monit. Assess.* **1999**, *54*, 47–68. [CrossRef]
4. Ridall, A.; Ingels, J. Suitability of free-living marine nematodes as bioindicators: Status and future considerations. *Front. Mar. Sci.* **2021**, *8*, 685327. [CrossRef]
5. Itaoka, M.; Tamai, K. Effect of eutrophication on the structure of meiobenthic communities in Hiroshima Bay. *Benthos Res.* **1993**, *1993*, 19–28. [CrossRef]
6. Kazemi-Dinan, A.; Schroeder, F.; Peters, L.; Majdi, N.; Traunspurger, W. The effect of trophic state and depth on periphytic nematode communities in lakes. *Limnologica* **2014**, *44*, 49–57. [CrossRef]
7. Ristau, K.; Spann, N.; Traunspurger, W. Species and trait compositions of freshwater nematodes as indicative descriptors of lake eutrophication. *Ecol. Indic.* **2015**, *53*, 196–205. [CrossRef]
8. Höckelmann, C.; Moens, T.; Jüttner, F. Odor compounds from cyanobacterial biofilms acting as attractants and repellents for free-living nematodes. *Limnol. Oceanogr.* **2004**, *49*, 1809–1819. [CrossRef]
9. Gaudes, A.; Sabater, S.; Vilalta, E.; Muñoz, I. The nematode community in cyanobacterial biofilms in the river Llobregat, Spain. *Nematology* **2006**, *8*, 909–919.
10. Ristau, K.; Traunspurger, W. Relation between nematode communities and trophic state in southern Swedish lakes. *Hydrobiologia* **2011**, *663*, 121–133. [CrossRef]

11. Wu, J.H.; Liang, Y.L.; Sun, X.D. Newly recorded species of free-living nematodes from china (Chromadorida Enoplida & Araeolaimida). *Acta Hydrobiol. Sin.* **1997**, *21*, 312–321, (In Chinese with English Abstract).
12. Wu, J.H.; Liang, Y.L. A comparative study of benthic nematodes in two Chinese lakes with contrasting sources of primary production. *Hdrobiologia* **1999**, *411*, 31–37. [CrossRef]
13. Wu, J.H.; Fu, C.Z.; Liang, Y.L.; Chen, J.K. Distribution of the meiofauna community in a eutrophic shallow lake of China. *Arch. Hydrobiol.* **2004**, *159*, 555–575. [CrossRef]
14. Huang, Y.; Wu, X.Q.; Wang, X.Y.; Qi, G.L.; Cheng, B. Studies on the Meiofauna community in Dongchang Lake of Liaocheng. *J. Liaocheng Univ. (Nat. Sci.)* **2008**, *21*, 68–72+78, (In Chinese with English Abstract).
15. Li, J.J.; Huang, Y. Studies on the Meiofauna Community in Weishan Lake. *J. Liaocheng Univ. (Nat. Sci.)* **2009**, *22*, 51–54, (In Chinese with English Abstract).
16. Yan, F.G. *Community Characteristics and Ecological Role of Meiofaunan in Shallow Lakes*; Institute of Hydrobiology, Chinese Academy of Sciences: Wuhan, China, 2012; (In Chinese with English Abstract).
17. Yan, F.G.; Wang, H.J.; Wang, H.Z.; Zhang, T.L.; Li, W. Community characteristics of Meiofaunan in an algae-dominated shallow lakes, with discussion on the ecological role. *Acta Hydrobiol. Sin.* **2010**, *34*, 634–638, (In Chinese with English Abstract). [CrossRef]
18. Schratzberger, M.; Ingels, J. Meiofauna matters: The roles of meiofauna in benthic ecosystems. *J. Exp. Mar. Biol. Ecol.* **2018**, *502*, 12–25. [CrossRef]
19. Strayer, D.L. The benthic micrometazoans of Mirror lake. New Hampshire. *Arch. Hydrobiol. Suppl.* **1985**, *72*, 287–426.
20. Tinson, S.; Laybourn-Parry, J. The distribution and abundance of benthic cyclopoid copepods in Esthwaite Water, Cumbria. *Hydrobiologia* **1986**, *131*, 225–234. [CrossRef]
21. Traunspurger, W. Distribution of benthic nematodes in the littoral of an oligotrophic lake (Konigssee, National Park Berchtesgaden, FRG). *Arch. Hydrobiol.* **1996**, *135*, 393–412. [CrossRef]
22. Traunspurger, W. Distribution of benthic nematodes in the littoriprofundal and profundal of an oligotrophic lake (Konigssee, National Park Berchtesgaden, FRG). *Arch. Hydrobiol.* **1996**, *135*, 557–575. [CrossRef]
23. Bergtold, M.; Traunspurger, W. The benthic community in the profundal of Lake Brunnsee: Seasonal and spatial patterns. *Arch. Hydrobiol.* **2004**, *160*, 527–554. [CrossRef]
24. Michiels, I.C.; Traunspurger, W. A three years study of seasonal dynamics of a zoobenthos community in a eutrophic lake. *Nematology* **2004**, *6*, 655–669. [CrossRef]
25. Michiels, I.C.; Traunspurger, W. Seasonal variation of biodiversity and assemblage structure in freshwater nematodes. *Arch. Hydrobiol.* **2005**, *163*, 183–194. [CrossRef]
26. Schroeder, F.; Traunspurger, W.; Pettersson, K.; Peters, L. Temporal changes in periphytic meiofauna in lakes of different trophic states. *J. Limnol.* **2012**, *71*, 216–227. [CrossRef]
27. Schroeder, F.; Peters, L.; Traunspurger, W. Temporal variations in epilithic nematode assemblages in lakes of different productivities. *Fundam. Appl. Limnol.* **2012**, *181*, 143–157. [CrossRef]
28. Kurashov, E.A. The role of meiobenthos in lake ecosystems. *Aquat. Ecol.* **2022**, *36*, 447–463. [CrossRef]
29. Ristau, K.; Faupel, M.; Traunspurger, W. The effects of nutrient enrichment on a freshwater meiofaunal assemblage. *Freshw. Biol.* **2012**, *57*, 824–834. [CrossRef]
30. Zhuang, C.; Ouyang, Z.; Xu, W.; Bai, Y.; Zhou, W. Impacts of human activities on the hydrology of Baiyangdian Lake, China. *Environ. Earth Sci.* **2011**, *62*, 1343–1350. [CrossRef]
31. Sun, X.Q.; Li, H.C.; Liao, C.Y.; Deng, J.Z.; Sun, H.M.; Yan, M.X.; Wang, H.W. Investigation of benthic invertebrates and analysis of water environment in Baiyangdian Lake. *Hebei Fish.* **2019**, *308*, 40–44+50, (In Chinese with English Abstract).
32. Liu, J.G.; Zhao, D.D.; Ye, B. Eeological aributes, restorlion, and prteeion of the Baivangdian in Xiong'an New Area. *Aela Eeologica Sin.* **2019**, *39*, 3019–3025, (In Chinese with English Abstract).
33. Huang, X.; Wang, F.; Wu, D.; Kong, F.Q.; Wang, Q. The community structure of macrobenthos and its response to environmental factors in Baiyangdian Lake. *Hebei Fish.* **2023**, *12*, 33–40, (In Chinese with English Abstract). [CrossRef]
34. Semprucci, F.; Losi, V.; Moreno, M. A review of Italian research on free-living marine nematodes and the future perspectives on their use as Ecological Indicators (EcoInds). *Mediterr. Mar. Sci.* **2015**, *16*, 352–365. [CrossRef]
35. The State Environmental Protection Administration. *The Water and Wastewater Monitoring Analysis Method Editorial Board. Water and Wastewater Monitoring Analysis Method*, 4th ed.; China Environmental Science Press: Beijing, China, 2002; pp. 38–47.
36. Wang, M.C.; Liu, X.Q.; Zhang, J.H. Evaluate method and classification standard on lake eutrophication. *Environ. Monit. China* **2002**, *5*, 47–49, (In Chinese with English Abstract).
37. Zhao, Q. *Traceable Model of Water Quality and the Control Countermeasures of Eutrophication in Shihou Lake*; China University of Geosciences Beijing: Beijing, China, 2012; (In Chinese with English Abstract).
38. Guo, Y.Q.; Zhang, Z.N.; Mu, F.H. Large-scale Patterns of Meiofaunal Abundance in the Bohai Sea. *Acta Ecol. Sin.* **2002**, *22*, 1463–1469.
39. Cao, Y.K. *Study on the Biodiversity and Taxonomy of Free-Living Marine Nematodes in the Xiamen Island, China*; Jimei University: Xiamen, China, 2012; (In Chinese with English Abstract).
40. Somerfield, P.J.; Warwick, R.M. *Meiofauna in Marine Pollution Monitoring Programmes: A Laboratory Manual*; Ministry of Agriculture, Fisheries and Food, Directorate of Fisheries Research Technical Series: London, UK, 1996.

41. Juario, J.V. Nematode species composition and seasonal fluctuation of a sublittoral meiofauna community in the German Bight. *Veröff. Inst. Meeresforsch. Bremerh.* **1975**, *15*, 283–337.
42. Zhang, Z.N.; Zhou, H.; Yu, Z.Z.; Han, J. Abundance and biomass of the benthic in the northern soft-bottom of the Jiaozhou Bay. *Oceanol. Limnol. Sin.* **2001**, *32*, 139–147, (In Chinese with English Abstract).
43. Liu, H.B. *Studies on Community Structure and Biodiversity of Meiofauna in Taiping Bay, Qingdao, China*; Ocean University of China: Qingdao, China, 2007.
44. Xu, M.; Liu, Q.H.; Zhang, Z.N.; Liu, X.S. Response of free-living marine nematodes to the southern Yellow Sea cold water mass. *Mar. Pollut. Bull.* **2016**, *105*, 58–64. [CrossRef] [PubMed]
45. Widbom, B. Determination of average individual dry weights and ash-free dry weights in different sieve fractions of marine meiofauna. *Mar. Biol.* **1984**, *84*, 101–108. [CrossRef]
46. Zhao, M.H.; Liu, X.S. Taxa composition and distribution patterns of meiofauna in the intertidal zones of Shandong Peninsula. *J. Liaocheng Univ. (Nat. Sci.)* **2021**, *34*, 100–110, (In Chinese with English Abstract).
47. Danovaro, R.; Gambi, C.; Lampadariou, N.; Tselepides, A. Deep-sea nematode biodiversity in the Mediterranean basin: Testing for longitudinal, bathymetric and energetic gradients. *Ecography* **2008**, *31*, 231–244. [CrossRef]
48. Van Den Hoogen, J.; Geisen, S.; Routh, D.; Ferris, H.; Traunspurger, W.; Wardle, D.A.; De Goede, R.G.M.; Adams, B.J.; Ahmad, W.; Andriuzzi, W.S.; et al. Soil nematode abundance and functional group composition at a global scale. *Nature* **2019**, *572*, 194–198. [CrossRef]
49. Strayer, D.; Likens, G.E. An energy budget for the zoobenthos of Mirror Lake, New Hampshire. *Ecology* **1986**, *67*, 303–313. [CrossRef]
50. Traunspurger, W. The biology and ecology of lotic nematodes. *Freshw. Biol.* **2000**, *44*, 29–45. [CrossRef]
51. Traunspurger, W. Nematoda. In *Encyclopedia of Inland Waters*; Likens, G.E., Ed.; Academic: London, UK; Boston, MA, USA, 2009; pp. 372–383.
52. Traunspurger, W.; Michiels, I.C.; Abebe, E. Composition and distribution of free-living freshwater nematodes: Global and local perspectives. In *Freshwater Nematodes: Ecology and Taxonomy*; Eyualem Abebe, A., Andràssy, I., Traunspurger, W., Eds.; CABI: Wallingford, UK, 2006; pp. 46–76.
53. Giere, O. *Meiobenthology. The Microscopic Motile Fauna of Aquatic Sediments*; Springer: Berlin/Heidelberg, Germany, 2009; 527p.
54. Wu, J.H. *Studies of Freshwater and Soil Nematodes of China*; Institute of Hydrobiology, Chinese Academy of Sciences: Wuhan, China, 1999; (In Chinese with English Abstract).
55. Wieser, W. Benthic studies in Buzzards Bay II. The Meiofauna 1. *Limnol. Oceanogr.* **1960**, *5*, 121–137. [CrossRef]
56. Traunspurger, W.; Wilden, B.; Majdi, N. An overview of meiofaunal and nematode distribution patterns in lake ecosystems differing in their trophic state. *Hydrobiologia* **2020**, *847*, 2665–2679. [CrossRef]
57. Peters, L.; Traunspurger, W. Species distribution of free-living nematodes and other meiofauna in littoral periphyton communities of lakes. *Nematology* **2005**, *7*, 267–280. [CrossRef]
58. Thiel, H.J. *Benthos in Upwelling Regions. Upwelling Ecosystems*; Springer: Berlin/Heidelberg, Germany, 1978; pp. 124–138.
59. Josefson, A.B.; Widbom, B. Differential response of benthic macrofauna and meiofauna to hypoxia in the Gullmar Fjord basin. *Mar. Biol.* **1988**, *100*, 31–40. [CrossRef]
60. Bruce, C.C. Freshwater Meiofauna: Biology and Ecology. *Q. Rev. Biol.* **2002**, *77*, 469.
61. Kreuzinger-Janik, B.; Schroeder, F.; Majdi, N.; Traunspurger, W. Depth-Related Effects on a Meiofaunal Community Dwelling in the Periphyton of a Mesotrophic Lake. *PLoS ONE.* **2018**, *10*, e0137793. [CrossRef] [PubMed]
62. Abebe, E.; Mees, J.; Coomans, A. Nematode communities of Lake Tana and other inland water bodies of Ethiopia. *Hydrobiologia* **2001**, *462*, 41–73. [CrossRef]
63. Michiels, C.I.; Traunspurger, W. Benthic community patterns and the composition of feeding types and reproductive modes in freshwater nematodes. *Nematology* **2005**, *7*, 21–36. [CrossRef]
64. Liu, X.S.; Xu, M.; Zhang, J.H.; Mou, G.; Liu, D.; Li, X. Abundance and biomass of deep-sea meiofauna in the northern South China Sea. *J. Trop. Oceanogr.* **2014**, *33*, 52–59.
65. Cai, L.Z.; Fu, S.J.; Yang, J.; Zhou, X.P. Distribution of meiofaunal abundance in relation to environmental factors in Beibu Gulf, South China Sea. *Acta Oceanol. Sin.* **2012**, *31*, 92–103. [CrossRef]
66. Pavlyuk, O.N.; Trebukhova, J.A. Meiobenthos in Nha Trang Bay of the South China Sea (Vietnam). *Ocean Sci. J.* **2006**, *41*, 139–148. [CrossRef]

water

MDPI

Article

Limnological Characteristics and Relationships with Primary Productivity in Two High Andean Hydroelectric Reservoirs in Ecuador

Gonzalo Sotomayor [1,*], Andrés Alvarado [2,3], Jorge Romero [4], Carlos López [5], Marta Aguilar [6], Marie Anne Eurie Forio [7] and Peter L. M. Goethals [7]

1 Centro de Investigaciones Aplicadas y Mejoramiento de la Productividad Empresarial (SAPIENS), Facultad de Ingeniería, Universidad Hemisferios, Paseo de la Universidad, 300, Quito 170147, Ecuador
2 Departamento de Recursos Hídricos y Ciencias Ambientales, Universidad de Cuenca, Av. 12 de abril s/n, Cuenca 010150, Ecuador; andres.alvarado@ucuenca.edu.ec
3 Facultad de Ingeniería, Universidad de Cuenca, Av. 12 de abril s/n, Cuenca 010150, Ecuador
4 Instituto de Estudios del Régimen Seccional del Ecuador (IERSE), Facultad de Ciencia y Tecnología, Universidad del Azuay, Av. 24 de Mayo 7-77, Cuenca 010150, Ecuador; jaromero@uazuay.edu.ec
5 Escuela Superior Politécnica del Litoral (ESPOL), Centro de Agua y Desarrollo Sustentable, Guayaquil 2019, Ecuador; cllopez@espol.edu.ec
6 Unidad de Gestión Socioambiental, Empresa Electro Generadora del Austro (Elecaustro S. A.), Av. 12 de abril s/n, Cuenca 010150, Ecuador; marta.aguilar@elecaustro.gob.ec
7 Department of Animal Sciences and Aquatic Ecology, Faculty of Bioscience Engineering, Ghent University, Coupure Links 653, 9000 Ghent, Belgium; marie.forio@ugent.be (M.A.E.F.); peter.goethals@ugent.be (P.L.M.G.)
* Correspondence: gonzalo.sotomayor-v@uhemisferios.edu.ec

Abstract: Studies on limnology are essential to reservoir management; nevertheless, few are known about the limnological features of the Andean reservoirs in Ecuador. To overcome this limitation in the information, from December 2018 to December 2019, the limnological characteristics of El Labrado and Chanlud reservoirs in the Machángara river basin (Ecuador south) were examined. Using the light/dark bottles technique, the primary productivity (PP) of phytoplankton was studied in conjunction with (1) vertical profiles of oxygen concentrations, water temperature, nitrogen, phosphorus, alkalinity, and heterotrophic bacteria; (2) Secchi disk transparency; and (3) meteorological factors such as wind force, precipitation, and water level. Data indicate that both reservoirs are polymictic, with alkaline waters, low nutrients, and low PP rates. Despite this, a principal component analysis revealed that Chanlud exhibits higher nitrogen, alkalinity, heterotrophic bacteria, and PP values. In two approaches through multiple linear regression analysis, each per reservoir, the PP was explained mainly by water temperature, depth, light, heterotrophic bacteria, and meteorological parameters. The low concentrations of nutrients and the low residency time explain the low PP values. Likewise, the altitudinal factor (i.e., both reservoirs are 3400 m above sea level) and the low human perturbations in surrounding reservoir zones play a crucial role in explaining their poor PP. Notwithstanding the low metabolic rates, clear seasonal trends were observed in both reservoirs; the lowest PP rates occurred during the cold season. To our knowledge, this is the first limnological study of high Andean reservoirs in Ecuador. These findings should be part of Andean reservoir management protocols, contributing significantly to local conservation efforts. Additionally, they could be extrapolated as a frame of reference to similar eco-hydrological systems.

Keywords: limnological features; primary productivity; high Andean reservoirs

1. Introduction

Reservoir construction is one of the most remarkable human activities in modifying freshwater ecosystems. It is a widespread practice worldwide that involves utilizing rivers by building a series of reservoirs [1]. In this sense, dams are infrastructures commonly related to reservoirs. Dams/reservoirs are primarily designed for hydroelectric power generation [2] and other uses such as irrigation, flood control, water supply, improved navigation, fish culture, recreation, or some combination. However, damming water courses exerts negative effects; dams/reservoirs can adversely affect the structure and functioning of aquatic ecosystems [3], downstream [4] and in situ. Several problems arise in situ with the slowing down of running water added to inadequate land use management of the surrounding areas of the reservoirs; the best known are eutrophication [5,6], the loss of biodiversity [7], and greenhouse gas emissions [8], among others.

Limnology is a comprehensive science that deals with water systems and the surrounding land. It considers physical, chemical, geological, and biological features to provide conceptual models of freshwater ecosystems and fundamental information needed to determine causes and potential solutions to environmental stresses [9]. Therefore, limnological characterization is a key tool for designing reservoirs' conservation and management plans [10,11] in response to, for example, ongoing climate change [12] or catchment disturbances [13]. Particularly in a region such as the Andes, whose ranges extend along western South America, from south Venezuela to Tierra del Fuego, with irregular topology [14], where multiple stressors and increasing water demands impose unprecedented stress on freshwater ecosystems, the inclusion of a limnological framework of research is crucial. The ability to assess the response of Andean freshwaters to anthropogenic stressors requires knowledge of baseline limnological conditions [15]. However, with a few notable exceptions, limnological surveys in the tropical Andes reservoirs are rare [16–21]. Except in Brazil [22–26], knowledge about neotropical reservoirs is extremely limited, and this trend is more significant for the high Andean region. In Ecuador, where many reservoirs have been implemented during the last two decades, limnological studies of these artificial ecosystems are non-existent, and no information is available. Up to now, only water quality has been analyzed for Ecuadorian reservoirs from the point of view of compliance with the limits required by Ecuadorian environmental legislation (gray literature), and most limnological research on lentic ecosystems is focused only on natural lakes [27]. Thus, although the overall principles and controlling factors that govern lakes and reservoirs are the same, the premise to consider is that reservoirs were made artificially, and their dynamics are not controlled naturally, i.e., they exhibit many differences relative to natural lakes; consequently, specific studies must be performed to evaluate and characterize reservoirs [28].

Our paper describes, for the first time, the limnological features of two adjacent high hydroelectric Andean reservoirs (i.e., Chanlud and El Labrado) in austral Ecuadorian Andes and their seasonal variations. This study aims to add to the understanding of the limnological characteristics of tropical Andean reservoirs and contribute to management efforts. The results will also form an important baseline for information on assessed reservoirs. We estimate the primary production of both reservoirs concerning selected environmental parameters. We addressed three research questions/goals: (i) What are the differences in limnological features between reservoirs? (ii) What are the main driving factors for primary productivity in the studied reservoirs? Furthermore, (iii) we aimed to study the seasonal limnological variability of the assessed reservoirs.

2. Materials and Methods

2.1. Study Area

The Andes in Ecuador are divided into eastern and western ranges. The second divides the Pacific and Atlantic slopes [29]. The studied reservoirs, Chanlud and El Labrado, are located on the Atlantic slopes in the upper area of the Machángara river basin, which belongs to the Paute river basin (Figure 1). The total area of the Machángara catchment is 323.55 km^2, and the altitude range is 2424 to 4424 m above sea level (m a.s.l.); that is, this is a typical hydrologic system of the Andean páramo ecosystem of South America [30]. The average annual precipitation varies between 877 mm and 363 mm per year, while the average annual temperature fluctuates between 16.0 °C and 9.0 °C in the lower and upper areas, respectively [31]. Two seasons are present during the year: a season of precipitation from the middle of February to the beginning of July and a dry season during the rest of the year. The average flow of the Machángara river measured from 1964 to 2010 was 8.4 $m^3 s^{-1}$, and this river is used for domestic and industrial purposes, agricultural irrigation, and hydropower generation. Currently, 60% of Cuenca's drinking water (Ecuador's third largest city) comes from the Machángara hydrological system.

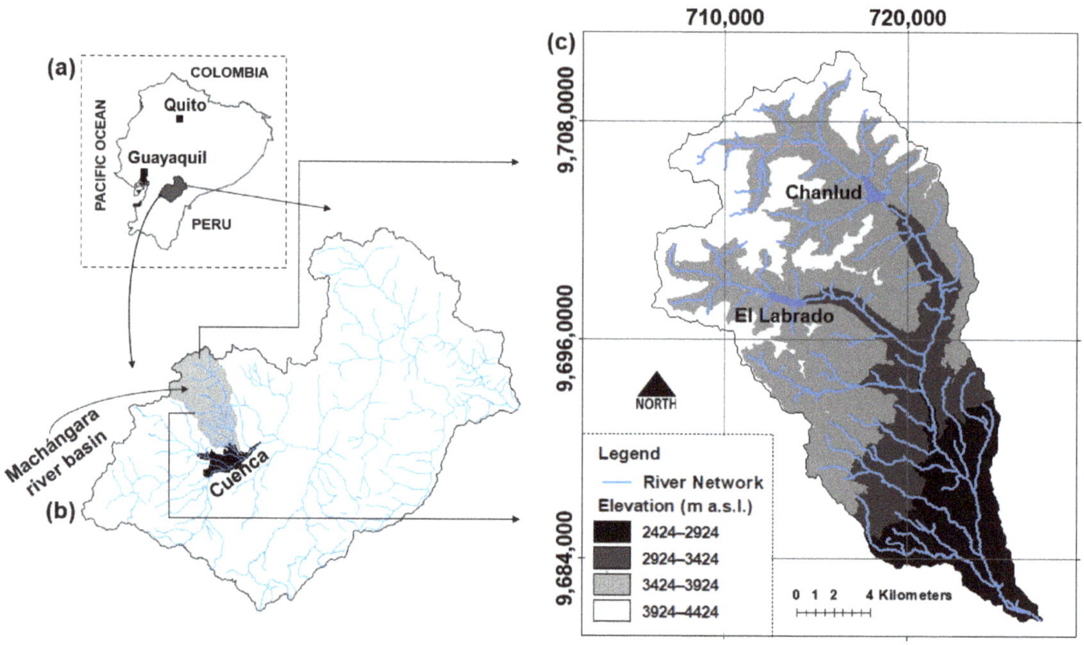

Figure 1. (**a**) Location of the Paute river basin in continental Ecuador and its largest city (Cuenca); (**b**) location of the Machángara river basin in the Paute river basin and (**c**) elevation distribution and the river network in the Machángara river basin. Coordinate system: WGS84 UTM 17S; coordinate units: meters.

Regarding the reservoirs Chanlud and El Labrado, the former is in the Machángara Alto River sub-basin, and the latter is in the Chulco river sub-basin. Both reservoirs form the Machángara River Hydroelectric Complex. The Chanlud reservoir became operational in 1992 and is located at 3464 m a.s.l. (at their centroid point); it has a maximum depth of 40 m, a storage capacity of 17 hm^3, and a regulated discharge of about 4.8 $m^3 s^{-1}$. The reservoir area is ~0.69 km^2, and its shoreline is 5.6 km. El Labrado reservoir became operational in 1972 and is located at 3418 m a.s.l. (at their centroid point); it has a maximum depth of 14 m, a storage capacity of 6.15 hm^3, and a regulated discharge of about 2.4 $m^3 s^{-1}$. The reservoir area is ~0.58 km^2, and its shoreline is 5.1 km [32]. In 1985, the national environmental authorities declared the upper area of the watershed of Machángara a protected forest (i.e., Machángara–Tomebamba protected area), so human activities such as agriculture, etc., were restricted in the surrounding areas of both studied reservoirs (for Chanlud, 0.3% of their contributing basin is anthropized, and for El Labrado, the percentage is 0.5%). However, the middle and lower zones of the Machángara watershed continued to deteriorate [33].

2.2. Sampling Design

For both reservoirs, sampling campaigns from December 2018 to October 2019, and December 2019, were performed in their deepest part (previously determined through bathymetry) between 8.30 and 16.30 h, with a spatial replicability of one site per reservoir. Vertical samples were taken every 5 and 3 m from surface to bottom in Chanlud and El Labrado, respectively, using a Van Dorn sampler (2L). The different intervals of vertical sample design were due to the significant dissimilarities in depth and their variability for both reservoirs. For Chanlud, the range of depth was 30.7–39.5 m, while for El Labrado, it was 8.4–14.1 m. Vertical sampling for El Labrado every 5 m was assumed to have a low number of samples, which would have negative implications for integrated analysis based on the scalar vector of depth. Therefore, vertical sampling was performed for this reservoir every 3 m of depth. Due to the water level variability of both reservoirs, their depths were first measured per visit using sonar (Speedtech® depth-mate portable sounder). Thus, the number of vertical samples per reservoir was established as a function of depth during the start of the field visit. For each water sample at each depth, measurements of alkalinity (mg L^{-1} of $CaCO_3$), orthophosphate (mg L^{-1} of PO_4^{3-}), nitrite (mg L^{-1} of NO_2^--N), nitrate (mg L^{-1} of NO_3^--N), heterotrophic bacteria (bacteria 100^{-1} mL^{-1}), and ammonium (mg L^{-1} of NH_4^+-N) were performed. Orthophosphate, nitrite, nitrate, and ammonium were obtained using a modular system for performing continuous flow analysis (OI Analytical—Flow Solution® FS 3100), with the detection limits being 0.001, 0.0005, 0.0005, and 0.0005 mg L^{-1}, respectively. Alkalinity was determined by a titrimetric approach, SM 2320 B method [34], with a detection limit of 8 mg L^{-1}. Heterotrophic bacteria were estimated by the spread plate method, in which the water sample was mixed for 5 min at low speed before serial dilution. Samples were spread in duplicate on a pre-cooled agar plate consisting of a glucose–nitrogen minimal medium with a 0.2% *w*/*v* casamino acids supplement. Plates were incubated at 22 °C for 3, 5, and 7 days and at 37 °C for 1, 3, and 5 days [35].

Vertical profiles of water temperature (°C) and dissolved oxygen (mg L^{-1} of O_2) were performed using EXO2 sonde [36] at one-second intervals. Light penetration in the water column was determined with a standard Secchi disk (Z_{sd}) (25 cm in diameter) in meters. The depth (m) of the euphotic zone (Z_{eu}) was assumed equivalent to 1% of the surface light level and was estimated by multiplying the Secchi-disk depth (Z_{sd}) by a factor of 2.7 [37,38]. The light vertical extinction coefficient (E_V) was calculated using this equation: $E_V{}^n \times Z_{sd}$ = constant, where n = 0.84, and the "constant" is 1.54 [39,40]. Likewise, an integrated water sample from the Z_{eu} was taken to determine chlorophyll-*a* concentration (µg L^{-1}) in the laboratory through spectrophotometric analyses of acetone extracts [41], with a detection limit of 0.023 µg L^{-1}.

In this context, photosynthesis and phytoplankton are the prime components of aquatic primary production, and chlorophyll-*a* is a fundamental indicator of phytoplankton abundance [42]. Herein, primary productivity (PP) was obtained by measuring oxygen concentrations in dark and clear bottles incubated in situ at various depths (i.e., every 5 and 3 m for the Chanlud and El Labrado reservoirs, respectively) throughout the water column, using a line of supports placed on an anchor located in the deepest part of each reservoir. Thus, the number of incubation levels was established as a function of the depth for each reservoir per visit. Each bottle was filled with water from the corresponding depth. Lightproof bags were used to store dark bottles wrapped in aluminum foil. A wood wire frame with an attached rope held the bottles vertically at 25 cm to prevent self-shading. Before placing the bottles on the supports and starting the incubation, the initial dissolved oxygen concentration (initial O_2) in each was measured using a HANNA oximeter (model HI 9146). Another measure of oxygen was performed after the incubation (final O_2). Calculations of gross primary production (GPP) and net primary production (NPP) and respiration (R) rates were based on the changes in the oxygen content in the light and dark bottles. The initial O_2 concentration (c1) could be expected to decrease to a lower value by R in dark bottles (c2) and increase to a higher concentration (c3) in clear bottles, according to the difference between photosynthetic production and respiration [43]. The difference (c1 − c2) represents R activity per unit volume during the incubation time interval (Δt) (for both bottles, the average Δt was 5:56 h, ± 0:52 standard deviation). The difference (c3 − c1) is equal to the NPP, and the sum (c3 − c1) + (c1 − c2) = (c3 − c2) corresponds to the GPP [40]. To obtain the PP (i.e., GPP + NPP) and R rates per area and time units (mg C m^{-2} d^{-1}) for each sampling month, vertical integrations were performed through the trapezoidal method [44–46]. Thus, since R, NPP, and GPP are vectors, the method calculated the integrated function for them concerning the scalar space specified by the depth (i.e., 0, 5, 10. . ., and 0, 3, 6. . ., m for the Chanlud and El Labrado reservoirs, respectively). The trapezoidal rule is a method that approximates integration over an interval by breaking the area under the curve into trapezoids with more easily comparable areas [47]. Herein, the trapezoidal vertical integrations were carried out using the "trapz" function of MATLAB® [48]. For the PP, the integrations were calculated on the Z_{eu}, and the day length was assumed to be 12 h. The integral function for R was calculated for the entire water column over 24 h, assuming that the R rate at night was the same as during the day [39,49].

Finally, meteorological parameters such as precipitation (mm) and wind force (m s^{-1}) were considered. These variables were obtained using two meteorological stations located in the proximity of each dam wall (for Chanlud, N −2°40′45″ E −79°2′1″ and for El Labrado, N −2°43′44″ E −79°4′22″). Sub-daily data were recorded from the meteorological stations during the study period, and they were aggregated into daily datasets and finally to a monthly frequency (i.e., accumulated for precipitation and averaged for wind force).

2.3. *Data Analysis*

Considering the varying number of records for each parameter in the two reservoirs (e.g., numerous monthly records for water temperature versus only one per month for primary productivity), the data analysis relied on a single monthly value for each parameter and reservoir. An aggregate process was carried out for the parameters with multiple values based on central tendency measures. To perform this, the normality of each dataset per parameter and month was checked using the Shapiro–Wilk (S-W) test [50], considering a 95% confidence level. For a particular parameter, if the S-W test suggested normality, their mean value was used for aggregating; otherwise, the median was used [51,52]. The S-W tests were performed through the shapiro.test() function in the R environment [53]. As a result, a matrix/database (X1) was developed for $n_{sp} = 2$ sampling points, one per reservoir, $n_{rep} = 24$ sampling replicates, 12 monthly records per reservoir that overall contained $n_{var} = 18$ variables, resulting in a total of $n_{obs} = 432$ observations ($n_{obs} = n_{var} \times n_{rep}$), which are represented by $X1_{i,j}$, with $i = 1, 2, \ldots, n_{var}$ and $j = 1, 2, \ldots, n_{rep}$. Once the matrix X1 was organized, a Pearson correlation analysis was performed for their n_{var} to exclude redundant information characterized by a positive or negative correlation magnitude above 0.75 [54]. This was achieved to minimize multicollinearity issues. Pearson correlation analysis was performed using the R environment's cor() function [53]. Likewise, parameters with weak chemical signals were excluded from the data analysis. After the matrix was stripped of redundant and weak signals (i.e., X2), a range scaling process [55] was used to standardize its associated distribution ($X2_z$). Then, a principal component analysis was performed for $X2_z$.

2.3.1. Principal Component Analysis (PCA)

The PCA is an ordination method where the original data matrix (herein $X2_z$) is reduced to parts A_L (loadings) and U_S (scores). A_L indicates how much an original variable is "loaded into" a principal component (PC), and U_S are the coordinates of one replicate in the new system [56,57]. Linear combinations of A_L and U_S reproduce the matrix $X2_z$ as new synthetic variables that are non-correlated between them, representing a certain quantity of variables of the $X2_z$ and explaining their variance [58]. A critical prior task of a PCA is selecting an optimal number of PCs [59]. Herein, the Average Eigenvalue Criterion

(AEC) was used for this purpose through the Venetian blinds cross-validation method. AEC is based on eigenvalues and only accepts components with an eigenvalue larger than the average eigenvalue as significant [59]. The Venetian blinds cross-validation method is based on splits of observed data [60–62]. In this research, the n_{rep} was split into five splits [63] for cross-validation (i.e., four groups for model training and the fifth group for validation). To perform the ordination PCA process, a quantitative response vector was loaded to $X2_z$, corresponding to n_{sp} (i.e., 1, 2 reservoirs) to assess the potential differences between studied reservoirs. The response vector does not affect PCA calculation, but it allows for a visual differentiation of samples/replicates during interpretation. Within this framework, through A_L, PCA identifies the most informative original variables that explain the differences between both reservoirs. For this purpose, the "cut-off rule" criterion was applied [64], which regards $|\text{loadings}| > 0.25$ as being significant [65]. Furthermore, Fisher's least significant difference (LSD) test was used to calculate/visualize intervals around the means of the most significant variables (i.e., $|\text{loadings}| > 0.25$). These intervals were constructed in such a way that if two means were the same, their intervals would overlap 95.0% of the time; on the contrary, any pair of intervals that did not overlap vertically corresponded to a pair of means that had a statistically significant difference [66]. The PCA was implemented with MATLAB® using the PCA toolbox version 1.3 [59], and the LSD test was calculated using the LSD.test() function in the R® package "agricolae" [67].

2.3.2. Multiple Linear Regression Analysis (MLRA)

Finally, to identify the informative parameters that explain the variability of the PP (GPP + NPP), using $X2_z$, two multiple linear regression analyses (MLRA) were performed, each one per reservoir (i.e., $MLRA_{Ch}$ and $MLRA_L$). In each MLRA, the PP was the dependent variable, and the rest of parameters were independent ones [68]. Backward stepwise selection (BSS) (i.e., a stepwise selection method) was applied to identify the independent variables with statistic effects on the dependent variable. Beginning with a model that includes all variables, the BSS removes variables one at a time if they are not statistically significant (i.e., variables are removed from the model at a given step if their p values are greater than the P-to-Remove value, herein fixed as 0.05) [69]. Thus, the most informative explanatory variables to explain the variability of PP were identified at the end of the process [70]. To evaluate each MLRA, the Adjusted R-squared (Adj—R^2) statistic was used. Also, the Durbin–Watson (DW) [71] statistic test was implemented for the residual and serial correlation assessment. Finally, an analysis of variance was performed for each MLRA. Of particular interest were the F-tests and their associated p-values, which test the statistical significance of the fitted model. A small p value (less than 0.05) indicated a significant relationship between PP and the independent variables. Both MLRAs were performed using the lm() function in the R environment [53].

3. Results

After Pearson correlation analysis, eleven variables were chosen for the planned statistical protocol: dissolved oxygen (O_2), water temperature (WT), nitrate (NO_3^--N), alkalinity ($CaCO_3$), heterotrophic bacteria (HB), wind force (WF), Secchi disk (Z_{sd}), maximum depth (D_{max}), primary productivity (PP), chlorophyll-*a* (Chl) and precipitation. The excluded variables were the euphotic zone (Z_{eu}), light vertical extinction coefficient (E_v), respiration (R), and gross and net primary productivity (GPP and NPP, respectively). Orthophosphates (PO_4^3), nitrites (NO_2^--N), and ammonium (NH_4^+-N) were not detected in both reservoirs; only trace values were registered for them. Therefore, these parameters were not considered for the statistical analysis. A summary of assessment parameters for both reservoirs is given in Table 1 (for most cases, henceforth, the mean is termed $\overline{x}$).

Table 1. Summary of assessment parameters for both reservoirs. The first ten columns show the means ± standard deviations. For the last nine columns, there are unique values because of the nature of the variables. Values highlighted in gray come from datasets with non-normal distributions for which the median instead of the mean value was used for data analysis. WT = water temperature, O_2 = dissolved oxygen, $CaCO_3$ = alkalinity, NO_3^--N = nitrate, NO_2^--N = nitrite, NH_4^+-N = ammonium, PO_4^{3-} = orthophosphate, HB = heterotrophic bacteria, WF = wind force, PPt = precipitation, E_v = light vertical extinction coefficient, D_{max} = maximum depth, Z_{sd} = Secchi disk, Z_{eu} = euphotic zone, Chl = chlorophyll-α, GPP = gross primary productivity, NPP = net primary productivity, PP = primary productivity, and R = respiration.

		WT (°C)	O_2	$CaCO_3$	NO_3^--N (mg L^{-1})	NO_2^--N	NH_4^+-N	PO_4^{3-}	HB (bac. 100^{-1} ml^{-1})	WF (m s^{-1})	PPt (mm)	E_v (–)	D_{max}	Z_{sd} (m)	Z_{eu}	Chl (µg L^{-1})	GPP	NPP (mg C m^{-2} d^{-1})	PP	R
Chanlud reservoir	dec-18	11.92 ± 0.47	6.09 ± 0.28	44.38 ± 2.17	0.11 ± 0.16	0.01 ± 0.01	0.03 ± 0.02	0.01 ± 0.00	3.43 ± 2.37	0.84 ± 0.13	0.12 ± 0.19	17.74	32.70	6.57	17.74	1.60	11.67	8.14	19.81	9.73
	jan-19	11.80 ± 0.70	6.26 ± 0.20	48.97 ± 17.66	0.04 ± 0.02	0.01 ± 0.00	0.04 ± 0.02	0.01 ± 0.01	9.67 ± 8.59	0.90 ± 0.21	0.16 ± 0.21	16.07	30.70	5.95	16.07	6.24	10.25	9.97	20.22	0.56
	feb-19	11.91 ± 0.87	6.61 ± 0.21	42.12 ± 4.9	0.10 ± 0.16	0.03 ± 0.07	0.05 ± 0.01	0.00 ± 0.00	0.50 ± 1.07	0.68 ± 0.20	0.23 ± 0.26	12.96	35.10	4.80	12.96	0.80	4.95	3.17	8.12	3.57
	mar-19	11.49 ± 0.62	6.19 ± 0.35	42.26 ± 3.24	0.02 ± 0.01	0.00 ± 0.00	0.06 ± 0.01	0.04 ± 0.03	1.38 ± 2.50	0.61 ± 0.15	0.24 ± 0.32	12.83	38.10	4.75	12.83	0.53	12.38	10.29	22.67	6.99
	apr-19	11.21 ± 0.30	5.80 ± 0.72	42.39 ± 2.29	0.01 ± 0.01	0.06 ± 0.03	0.04 ± 0.02	0.01 ± 0.01	3.50 ± 5.81	0.51 ± 0.20	0.16 ± 0.22	11.29	39.10	4.18	11.29	0.53	13.65	8.02	21.67	12.75
	may-19	11.15 ± 0.62	6.06 ± 0.43	39.83 ± 3.06	0.03 ± 0.01	0.02 ± 0.01	0.04 ± 0.01	0.03 ± 0.00	4.29 ± 7.93	0.64 ± 0.22	0.17 ± 0.20	14.93	38.10	5.53	14.93	2.67	10.47	6.81	17.28	9.37
	jun-19	8.97 ± 0.46	7.80 ± 0.17	35.28 ± 2.62	0.07 ± 0.05	0.01 ± 0.00	0.05 ± 0.01	0.00 ± 0.00	42.50 ± 45.53	0.85 ± 0.32	0.26 ± 0.28	12.47	38.40	4.62	12.47	2.41	8.06	4.32	12.38	8.13
	jul-19	8.97 ± 0.38	7.77 ± 0.16	33.63 ± 1.70	0.03 ± 0.02	0.01 ± 0.00	0.05 ± 0.01	0.00 ± 0.00	63.86 ± 94.36	0.84 ± 0.20	0.19 ± 0.23	11.53	38.50	4.27	11.53	1.17	4.12	3.85	7.97	0.53
	aug-19	7.96 ± 0.31	7.12 ± 0.10	37.03 ± 2.62	3.51 ± 2.79	0.02 ± 0.02	0.07 ± 0.02	0.01 ± 0.00	29.86 ± 18.22	1.00 ± 0.25	0.16 ± 0.22	12.64	38.75	4.68	12.64	0.53	6.64	5.44	12.08	2.74
	sep-19	9.14 ± 1.08	7.17 ± 0.11	40.58 ± 5.07	4.24 ± 2.81	0.01 ± 0.00	0.06 ± 0.01	0.01 ± 0.00	76.57 ± 34.80	0.84 ± 0.08	0.09 ± 0.14	16.36	34.50	6.06	16.36	1.34	8.75	7.41	16.16	2.72
	oct-19	9.87 ± 0.84	7.11 ± 0.29	40.91 ± 1.16	0.06 ± 0.06	0.01 ± 0.00	0.05 ± 0.01	0.01 ± 0.00	13.29 ± 11.25	0.75 ± 0.16	0.23 ± 0.26	13.64	39.50	5.05	13.64	0.41	8.85	5.64	14.49	9.17
	dec-19	11.87 ± 0.42	6.40 ± 0.52	39.51 ± 2.32	0.02 ± 0.01	0.01 ± 0.00	0.03 ± 0.01	0.00 ± 0.00	20.71 ± 19.20	0.87 ± 0.17	0.19 ± 0.21	14.90	38.00	5.52	14.90	1.84	19.45	12.96	32.41	22.44
El Labrado reservoir	dec-18	11.92 ± 0.14	6.48 ± 0.01	29.78 ± 3.10	0.03 ± 0.00	0.00 ± 0.00	0.04 ± 0.03	0.02 ± 0.02	3.67 ± 2.08	0.84 ± 0.25	0.10 ± 0.11	17.82	8.40	6.60	17.82	1.34	2.04	1.38	3.42	1.36
	jan-19	12.02 ± 0.28	6.52 ± 0.09	29.23 ± 1.25	0.04 ± 0.03	0.01 ± 0.01	0.03 ± 0.00	0.01 ± 0.01	2.25 ± 2.22	0.90 ± 0.29	0.14 ± 0.18	14.31	10.10	5.30	14.31	2.67	1.90	1.72	3.62	0.39
	feb-19	12.23 ± 0.41	6.60 ± 0.07	28.58 ± 1.81	0.11 ± 0.04	0.00 ± 0.00	0.03 ± 0.00	0.0 ± 0.00	6.20 ± 5.31	0.68 ± 0.28	0.21 ± 0.22	16.47	14.10	6.10	16.47	0.36	2.05	1.32	3.37	1.46
	mar-19	12.93 ± 0.68	6.52 ± 0.16	28.19 ± 2.69	0.03 ± 0.02	0.00 ± 0.00	0.05 ± 0.02	0.03 ± 0.03	0.00 ± 0.00	0.61 ± 0.26	0.21 ± 0.24	17.36	10.80	6.43	17.36	0.27	3.46	1.88	5.34	3.20
	apr-19	12.90 ± 0.10	6.76 ± 0.06	29.70 ± 1.08	0.02 ± 0.00	0.07 ± 0.02	0.04 ± 0.01	0.01 ± 0.01	98.50 ± 77.29	0.51 ± 0.27	0.12 ± 0.16	16.39	11.30	6.07	16.39	0.00	3.58	3.02	6.60	1.12
	may-19	12.31 ± 0.38	6.78 ± 0.20	28.08 ± 1.76	0.02 ± 0.00	0.01 ± 0.01	0.03 ± 0.02	0.02 ± 0.01	1.25 ± 0.96	0.64 ± 0.33	0.15 ± 0.17	15.23	10.70	5.64	15.23	3.47	1.80	1.09	2.89	1.42
	jun-19	8.46 ± 0.13	7.36 ± 0.03	25.92 ± 1.76	0.04 ± 0.01	0.01 ± 0.00	0.06 ± 0.01	0.00 ± 0.00	44.00 ± 51.12	0.85 ± 0.59	0.20 ± 0.21	15.09	14.00	5.59	15.09	1.60	2.40	1.37	3.77	2.06
	jul-19	8.80 ± 0.10	7.17 ± 0.03	25.38 ± 1.08	0.02 ± 0.01	0.01 ± 0.00	0.05 ± 0.00	0.00 ± 0.00	0.25 ± 0.50	0.84 ± 0.34	0.10 ± 0.13	17.09	12.00	6.33	17.09	0.67	1.81	1.20	3.01	1.18
	aug-19	8.41 ± 0.12	8.00 ± 0.05	24.84 ± 1.25	0.02 ± 0.02	0.01 ± 0.00	0.06 ± 0.03	0.00 ± 0.00	26.25 ± 16.88	1.00 ± 0.52	0.12 ± 0.18	16.50	12.40	6.11	16.50	0.00	2.30	1.45	3.75	1.70
	sep-19	10.61 ± 0.10	7.35 ± 0.21	26.08 ± 1.04	0.35 ± 0.26	0.01 ± 0.00	0.03 ± 0.01	0.00 ± 0.00	30.00 ± 11.97	0.84 ± 0.18	0.06 ± 0.10	18.04	10.80	6.68	18.04	0.80	0.51	0.21	0.72	0.64
	oct-19	10.84 ± 0.19	7.06 ± 0.01	32.20 ± 5.61	0.03 ± 0.03	0.01 ± 0.00	0.05 ± 0.01	0.00 ± 0.00	2.25 ± 3.30	0.75 ± 0.30	0.26 ± 0.29	15.39	11.60	5.70	15.39	0.27	1.64	1.58	3.22	0.12
	dec-19	12.51 ± 0.06	6.56 ± 0.21	27.12 ± 1.71	0.01 ± 0.00	0.01 ± 0.00	0.04 ± 0.01	0.00 ± 0.00	18.25 ± 6.13	0.87 ± 0.25	0.23 ± 0.29	12.96	10.20	4.80	12.96	1.54	2.23	2.23	4.46	0.00

3.1. Principal Component Analysis (PCA)

Regarding the Average Eigenvalue Criterion (AEC), two components (PCs) were identified, i.e., the optimal number of PCs. The percentage of explained variance for these two components were 70.1% and 13.8% for PC1 and PC2, respectively. Figure 2 shows the score plot of the PCA of each monthly record of both studied reservoirs. A clear distinction between both reservoirs is observed.

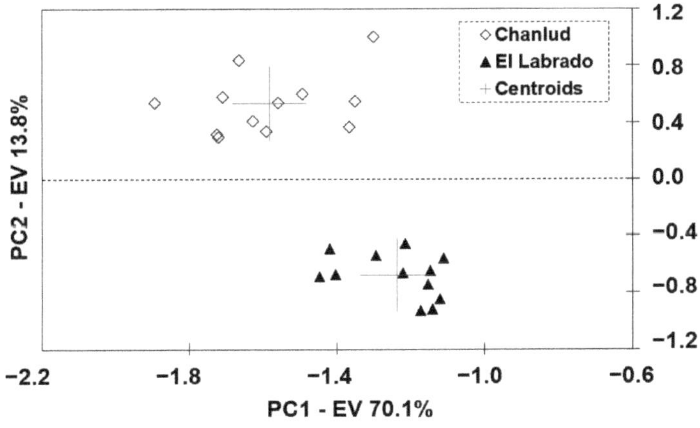

Figure 2. Score plot from the principal component analysis performed for the Chanlud and El Labrado reservoirs.

In Figure 2, the centroids of the coordinates of each reservoir are shown, (−1.59, 0.53) and (−1.24, −0.68) for Chanlud and El Labrado, respectively. The points for El Labrado are closer to their centroid than the points of the Chanlud reservoir, which are more dispersed from its centroid. This finding implies that fewer temporal (monthly) differences exist for El Labrado than for Chanlud. The average Euclidean distance between points belonging to the Chanlud reservoir and their centroid is 0.23 ± 0.13; meanwhile, for the points of El Labrado, it is 0.18 ± 0.07.

Eight informative variables were identified by the PCA (| loadings | > 0.25) that explain the variability of studied reservoirs (Table 2, Figure 3).

Table 2. Loading values for the principal components of the PCA model. Bold and italic values indicate a strong influence of the variables (i.e., | loadings | > 0.25). WT = water temperature, PPt = precipitation, D_{max} = maximum depth, Z_{sd} = Secchi disk, $CaCO_3$ = alkalinity, O_2 = dissolved oxygen, WF = wind force, PP = primary productivity, HB = heterotrophic bacteria, Chl = chlorophyll-*a*, and NO_3^--N = nitrate.

Parameter	PC1	PC2
WT	***−0.40***	−0.17
PPt	***−0.39***	0.06
D_{max}	***−0.38***	***0.54***
Z_{sd}	***−0.37***	***−0.40***
$CaCO_3$	***−0.36***	***0.40***
O_2	***−0.34***	−0.23
WF	***−0.27***	***−0.46***
PP	−0.23	***0.28***
HB	−0.13	−0.06
Chl	−0.12	−0.06
NO_3^--N	−0.06	0.07

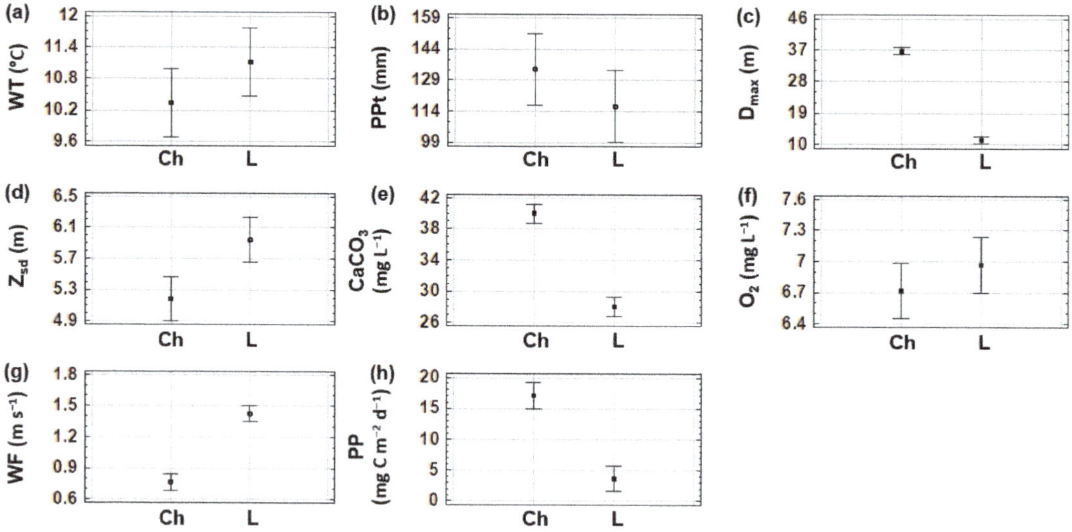

Figure 3. Means and Fisher's test-based intervals of the significant variables for studied reservoirs (Ch = Chanlud and L = El Labrado). (**a**) WT = water temperature, (**b**) PPt = precipitation, (**c**) D_{max} = maximum depth, (**d**) Z_{sd} = Secchi disk, (**e**) $CaCO_3$ = alkalinity, (**f**) O_2 = dissolved oxygen, (**g**) WF = wind force, and (**h**) PP = primary productivity. Mean values are depicted with a black point symbol.

Table 2 shows the informative variables for PC1 and PC2 based on the values of their loadings. Considering the percentage of explained variance for these two components (i.e., 70.1 and 13.8% for PC1 and PC2, respectively), the variables of PC1 have the most informative load to comparatively evaluate the studied reservoirs.

The mean values of the following parameters, WT (Figure 3a), Z_{sd} (Figure 3d), O_2 (Figure 3f), and WF (Figure 3g), were reported to be higher in El Labrado than Chanlud ($\overline{x}$ of WT expressed in °C is 11.1 for El Labrado and 10.3 °C for Chanlud; $\overline{x}$ of Z_{sd} is 5.9 m for El Labrado and 5.2 m for Chanlud; $\overline{x}$ of O_2 is 7.0 mg L^{-1} for El Labrado and 6.7 mg L^{-1} for Chanlud; and $\overline{x}$ of WF expressed in m s^{-1} is 1.4 for El Labrado and 0.8 for Chanlud). For precipitation (Figure 3b), D_{max} (Figure 3c), $CaCO_3$ (Figure 3e), and PP (Figure 3h), the inverse trend was observed ($\overline{x}$ of precipitation expressed in mm is 116.5 for El Labrado and 134.4 for Chanlud; $\overline{x}$ of D_{max} is 11.4 m for El Labrado and 36.7 m for Chanlud; $\overline{x}$ of $CaCO_3$ expressed in mg L^{-1} is 28.1 for El Labrado and 40.0 for Chanlud; and $\overline{x}$ of PP expressed in mg C m^{-2} d^{-1} is 3.7 for El Labrado and 17.1 for Chanlud). However, for the specific cases of WT (Figure 3a), precipitation (Figure 3b), and O_2 (Figure 3f), the intervals displayed around the mean based on Fisher's least significant difference (LSD) procedure overlap 95.0% of the time, indicating that nonsignificant differences exist between the means of those parameters. It is important to emphasize this finding because although WT, precipitation, and O_2 do not exhibit significant statistical differences between assessed reservoirs (Figures 3a, 3b, and 3f, respectively), they are parameters that, under the PCA, are key to explaining each reservoir's temporal variability.

Analysis of the Variables Identified by PCA as Informative to the Studied Reservoirs

With regard to meteorological parameters, for the Chanlud reservoir, the annual wind force (WF, Figure 4) ranged from 0.51 to 1.0 m s^{-1} ($\overline{x}$ = 0.8 ± 0.1 m s^{-1}). In many months, the WF was similar (range = 0.7–1, $\overline{x}$ = 0.9 ± 0.1 m s^{-1}). However, in the months of February, March, April, and May, the WF ranged from 0.5 to 0.7 m s^{-1}, with the mean value of 0.6 ± 0.1 m s^{-1} reported in those months. For El Labrado reservoir, the annual wind force (WF) ranged from 1.2 to 1.9 m s^{-1} ($\overline{x}$ = 1.4 ± 0.2 m s^{-1}) and was similar for most months (range = 1.2–1.4, $\overline{x}$ = 0.9 ± 0.1 m s^{-1}). However, a comparatively high WF was reported in June, July, and August and ranged from 1.6 to 1.9 m s^{-1} with a mean value of 1.7 ± 0.1 m s^{-1}. Regarding precipitation, in Chanlud, it ranges from 66.0 to 184.4 mm ($\overline{x}$ = 134.4 ± 35.8 mm), and for El Labrado, it ranges from 46.0 to 191.3 mm ($\overline{x}$ = 116.5 ± 44.9 mm). A seasonal trend was observed for El Labrado where for July, August, and September, the lowest precipitation was registered, $\overline{x}$ = 71 ± 22.8 mm (Figure 4).

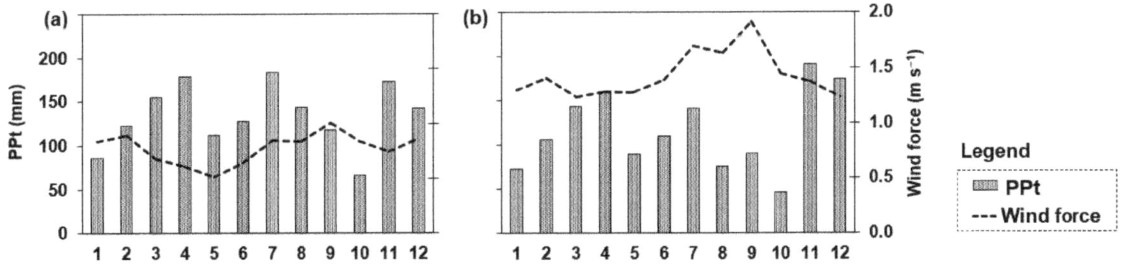

Figure 4. Monthly data of the cumulative precipitation (PPt) and the average wind force recorded in meteorological stations located in the proximity of the studied reservoirs: (**a**) Chanlud and (**b**) El Labrado in (1) December 2018 and (2) January, (3) February, (4) March, (5) April, (6) May, (7) June, (8) July, (9) August, (10) September, (11) October, and (12) December of 2019.

Vertical changes in the water temperature (WT) in the reservoir Chanlud are given in Figure 5. No evidence of consistent thermal stratification was observed. The coldest months were from June to October, and the mean temperature reported for this period was 9.0 ± 0.7 °C (Figure 5). The rest of the months reported a mean temperature of 11.6 ± 0.3 °C, i.e., the warmest period. The change in the temperature between the warmest period and the rest of the time is 2.6 °C. The maximum and minimum differences between the surface and the bottom were 2.9 and 0.9 °C in October and April, respectively. This emphasizes that mixing events are related to the warmest periods, and weak stratification occurs during colder times. Like WT, vertical profiles of dissolved oxygen (O_2) exhibit the highest values at the surface and the lowest O_2 concentrations at the bottom (Figure 5). A seasonal O_2 concentration is present, i.e., during the coldest period, a higher solubility of O_2 is notorious ($\overline{x}$ = 7.4 ± 0.4 mg L^{-1}), while, on the other hand, a lower O_2 solubility ($\overline{x}$ = 6.2 ± 0.3 mg L^{-1}) is associated with the warmest months. Water transparency by the Secchi disk (Z_{sd}) ranged between 4.2 and 6.6 m ($\overline{x}$ = 5.2 ± 0.8 m). The depth of the Chanlud reservoir ranged from 30.7 to 39.5 m, and the mean depth was reported as 36.8 ± 2.8 m. The euphotic zone (Z_{eu}) ranged from 13.0 to 18.0 m ($\overline{x}$ = 16.1 ± 2.0 m). The light vertical extinction coefficient (E_v) ranged from 0.2 to 0.3 ($\overline{x}$ = 0.24 ± 0.04) (these last two parameters were not significant regarding the PCA, namely, | loadings | < 0.25).

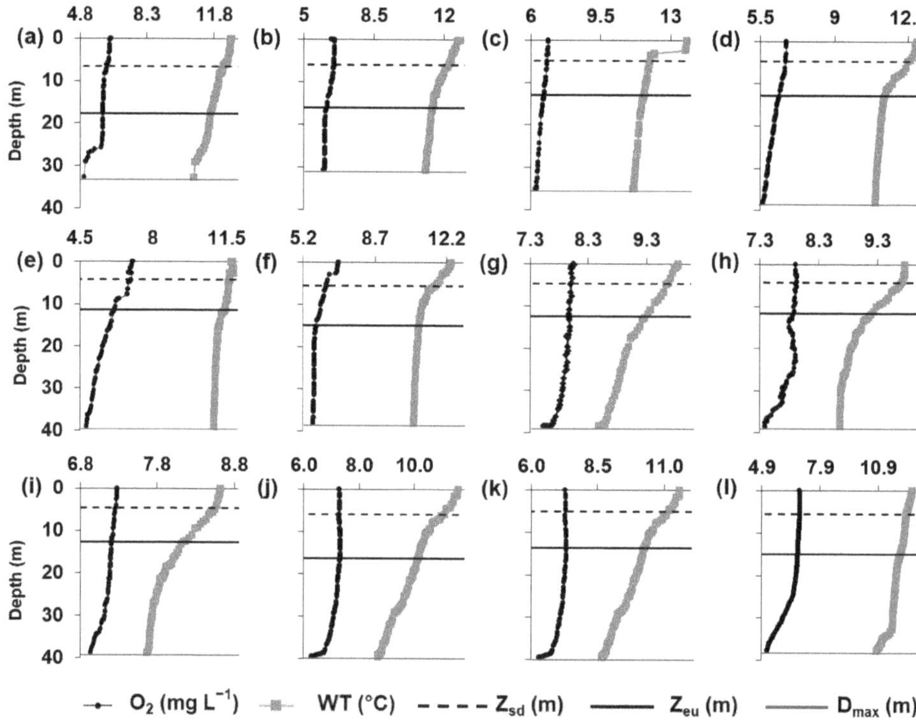

Figure 5. Monthly vertical profiles of dissolved oxygen (O_2) and water temperature (WT) of the Chanlud reservoir: depth of the Secchi disk (Z_{sd}), euphotic zone (Z_{eu}), and maximum depth (D_{max}) in (**a**) December 2018 and (**b**) January, (**c**) February, (**d**) March, (**e**) April, (**f**) May, (**g**) June, (**h**) July, (**i**) August, (**j**) September, (**k**) October, and (**l**) December of 2019.

The water temperature (WT) vertical profiles of El Labrado reservoir (Figure 6) show a typical curve, with the warmest water temperatures at the surface and the coldest at the bottom for most months. No thermic stratification is observed (with the exception of August, see Figure 6i). Evenly, in El Labrado reservoir, a clear thermic temporal trend is reflected, where, as in the case of Chanlud, June, July, August, September, and October are the coldest months ($\overline{x}$ = 9.4 ± 1.2 °C), and the rest of the months correspond with the warmest period ($\overline{x}$ = 12.4 ± 0.4 °C). There is a 3.0 °C difference between the warmest and coldest periods. The maximum and minimum differences between the surface and the bottom were 1.8 and 0.2 °C in March and December 2019, respectively. The vertical profiles of O_2 do not exhibit a trend regarding the depth gradient. Still, a seasonal O_2 concentration is present, where a higher solubility of O_2 is during the coldest period ($\overline{x}$ = 7.4 ± 0.4 mg L^{-1}). To the contrary, a lower O_2 solubility ($\overline{x}$ = 6.6 ± 0.1 mg L^{-1}) is associated with the warmest months. Z_{sd} ranged between 4.8 and 6.7 m ($\overline{x}$ = 6.0 ± 0.6 m), and the depth range values were between 8.4 and 14.1 ($\overline{x}$ = 11.4 ± 1.6 m). Z_{eu} was always greater than the maximum depth for El Labrado reservoir, and the E_v ranged from 0.2 to 0.3 ($\overline{x}$ = 0.2 ± 0.0.2).

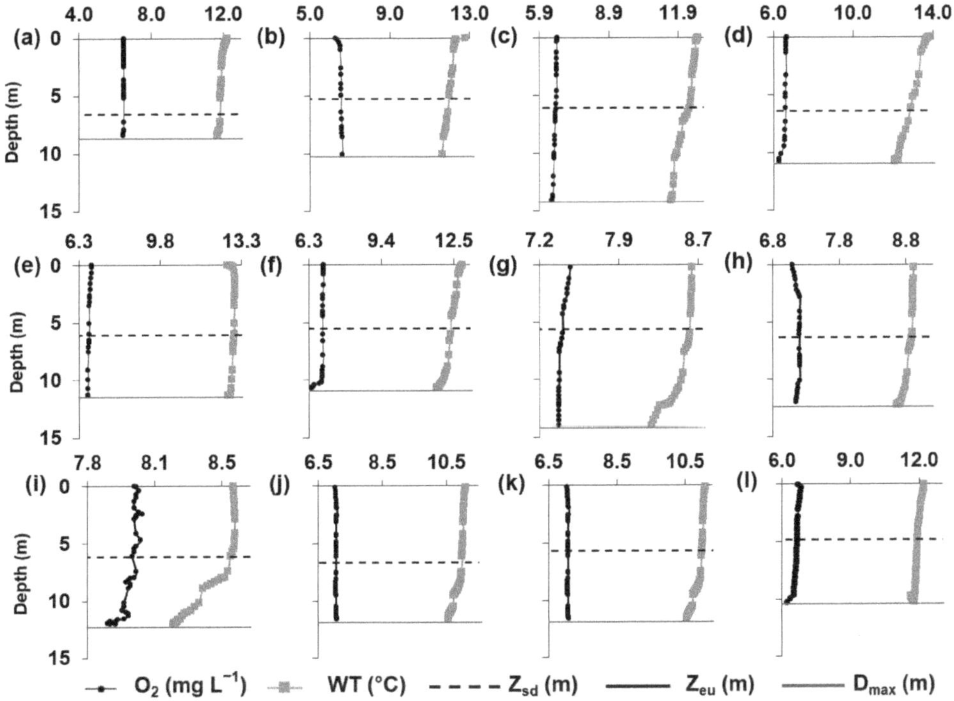

Figure 6. Monthly vertical profiles of dissolved oxygen (O_2) and water temperature (WT) of El Labrado reservoir: depth of the Secchi disk (Z_{sd}) and maximum depth (D_{max}) in (**a**) December 2018 and (**b**) January, (**c**) February, (**d**) March, (**e**) April, (**f**) May, (**g**) June, (**h**) July, (**i**) August, (**j**) September, (**k**) October, and (**l**) December of 2019.

The total alkalinity in Chanlud fluctuated between 32.4 and 85.0 mg L^{-1} of $CaCO_3$ ($\overline{x}$ = 40.6 ± 6.4 mg L^{-1}). No evidence of a trend regarding depth gradient exists for total alkalinity, and their maximum value was detected in January (i.e., 85.0 mg L^{-1} of $CaCO_3$). For El Labrado reservoir, the total alkalinity fluctuated between 23.8 and 38.8 mg L^{-1} of $CaCO_3$, with an overall mean of 27.9 ± 2.9 mg L^{-1}. No trend regarding depth gradient exists, and their maximum value was detected in October (i.e., 38.8 mg L^{-1} of $CaCO_3$).

Concerning primary productivity (PP), it was used as a single value in the PCA; however, in the current section, we approach this parameter according to its components (i.e., GPP + NPP = PP) to give the reader a more detailed description. Thus, the $\overline{x}$ gross primary productivity (GPP) value in the Chanlud reservoir was 9.9 ± 4.1 mg C m^{-2} d^{-1}, and for net primary productivity (NPP), the $\overline{x}$ = 7.2 ± 2.9 mg C m^{-2} d^{-1}. Higher values were found near the surface and decreased irregularly to the bottom. For the temporal component, a downward trend in GPP and NPP was observed during the coldest period (i.e., for the warm period, $\overline{x}_{GPP}$ = 11.8 ± 4.3 mg C m^{-2} d^{-1}; $\overline{x}_{NPP}$ = 8.5 ± 3.1 mg C m^{-2} d^{-1}; for the cold period, $\overline{x}_{GPP}$ = 7.3 ± 2.0 mg C m^{-2} d^{-1}; $\overline{x}_{NPP}$ = 5.3 ± 1.4 mg C m^{-2} d^{-1}) (see Figure 7a).

For El Labrado reservoir, the $\overline{x}_{GPP}$ = 2.1 ± 0.8 mg C m^{-2} d^{-1}, and for NPP, it was 1.5 ± 0.7 mg C m^{-2} d^{-1}. As in the case of the Chanlud reservoir, seasonality is essential to explain the trends in GPP and NPP—namely, for the warm period, $\overline{x}_{GPP}$ = 2.4 ± 0.8 mg C m^{-2} d^{-1}; $\overline{x}_{NPP}$ = 1.8 ± 0.7 mg C m^{-2} d^{-1}; for the cold period, $\overline{x}_{GPP}$ = 1.7 ± 0.8 mg C m^{-2} d^{-1}; $\overline{x}_{NPP}$ = 1.2 ± 0.5 mg C m^{-2} d^{-1} (Figure 7b).

Concerning the respiration (R) rates for the Chanlud reservoir, X_R = 7.4 ± 6.2 mg C m^{-2} d^{-1}. For the warm period, $\overline{x}_R$ = 9.4 ± 7.1 mg C m^{-2} d^{-1}, and for the cold period, $\overline{x}_R$ = 4.7 ±

3.8 mg C m^{-2} d^{-1}. The mean R rate in El Labrado reservoir was 1.2 0.9 ± 0.9 mg C m^{-2} d^{-1}. However, contrary to the Chanlud reservoir pattern, seasonality marginally influences R in this reservoir (i.e., for the warm period, $\bar{x}_R$ = 1.3 ± 1.0 mg C m^{-2} d^{-1} and for the cold period, $\bar{x}_R$ = 1.1 ± 0.8 mg C m^{-2} d^{-1}) (R was not significant regarding the PCA, i.e., | loadings | < 0.25).

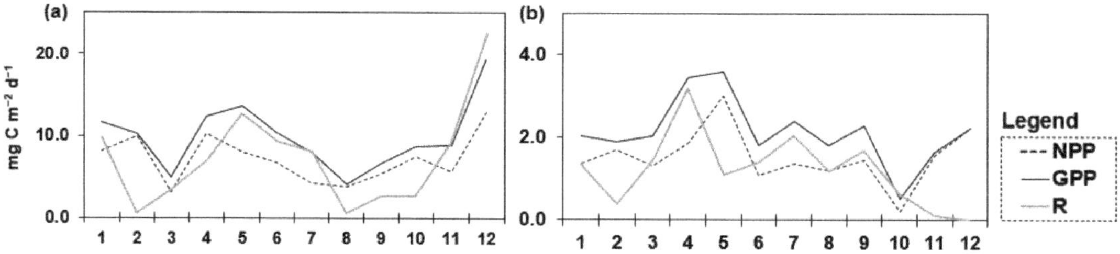

Figure 7. Gross primary productivity (GPP), net primary productivity (NPP), and respiration (R) rates for each month at the (**a**) Chanlud and (**b**) El Labrado reservoirs in (1) December 2018 and (2) January, (3) February, (4) March, (5) April, (6) May, (7) June, (8) July, (9) August, (10) September, (11) October, and (12) December of 2019.

3.2. Multiple Linear Regression Analysis (MLRA)

Concerning the individual MLRAs performed for each reservoir, the Adjusted R-Squared (Adj—R^2) values were 0.7 and 0.8 for the Chanlud ($MLRA_{Ch}$) and El Labrado ($MLRA_L$) regression models, respectively, which implies that the models, as fitted, very well explain the variability in PP in both cases. The Durbin–Watson (DW) statistic value was 2.1 in both cases, which is congruent with the absence of serial autocorrelation in the residuals at the 95.0% confidence level (the DW statistic becomes smaller as the serial correlation increases). The *p*-values of the analysis of variance of both cases were less than 0.05 (i.e., 0.04 for $MLRA_{Ch}$ and $MLRA_L$); therefore, there are statistically significant relationships between the independent variables and PP at the 95.0% confidence level in both models.

The output equation of the fitted model for the Chanlud reservoir is as follows:

$$PP_{Ch} = -5.39 + 0.87 \times WT + 3.87 \times D_{max} + 1.14 \times Chl + 1.66 \times Alkalinity + 0.75 \times HB + 1.70 \times WF$$

and for El Labrado reservoir, it is as follows:

$$PP_L = -0.26 - 0.31 \times O_2 + 0.12 \times WT + 0.12 \times HB + 0.09 \times Precipitation + 0.60 \times WF$$

Different planned approaches were designed for PCA and MLRAs in the current research; despite this, both analyses choose similar variables as informative to perform their corresponding modeling. For example, WT, precipitation, and O_2 are critical variables according to the PCA to describe the seasonal variability of both reservoirs and are equally chosen by MLRAs as crucial variables to explain the variability of the PP in the studied reservoirs. Also, D_{max}, $CaCO_3$, and WF are all critical variables identified by the PCA to perform a clear discrimination between both reservoirs. MLRAs equally choose them as the most informative variables to explain the variability of the PP in the evaluated reservoirs.

Nevertheless, notwithstanding these similarities, each reservoir itself has its own set of descriptive parameters to explain the variability of its corresponding PP trends. Hence, checking the output MLRA equations shows that Chanlud is more complex than El Labrado, not only due to the number of variables but also due to their nature. For example, the hydraulic and chemical components in Chanlud are more critical than in El Labrado; conversely, the metrological component in El Labrado is more relevant than in Chanlud. Orographic factors (the almost complete absence of hills around Chanlud, which is opposite El Labrado) and the critical size difference between the contributing hydrological areas of both reservoirs (for Chanlud, the contributing hydrological area is 85.5 km^2, and for El Labrado, it is 42.0 km^2) could explain these differences in terms of complexity between both multiple regression analyses ($MLRA_{Ch}$ and $MLRA_L$).

On the other hand, in both MLRAs, some variables identified as not significant by the PCA (i.e., | loadings | < 0.25) were chosen as essential to explain the variability of the PP (i.e., heterotrophic bacteria and chlorophyll-α, HB and Chl, respectively). Thus, for Chanlud, values of HB were found between 0 and 250 with a mean value of 21.6 ± 39.2 bacteria 100^{-1} mL^{-1}. The highest values were found in July and September (250 and 141 bacteria 100^{-1} mL^{-1}, respectively). In the case of El Labrado reservoir, the heterotrophic bacteria were found in a range of 0 to 190 bacteria 100^{-1} mL^{-1} with a mean value of 19.5 ± 36.8 bacteria 100^{-1} mL^{-1} (Table 1). The highest values were found in April and June (190 and 120 bacteria 100^{-1} mL^{-1}, respectively). Both reservoirs showed non-trends in HB concerning depth gradients. The chlorophyll-*a* (Chl) measurements in the Chanlud reservoir ranged from 0.4 to 6.2, with a mean of 1.7 ± 1.6 μg L^{-1}, while it ranged from 0 to 3.5, with a mean of 1.1 ± 1.1 μg L^{-1} in El Labrado reservoir (Table 1).

4. Discussion

In Ecuador, there are no previous studies that address the issue of limnology in reservoirs, and the availability of baseline data are restricted only to lakes [27,72–76]. This present paper is a pioneer effort to address the study of limnological aspects in high Andean reservoirs in Ecuador and one of the few references to tropical Andean reservoirs [16,21].

Both statistical approaches used in this research, i.e., the principal components analysis (PCA) and multiple linear regression analysis (MLRA), gave relevant results in selecting appropriate explanatory variables representative to (i) explain the seasonal variability of each reservoir, the (ii) potential differences between the two studied reservoirs, and (iii) the variability of the PP in each reservoir. Furthermore, the high values of their performance statistics validate the scientific reliability of the outputs.

For both reservoirs, the monthly mean values of dissolved oxygen and water temperature followed temporal patterns previously reported for reservoirs and lakes; namely, the higher the temperature, the lower the diffusion of dissolved oxygen and vice versa [77,78]. Additionally, both variables exhibited a downward trend regarding depth; however, no thermal stratification was observed, which implies frequent mixing events, probably due to strong winds (wind force was a variable that both PCA and MLRAs chose as informative) and the lack of hills or mountains around the reservoirs [21], mainly in Chanlud. The downward pattern of dissolved oxygen regarding depth was weaker in El Labrado than in Chanlud; however, no statistically significant differences existed for this parameter between both reservoirs (Figure 3f). This last finding can be partially explained, considering that wind force is significantly higher in El Labrado than in the Chanlud reservoir (Figure 3g). Similar to most tropical lakes, the thermal stratification of the water column in both reservoirs is typically weak due to limited seasonal variation in temperature [79]. Thus, in tropical Andean reservoirs, the high and cold inflows during the wet season homogenize the water column below a shallow surface mixed layer; in dry seasons, warmer inflows enter at intermediate depths, favoring the development of a thick metalimnion with sharper temperature gradients at its top and base [20,77,80,81]. Our results did not show this seasonal trend. Other studies in the lakes of Cajas National Park, near the area of this current research, identified the same colder period; however, contrary to this study, the authors linked this period with the thermal stratifications of some of the studied lakes [76]. The differences in the flushing time of water bodies and the effects of riverine inflows at the studied sites could explain this non-congruence between both studies.

Concerning nutrients, only nitrates were representative of signal chemical detection. Nitrates were most notorious in the Chanlud reservoir (i.e., the mean of nitrates for Chanlud is 0.56 ± 1.52 mg L^{-1}, and for El Labrado, it is 0.06 ± 0.12 mg L^{-1}). We interpret the increased availability of nitrates in Chanlud as mainly the result of the notoriously larger size of their contributing hydrological area relative to El Labrado. Thus, higher rates of organic material production in the terrestrial part of the catchment subsequently increase the leaching of nitrates into aquatic systems [82,83]. However, despite the nitrate concentrations detected, their values were generally consistent with very low concentrations, which is congruent with the fact that both PCA and MLRAs considered nitrates a nonsignificant variable. The rest of the nutrients (i.e., nitrites, ammonium, and orthophosphates) presented weak chemical signals (Table 1). Two explanations are provided in this regard: (i) In tropical regions, in the first rainy season, many nutrients are carried into reservoirs by the first precipitations. A peak in nutrient input can often be attributed to surface runoff from nutrient-rich soils [84]. However, most of these nutrients will not remain in the system for long if the residence time of the water is short, i.e., if there is a high flushing rate, as is the case in many reservoirs. (ii) There is an almost complete absence of human populations in the region's reservoirs [85]; i.e., the contributing hydrological areas for both reservoirs are highly conserved. Their landscapes are conformed by pristine ecosystems (for Chanlud, 0.3% of their contributing basin is anthropized, and for El Labrado, the percentage is 0.5%) [86].

Regarding alkalinity, the current outputs are congruent with other Andean lentic systems, such as La Brava and La Punta lakes in northern Chile, where very similar values were reported [87]. Thus, the current results suggest that for both reservoirs, the waters are slightly alkaline, especially Chanlud (Figure 3e). In this context, the more alkaline a lake is, the greater the concentration of carbonates [88], which has been reported as a promoter of the growth of many groups of phytoplankton [89,90]. This is consistent with this study's findings, where a congruence between trends in alkalinity and primary productivity (Figure 3e,h) was observed for both reservoirs. Furthermore, concerning the $MLRA_{Ch}$, alkalinity is an important explanatory variable to explain the variability of primary productivity.

For water transparency (Z_{sd}), there was a difference between the monthly measurements of both reservoirs. El Labrado reservoir contained the most transparent waters (Figure 3d), which could be explained by the lowest nutrient concentration and primary productivity (Figure 3h) in this reservoir compared to Chanlud. Furthermore, this finding could be explained by the contributing hydrological areas of both reservoirs (86.8 and 40.1 km^2 for Chanlud and El Labrado, respectively); i.e., the greater the contribution area, the greater the runoff area, and therefore, the greater the amounts of total solids going into the reservoir. Despite this, the reservoirs' current water transparency values were slightly lower than reported in the lakes of the Cajas National Park, Z_{sd} $\overline{x}$ = 6.7 m [76].

Regarding the primary productivity (PP) values, $PP_{Chanlud} > PP_{El\ Labrado}$ (Figure 3h). The outputs for PP correspond with low ecosystem metabolic activity and a low loading nutrient rate. In general, in lakes and reservoirs, it has been reported that low PP occurs when (i) human activities in the surrounding areas are scarce, (ii) as a product of wind stress, (iii) when there is no stability concerning solar radiation, and (iv) as a product of non-effective nutrient recycling [91–93]. This latter factor is congruent with findings for twenty-four tropical high-altitude lakes in southern Ecuador [94]. In this study, the authors found that lower phosphate concentrations explain the low productivity of the studied lakes, which is consistent with our findings in the studied reservoirs, where the phosphates (among other nutrients) were undetectable. In addition to these four factors, two more explanatory elements must be included for this current study case: (v) the relatively short water residence times and (vi) the fact that both reservoirs are above 3400 m a.s.l. Hence, despite the low PP values obtained, the PP method efficiently captured increased and decreased dissolved oxygen data in the clear and dark bottles. It is necessary to consider that current PP datasets are expected to be typical of the Andean lentic water bodies, mainly due to their pristine conditions and altitudinal factors The latter is congruent with findings for northern Sweden lakes, where the PP decreases as altitude increases [95]. Unfortunately, very few studies on similar reservoirs in physiographic terms to those studied in this research have been conducted in order to be able to perform a comparative analysis (mainly in the Andean region). However, some findings about metabolic rates in this current research are congruent with other studies. For example, in the case of Lake Monte Alegre (south-eastern Brazil), a study reported that the higher PP in the water column occurred in the transition periods when thermal stratification was unstable and lower PP occurred in the cold season (frequent mixing) [96]. Both findings are congruent with the outputs of the two studied reservoirs. In Chanlud, the peak of PP is just after the coldest period, and for El Labrado, it is before it (Figure 7); i.e., timing/season is essential. Thus, for Chanlud, during the coldest months, the $\bar{x}_{PP}$ = 12.6 mg C m^{-2} d^{-1}, and in the warmest period, the $\bar{x}_{PP}$ = 20.3 mg C m^{-2} d^{-1}; for El Labrado, for the coldest months, $\bar{x}_{PP}$ = 2.9 mg C m^{-2} d^{-1}, and for the warmest period, $\bar{x}_{PP}$ = 4.2 mg C m^{-2} d^{-1}. These results are consistent with more studies carried out in other lakes and reservoirs, where the same PP and water temperature relationship was observed [97,98] (herein, the MLRAs chose water temperature as one of the critical parameters to explain the variability of PP in both reservoirs). In this context, the PP in the studied reservoirs is subject to multifactorial regulation. Besides water temperature, meteorological (i.e., wind force, precipitation) and biological (i.e., heterotrophic bacteria) factors describe the PP and their variability. This is like other studies, where there is evidence that a set of variables (like the ones reported here) have been described as determining factors in regulating biological processes in the lakes of the Andes [73,99,100].

Following a statistically sound approach, this study aims to increase our understanding of the limnological characteristics of tropical high Andean reservoirs, which have yet to be studied in much detail. The results will also form an important baseline for information to determine future changes that might take place in them, such as climatic changes, which are predicted to be dramatic at high latitudes [101], and local stressors. Also, this research identifies significant and nonsignificant descriptive variables to describe (i) the variability of each reservoir and discriminatory factors between them and (ii) the primary productivity of the studied reservoirs. Both findings have the potential to reduce the number of variables to be monitored in future similar research and, consequently, the monitoring time and related monetary expenses.

5. Conclusions

Some differences are evident between both reservoirs in aspects related to physicochemical (i.e., alkalinity, light penetration) and biological factors (i.e., primary productivity). Chanlud exhibits a more photosynthetically efficient euphotic zone. Although the primary productivity of Chanlud was higher than El Labrado, the results for both reservoirs correspond to low metabolic rates. This was expected since primary productivity rates are linked to the intrinsic conditions of reservoirs, that is, high elevation, low nutrient recycling, low temperatures, wind stress, and short retention times. The method used to estimate the metabolic rates of the reservoirs was practical and provided representative and reliable information. This is validated by contrasting the vertical integrations of primary productivity with variables such as temperature and dissolved oxygen. Temperature, heterotrophic bacteria, and wind force are critical variables in both reservoirs since the results significantly influence primary productivity rates. The study's weaknesses include its reliance on accessible areas near dam walls. Incorporating new sampling sites associated with a horizontal zonation of reservoirs would benefit future evaluations. Also, more accurate methodologies to measure the primary productivity and respiration rates could be tested—for example, the oxygen isotope ($\delta^{18}O$) mass balance approach.

Author Contributions: Conceptualization, G.S.; Methodology, G.S., J.R., C.L., M.A.E.F. and P.L.M.G.; Software, G.S.; Formal analysis, G.S.; Investigation, G.S. and J.R.; Data curation, G.S. and J.R.; Writing—original draft, G.S.; Writing—review & editing, G.S., A.A., C.L., M.A.E.F. and P.L.M.G.; Supervision, A.A., C.L., M.A., M.A.E.F. and P.L.M.G.; Project administration, A.A. and M.A.; Funding acquisition, A.A. and M.A. All authors have read and agreed to the published version of the manuscript.

Funding: The APC was kindly paid by the Universidad Hemisferios. Not specific grant number applicable.

Data Availability Statement: The data might be available directly from the Departamento de Recursos Hídricos y Ciencias Ambientales of the Universidad de Cuenca.

Acknowledgments: The authors would like to thank the company Electro Generadora del Austro (Elecaustro S.A.) and the University of Cuenca for funding this current study. The preparation of the manuscript took place under the framework of the project "Investigación biótica, metabólica y trófica de los embalses de Chanlud y El Labrado".

Conflicts of Interest: Author Marta Aguilar was employed by the company Empresa Electro Generadora del Austro (Elecaustro S. A.). The remaining authors declare that the research was conducted in the absence of any commercial or financial relationships that could be construed as a potential conflict of interest.

References

1. Rodgher, S.; Espíndola, E.L.; Rocha, O.; Fracácio, R.; Pereira, R.H.; Rodrigues, M.H. Limnological and ecotoxicological studies in the cascade of reservoirs in the Tietê River (São Paulo, Brazil). *Braz. J. Biol.* **2005**, *65*, 697–710. [CrossRef] [PubMed]
2. Zarfl, C.; Lumsdon, A.E.; Berlekamp, J.; Tydecks, L.; Tockner, K. A global boom in hydropower dam construction. *Aquat. Sci.* **2015**, *77*, 161–170. [CrossRef]
3. Bredenhand, E.; Samways, M.J. Impact of a dam on benthic macroinvertebrates in a small river in a biodiversity hotspot: Cape Floristic Region, South Africa. *J. Insect Conserv.* **2009**, *13*, 297–307. [CrossRef]
4. Nguyen, T.H.T.; Everaert, G.; Boets, P.; Forio, M.A.E.; Bennetsen, E.; Volk, M.; Hoang, T.H.T.; Goethals, P.L.M. Modelling tools to analyze and assess the ecological impact of hydropower dams. *Water* **2018**, *10*, 259. [CrossRef]
5. Margalef, R. *Limnología*; Omega: Barcelona, Spain, 1983.
6. Bengtsson, L.; Herschy, R.W.; Fairbridge, R.W. *Encyclopedia of Lakes and Reservoirs*; Springer: Dordrecht, The Netherland, 2012.
7. Simões, N.R.; Nunes, A.H.; Dias, J.D.; Lansac-Tôha, F.A.; Velho, L.F.M.; Bonecker, C.C. Impact of reservoirs on zooplankton diversity and implications for the conservation of natural aquatic environments. *Hydrobiologia* **2015**, *758*, 3–17. [CrossRef]
8. Song, C.; Gardner, K.H.; Klein, S.J.W.; Souza, S.P.; Mo, W. Cradle-to-grave greenhouse gas emissions from dams in the United States of America. *Renew. Sustain. Energy Rev.* **2018**, *90*, 945–956. [CrossRef]
9. Dodds, W.K.; Whiles, M.R. *Freshwater Ecology*; Academic Press: Cambridge, MA, USA, 2019.
10. Páez, R.; Ruiz, G.; Márquez, R.; Soto, L.M.; Montiel, M.; López, C. Limnological studies on a shallow reservoir in Western Venezuela (Tulé Reservoir). *Limnologica* **2001**, *31*, 139–145. [CrossRef]
11. Silva, R.F.C.M.; de Almeida, T.; Cicerelli, R.E.; Gomes, L.N.L. A spatiotemporal analysis of the physicochemical parameters after the operation of the Corumbá IV reservoir (Midwest Brazil) to support better management decision. *Environ. Monit. Assess.* **2021**, *193*, 247. [CrossRef] [PubMed]
12. Williamson, C.E.; Saros, J.E.; Vincent, W.F.; Smol, J.P. Lakes and reservoirs as sentinels, integrators, and regulators of climate change. *Limnol. Oceanogr.* **2009**, *54*, 2273–2282. [CrossRef]
13. Soares, M.C.S.; Marinho, M.M.; Huszar, V.L.M.; Branco, C.W.C.; Azevedo, S.M.F.O. The effects of water retention time and watershed features on the limnology of two tropical reservoirs in Brazil. *Lakes Reserv. Res. Manag.* **2008**, *13*, 257–269. [CrossRef]
14. Gansser, A. Facts and theories on the andes. *J. Geol. Soc.* **1973**, *129*, 93–131. [CrossRef]
15. Michelutti, N.; Tapia, P.M.; Labaj, A.L.; Grooms, C.; Wang, X.; Smol, J.P. A limnological assessment of the diverse waterscape in the Cordillera Vilcanota, Peruvian Andes. *Inland Waters* **2019**, *9*, 395–407. [CrossRef]
16. Aguilera, X.; Declerck, S.; De Meester, L.; Maldonado, M.; Ollevier, F. Tropical high Andes lakes: A limnological survey and an assessment of exotic rainbow trout (Oncorhynchus mykiss). *Limnologica* **2006**, *36*, 258–268. [CrossRef]
17. Cabrol, N.A.; Grin, E.A.; Chong, G.; Minkley, E.; Hock, A.N.; Yu, Y.; Bebout, L.; Fleming, E.; Häder, D.P.; Demergasso, C.; et al. The High-Lakes Project. *J. Geophys. Res. Biogeosci.* **2009**, *114*, G00D06. [CrossRef]
18. Fortner, S.K.; Mark, B.G.; McKenzie, J.M.; Bury, J.; Trierweiler, A.; Baraer, M.; Burns, P.J.; Munk, L. Elevated stream trace and minor element concentrations in the foreland of receding tropical glaciers. *Appl. Geochem.* **2011**, *26*, 1792–1801. [CrossRef]
19. Benito, X.; Fritz, S.C.; Steinitz-Kannan, M.; Tapia, P.M.; Kelly, M.A.; Lowell, T.V. Geo-climatic factors drive diatom community distribution in tropical South American freshwaters. *J. Ecol.* **2018**, *106*, 1660–1672. [CrossRef]
20. Marín-Ramírez, A.; Gómez-Giraldo, A.; Román-Botero, R. Variación estacional de la temperatura media y los flujos advectivos y atmosféricos de calor en un embalse tropical andino. *Rev. Acad. Colomb. Cienc. Exactas Físicas Nat.* **2020**, *44*, 360–375. [CrossRef]
21. Villabona-González, S.L.; Benjumea-Hoyos, C.A.; Gutiérrez-Monsalve, J.A.; López-Muñoz, M.T.; González, E.J. Main physicochemical and biological variables in the trophic state of five Colombian Andean reservoirs. *Rev. Acad. Colomb. Cienc. Exactas Físicas Nat.* **2020**, *44*, 344–359.
22. Tundisi, J.G.; Matsumura-Tundisi, T.; Calijuri, M.C. Limnology and management of reservoirs in Brazil. *Comp. Reserv. Limnol. Water Qual. Manag.* **1993**, *1000*, 25–55.
23. Tundisi, J.G.; Matsumura-Tundisi, T. Integration of research and management in optimizing multiple uses of reservoirs: The experience in South America and Brazilian case studies. *Hydrobiologia* **2003**, *500*, 231–242. [CrossRef]
24. Agostinho, A.A.; Gomes, L.C.; Latini, J.D. Fisheries management in brazilian reservoirs: Lessons from/for South America. *Interciencia* **2004**, *29*, 334–338.

25. Almeida, V.L.S.; Dantas, Ê.W.; Melo-Júnior, M.; Bittencourt-Oliveira, M.C.; Moura, A.N. Zooplanktonic community of six reservoirs in northeast Brazil. *Braz. J. Biol.* **2009**, *69*, 57–65. [CrossRef] [PubMed]
26. Tundisi, J.G. Reservoirs: New challenges for ecosystem studies and environmental management. *Water Secur.* **2018**, *4–5*, 1–7.
27. Steinitz-Kannan, M.; López, C.; Jacobsen, D.; Guerra, M.L. History of limnology in Ecuador: A foundation for a growing field in the country. *Hydrobiologia* **2020**, *847*, 4191–4206. [CrossRef]
28. Hwang, S.-J.; Kwun, S.-K.; Yoon, C.-G. Water quality and limnology of Korean reservoirs. *Paddy Water Environ.* **2003**, *1*, 43–52. [CrossRef]
29. Ulloa, C.; Jørgensen, P. *Arboles y Arbustos de los Andes del Ecuador*; Ediciones ABYA-YALA: Quito, Ecuador, 2002.
30. Buytaert, W.; Célleri, R.; De Bièvre, B.; Cisneros, F.; Wyseure, G.; Deckers, J.; Hofstede, R. Human impact on the hydrology of the Andean páramos. *Earth-Sci. Rev.* **2006**, *79*, 53–72. [CrossRef]
31. Buytaert, W.; Celleri, R.; Willems, P.; De Bièvre, B.; Wyseure, G. Spatial and temporal rainfall variability in mountainous areas: A case study from the south Ecuadorian Andes. *J. Hydrol.* **2006**, *329*, 413–421. [CrossRef]
32. Matute-Pinos, V.O. Análisis de Factibilidad de Generación Eléctrica a pie de la Presa de Chanlud. Bachelor's Thesis, Universidad de Cuenca, Cuenca, Ecuador, 2014.
33. Herrera, I.A.; Burneo, P. Environmental flow assessment in Andean rivers of Ecuador, case study: Chanlud and El Labrado dams in the Machángara River. *Ecohydrol. Hydrobiol.* **2017**, *17*, 103–112. [CrossRef]
34. American Public Health Association. *Standard Methods for the Examination of Water and Wastewater*, 20th ed.; American Public Health Association: Washington, DC, USA, 1998; Volume 34.
35. Massa, S.; Caruso, M.; Trovatelli, F.; Tosques, M. Comparison of plate count agar and R2A medium for enumeration of heterotrophic bacteria in natural mineral water. *World J. Microbiol. Biotechnol.* **1998**, *14*, 727–730. [CrossRef]
36. Snazelle, T. *Evaluation of Xylem EXO Water-Quality Sondes and Sensors*; USGS: Reston, VA, USA, 2015.
37. de Castro Medeiros, L.; Mattos, A.; Lürling, M.; Becker, V. Is the future blue-green or brown? The effects of extreme events on phytoplankton dynamics in a semi-arid man-made lake. *Aquat. Ecol.* **2015**, *49*, 293–307. [CrossRef]
38. Campos-Silva, J.V.; Peres, C.A.; Amaral, J.H.F.; Sarmento, H.; Forsberg, B.; Fonseca, C.R. Fisheries management influences phytoplankton biomass of Amazonian floodplain lakes. *J. Appl. Ecol.* **2021**, *58*, 731–743. [CrossRef]
39. Dokulil, M.; Silva, I.; Bauer, K. An assessment of the phytoplankton biomass and primary productivity of Parakrama Samudra, a shallow man-made lake in Sri Lanka. In *Limnology of Parakrama Samudra—Sri Lanka*; Springer: Dordrecht, The Netherland, 1983; pp. 49–76.
40. Amarasinghe, P.B.; Vijverberg, J. Primary production in a tropical reservoir in Sri Lanka. *Hydrobiologia* **2002**, *487*, 85–93. [CrossRef]
41. Wetzel, R.G.; Likens, G.E. *Limnological Analyses*, 3rd ed.; Springer Science+Business Media, LLC: New York, NY, USA, 1993.
42. Egna, H.S.; Boyd, C.E. *Dynamics of Pond Aquaculture*; CRC Press: Boca Raton, FL, USA, 1997.
43. Villarreal-Quintero, E.Y.; Alcocer, J.; Oseguera, L.A. Producción primaria en un lago oligotrófico tropical. In *Estado Actual del Conocimiento del Ciclo del Carbono y sus Interacciones en México: Síntesis a 2013*; Programa Mexicano del Carbono A.C.: Texcoco, Mexico, 2014; pp. 322–327.
44. Lehman, J.T. Hypolimnetic metabolism in Lake Washington: Relative effects of nutrient load and food web structure on lake productivity. *Limnol. Oceanogr.* **1988**, *33*, 1334–1347. [CrossRef]
45. Sterner, R.W. In situ-measured primary production in Lake Superior. *J. Great Lakes Res.* **2010**, *36*, 139–149. [CrossRef]
46. Lee, H.; Shum, C.K.; Tseng, K.H.; Guo, J.Y.; Kuo, C.Y. Present-day lake level variation from envisat altimetry over the northeastern qinghai-tibetan plateau: Links with precipitation and temperature. *Terr. Atmos. Ocean. Sci.* **2011**, *22*, 169–175. [CrossRef]
47. Larson, R.; Edwards, B.H.; Falvo, D.C. *Calculus I with Precalculus—A One-Year Course*, 3rd ed.; Brooks/Cole, Cengage Learning: Pacific Grove, CA, USA, 2011.
48. Hanselman, D.; Littlefield, B. *Mastering MATLAB®*; Pearson Education Limited: London, UK, 2012.
49. Wondie, A.; Mengistu, S.; Vijverberg, J.; Dejen, E. Seasonal variation in primary production of a large high altitude tropical lake (Lake Tana, Ethiopia): Effects of nutrient availability and water transparency. *Aquat. Ecol.* **2007**, *41*, 195–207. [CrossRef]
50. Shapiro, S.S.; Wilk, M.B. An Analysis of Variance Test for Normality (Complete Samples). *Biom. Trust.* **1965**, *52*, 591–611. [CrossRef]
51. Anderson, T.; Finn, J. *The New Statistical Analysis of Data*; Springer: New York, NY, USA, 1996.
52. Helsel, D.R.; Hirsch, R.M.; Ryberg, K.R.; Archfield, S.A.; Gilroy, E.J. Statistical Methods in Water Resources. In *Book 4, Hydrologic Analysis and Interpretation*; U.S. Geological Survey: Reston, VA, USA, 2020; p. 458.
53. R Core Team. *Package 'stats'*; R Core Team: Vienna, Austria, 2016.
54. Shen, C.; Shi, Y.; Fan, K.; He, J.-S.; Adams, J.M.; Ge, Y.; Chu, H. Soil pH dominates elevational diversity pattern for bacteria in high elevation alkaline soils on the Tibetan Plateau. *FEMS Microbiol. Ecol.* **2019**, *95*, fiz003. [CrossRef] [PubMed]
55. Frank, I.E.; Todeschini, R. *The Data Analysis Handbook*; Elsevier Science B.V.: Amsterdam, The Netherlands, 1994; Volume 14.
56. Zupan, J. *PCs for Chemists*; Elsevier Science Publishers B.V.: Amsterdam, The Netherlands, 1990; Volume 5.
57. Varmuza, K.; Filzmoser, P. *Introduction to Multivariate Statistical Analysis in Chemometrics*; Taylor & Francis Group, LLC.: Abingdon, UK, 2010; Volume 64.
58. Einax, J.W.; Zwanziger, H.W.; Geiss, S. *Chemometrics in Environmental Analysis*; Wiley-VCH Verlag GmbH: Hoboken, NJ, USA, 1997.

59. Ballabio, D. A MATLAB toolbox for Principal Component Analysis and unsupervised exploration of data structure. *Chemom. Intell. Lab. Syst.* **2015**, *149*, 1–9. [CrossRef]
60. Ballabio, D.; Consonni, V. Classification tools in chemistry. Part 1: Linear models. PLS-DA. *Anal. Methods* **2013**, *5*, 3790–3798. [CrossRef]
61. Mevik, B.; Wehrens, R. Introduction to the pls Package. In *Help Section of the "Pls" Package of R Studio Software*; R Core Team: Vienna, Austria, 2015; Section 7; pp. 1–23.
62. Rácz, A.; Bajusz, D.; Héberger, K. Modelling methods and cross-validation variants in QSAR: A multi-level analysis. *SAR QSAR Environ. Res.* **2018**, *29*, 661–674. [CrossRef] [PubMed]
63. Rani, M.; Marchesi, C.; Federici, S.; Rovelli, G.; Alessandri, I.; Vassalini, I.; Ducoli, S.; Borgese, L.; Zacco, A.; Bilo, F.; et al. Miniaturized near-infrared (MicroNIR) spectrometer in plastic waste sorting. *Materials* **2019**, *12*, 2740. [CrossRef] [PubMed]
64. Peres-Neto, P.R.; Jackson, D.A.; Somers, K.M. Giving meaningful interpretation to ordination axes: Assessing loading significance in principal component analysis. *Ecology* **2003**, *84*, 2347–2363. [CrossRef]
65. Chatfield, C.; Collins, A.J. *Introduction to Multivariate Analysis*, 1st ed.; Chapman and Hall: London, UK, 1980.
66. Williams, L.J.; Abdi, H. Fisher's Least Significant Difference (LSD) Test. In *Encyclopedia of Research Design*; Salkind, N., Ed.; Sage: Thousand Oaks, CA, USA, 2010; pp. 840–853.
67. Mendiburu, F. *Agricolae: Statistical Procedures for Agricultural Research, R Package Version 1.3-2*; R Core Team: Vienna, Austria, 2019; Volume 1.
68. Montgomery, D.; Peck, E.A.; Geoffrey, V. *Introduction to Linear Regression Analysis*, 5th ed.; John Wiley & Sons: Hoboken, NJ, USA, 2012.
69. Rawlings, J.O.; Pantula, S.G.; Dickey, D.A. *Applied Regression Analysis: A Research Tool*, 2nd ed.; Springer: New York, NY, USA, 1998.
70. Zuur, A.F.; Ieno, E.N.; Walker, N.J.; Saveliev, A.A.; Smith, G.M. *Mixed Effects Models and Extensions in Ecology with R*; Springer Science+Business Media: New York, NY, USA, 2009; Volume 574.
71. Durbin, J.; Watson, G.S. Testing for serial correlation in least squares regression. III. *Biometrika* **1971**, *58*, 1–19. [CrossRef]
72. Steinitz-Kannan, M. Comparative Limnology of Ecuadorian Lakes: A Study of Species Number and Composition of Plankton Communities of the Galapagos Islands and the Equatorial Andes. Ph.D. Thesis, Ohio State University, Columbus, OH, USA, 1979.
73. Miller, M.C.; Kannan, M.; Colinvaux, P.A. Limnology and primary productivity of Andean and Amazonian tropical lakes of Ecuador. *Int. Ver. Theor. Angew. Limnol. Verhandlungen* **1984**, *22*, 1264–1270. [CrossRef]
74. Gunkel, G. Limnology of an Equatorial High Mountain Lake in Ecuador, Lago San Pablo. *Limnologica* **2000**, *30*, 113–120. [CrossRef]
75. Gunkel, G.; Beulker, C. Limnology of the crater lake Cuicocha, Ecuador, a cold water tropical lake. *Int. Rev. Hydrobiol.* **2009**, *94*, 103–125. [CrossRef]
76. Van Colen, W.R.; Mosquera, P.; Vanderstukken, M.; Goiris, K.; Carrasco, M.C.; Decaestecker, E.; Alonso, M.; León-Tamariz, F.; Muylaert, K. Limnology and trophic status of glacial lakes in the tropical Andes (Cajas National Park, Ecuador). *Freshw. Biol.* **2017**, *62*, 458–473. [CrossRef]
77. Gómez-Giraldo, A.; Román-Botero, R.; Toro, F.M. Seasonal evolution of the thermal structure of a tropical reservoir. In Proceedings of the 16th Workshop on Physical Processes in Natural Waters, Queensland, Australia, 8–11 July 2013; pp. 2–3.
78. Zhang, Y.; Wu, Z.; Liu, M.; He, J.; Shi, K.; Zhou, Y.; Wang, M.; Liu, X. Dissolved oxygen stratification and response to thermal structure and long-term climate change in a large and deep subtropical reservoir (Lake Qiandaohu, China). *Water Res.* **2015**, *75*, 249–258. [CrossRef] [PubMed]
79. Lewis, W.M. Biogeochemistry of tropical lakes. *Int. Ver. Theor. Angew. Limnol. Verhandlungen* **2010**, *30*, 1595–1603. [CrossRef]
80. Posada-Bedoya, A.; Gómez-Giraldo, A.; Román-Botero, R. Effects of riverine inflows on the climatology of a tropical Andean reservoir. *Limnol. Oceanogr.* **2021**, *66*, 3535–3551. [CrossRef]
81. Román-Botero, R.; Gómez-Giraldo, A.; Toro-Botero, M. Efecto estacional de los afluentes en la estructura térmica de un pequeño embalse neotropical, La Fe—Colombia. *Dyna* **2013**, *80*, 152–161.
82. Hope, D.; Billett, M.F.; Cresser, M.S. A review of the export of carbon in river water: Fluxes and processes. *Environ. Pollut.* **1994**, *84*, 301–324. [CrossRef] [PubMed]
83. Tranvik, L.; Jansson, M. Climate change—Terrestrial export of organic carbon. *Nature* **2002**, *415*, 861–862. [CrossRef]
84. Thomas, S.; Cecchi, P.; Corbin, D.; Lemoalle, J. The different primary producers in a small African tropical reservoir during a drought: Temporal changes and interactions. *Freshw. Biol.* **2000**, *45*, 43–56. [CrossRef]
85. Gunkel, G.; Lange, U.; Walde, D.; Rosa, J.W.C. The environmental and operational impacts of Curuá-Una, a reservoir in the Amazon region of Pará, Brazil. *Lakes Reserv. Res. Manag.* **2003**, *8*, 201–216. [CrossRef]
86. Ministerio del Ambiente del Ecuador (MAE). *Sistema de Clasificación de Ecosistemas del Ecuador Continental*; Subsecretaría de Patrimonio Natural—Proyecto Mapa de Vegetación: Quito, Ecuador, 2013.
87. Salazar, R.N.; Aguirre, C.; Soto, J.; Salinas, P.; Salinas, C.; Prieto, H.; Paneque, M. Physicochemical parameters affecting the distribution and diversity of the water column microbial community in the high-altitude andean lake system of la brava and la punta. *Microorganisms* **2020**, *8*, 1181. [CrossRef]
88. Jones, B.E.; Grant, W.D.; Duckworth, A.W.; Owenson, G.G. Microbial diversity of soda lakes. *Extremophiles* **1998**, *2*, 191–200. [CrossRef]
89. Munawar, M. Limnological studies on freshwater ponds of Hyderabad—India. *Hydrobiologia* **1970**, *44*, 13–27. [CrossRef]

90. Hulyal, S.B.; Kaliwal, B.B. Dynamics of phytoplankton in relation to physico-chemical factors of Almatti reservoir of Bijapur District, Karnataka State. *Environ. Monit. Assess.* **2009**, *153*, 45–59. [CrossRef]
91. Lewis, W.M. Primary Production in the Plankton Community of a Tropical Lake. *Ecol. Monogr.* **1974**, *44*, 377–409. [CrossRef]
92. Kimmel, B.L.; Groeger, A.W. Factors controlling primary production in lakes and reservoirs: A perspective. *Lake Reserv. Manag.* **1984**, *1*, 277–281. [CrossRef]
93. Lee, Y.; Ha, S.Y.; Park, H.K.; Han, M.S.; Shin, K.H. Identification of key factors influencing primary productivity in two river-type reservoirs by using principal component regression analysis. *Environ. Monit. Assess.* **2015**, *187*, 213. [CrossRef] [PubMed]
94. Cartuche, A.; Guan, Z.; Ibelings, B.W.; Venail, P. Phytoplankton Diversity Relates Negatively with Productivity in Tropical High-Altitude Lakes from Southern Ecuador. *Sustainability* **2019**, *11*, 5235. [CrossRef]
95. Karlsson, J.; Jonsson, A.; Jansson, M. Productivity of high-latitude lakes: Climate effect inferred from altitude gradient. *Glob. Chang. Biol.* **2005**, *11*, 710–715. [CrossRef]
96. Feresin, E.G.; Arcifa, M.S.; da Silva, L.H.S.; Esguícero, A.L.H. Primary productivity of the phytoplankton in a tropical Brazilian shallow lake: Experiments in the lake and in mesocosms. *Acta Limnol. Bras.* **2010**, *22*, 384–396. [CrossRef]
97. Morin, A.; Lamoureux, W.; Busnarda, J. Empirical models predicting primary productivity from chlorophyll a and water temperature for stream periphyton and lake and ocean phytoplankton. *J. N. Am. Benthol. Soc.* **1999**, *18*, 299–307. [CrossRef]
98. Chattopadhyay, C.; Banerjee, T.C. Water Temperature and Primary Production in the Euphotic Zone of a Tropical Shallow Freshwater Lake. *Asian J. Exp. Sci* **2008**, *22*, 103–108.
99. Montecino, V. Primary productivity in South American temperate lakes and reservoirs. *Rev. Chil. Hist. Nat.* **1991**, *64*, 555–567.
100. Gunkel, G.; Casallas, J. Limnology of an equatorial high mountain lake—Lago San Pablo, Ecuador: The significance of deep diurnal mixing for lake productivity. *Limnologica* **2002**, *32*, 33–43. [CrossRef]
101. Herzog, S.K.; Martinez, R.; Jørgensen, P.M.; Tiesse, H. *Climate Change and Biodiversity in the Tropical Andes*; Inter-American Institute for Global Change Research (IAI): Montevideo, Uruguay; Scientific Committee on Problems of the Environment (SCOPE): Paris, France, 2011.

water

MDPI

Article

Environmental DNA Metabarcoding as a Promising Conservation Tool for Monitoring Fish Diversity in Dongshan Bay, China

Yanxu Zhang [1,2], Weiyi He [2], Lei Wang [2,3,4], Danyun Ou [2,3,4], Jinli Qiu [2], Weiwen Li [1,2,3,4,*] and Hao Huang [1,2,3,4,*]

1 College of Marine Living Resource Sciences and Management, Shanghai Ocean University, Shanghai 201306, China
2 Key Laboratory of Marine Ecological Conservation and Restoration, Third Institute of Oceanography, Ministry of Natural Resources, Xiamen 361005, China
3 Observation and Research Station of Island and Coastal Ecosystem in the Western Taiwan Straits, Ministry of Natural Resources, Xiamen 361005, China
4 Fujian Provincial Station for Filed Observation and Research of Island and Coastal Zone in the Zhangzhou, Xiamen 361005, China
* Correspondence: liweiwen@tio.org.cn (W.L.); huanghao@tio.org.cn (H.H.)

Abstract: Dongshan Bay is a typical subtropical semi-enclosed bay characterized by abundant fish resources. We aimed to assess fish diversity and its seasonal variation in Dongshan Bay and to provide a scientific basis for the sustainable management and conservation of the fishery's resources. In this study, we employed environmental DNA (eDNA) metabarcoding technology to analyze fish diversity in the bay during winter 2023 and summer 2024. A total of 76 fish species were detected across 12 sampling sites, with 43 species identified in summer and 45 species seen in winter. Overall, 13 species were detected in both the winter and summer. Non-significant differences were observed in Alpha diversity among the sampling sites. Fish species richness at the HXH2 site was the lowest among all the sampling sites for the reason that this sampling site was near to the effluent outlet of the Zhangzhou nuclear power plant and notably influenced by the thermal discharge. In general, fish diversity and abundance were higher in winter than in summer. RDA test analysis revealed that water temperature and dissolved oxygen were the primary environmental factors influencing fish distribution in summer. In winter, the influence of various factors is relatively balanced, with chlorophyll and Blue Green Algae Phycoerythrin (BGA PE) having a relatively greater impact than other factors. Our results offer valuable insights into enhancing fish diversity management in Dongshan Bay.

Keywords: eDNA; semi-enclosed bay; fish; biodiversity; thermal discharge

1. Introduction

Fish is the largest group of vertebrates, constituting nearly half of all vertebrate species, and plays an important role in global biodiversity protection [1]. Fish contribute substantially to human society by providing valuable fishery resources. Conducting comprehensive and systematic evaluations of biological and ecosystems is central to ecological research and serves as a vital step in safeguarding global biodiversity [2]. Fish diversity plays a pivotal role in biodiversity, while the assessment and monitoring of fish diversity serve as the foundation for ecosystem monitoring and health evaluations [3]. Traditional methods for fish diversity monitoring, such as electrofishing, netting, and trapping, typically involve collecting samples and then determining the abundance and biomass of fish through the morphological identification, counting, and weighing of the catch [4]. However, these conventional techniques are often environmentally invasive, time-consuming, labor-intensive, require advanced expertise in morphological identification, and face challenges in capturing low-density species [5]. The emergence of environmental DNA technology has introduced a novel approach to biodiversity monitoring [6]. Environmental DNA refers to the genetic material left behind by organisms in the environment, which can be derived from mitochondrial or nuclear DNA, including shed cells from tissues such as the intestines and skin, as well as bodily fluids such as urine, mucus, eggs, and sperm [7]. This technology enables researchers to detect the presence or recent presence of species without directly observing or capturing the organisms [5,8]. Environmental DNA metabarcoding allows for the identification of multiple target species in the environmental samples (such as water, sediments, or soil) by extracting DNA from these samples, using universal primers for the target taxa and performing PCR amplification using high-throughput sequencing [9]. This method does not require the capture of the organisms, and this non-invasive, efficient, and highly sensitive approach overcomes the limitations of traditional morphological surveys, presenting significant potential for biodiversity assessments [10].

Since DNA metabarcoding's initial application in 2008, when Ficetola et al. (2008) used environmental DNA technology to monitor the invasive *Rana catesbeiana* in ponds, eDNA metabarcoding has seen rapidly development [11]. It is widely used in fishery management and in fish diversity monitoring in both freshwater and marine ecosystems, with particularly widespread use in biodiversity studies [12]. Thomsen et al. employed eDNA technology to evaluate fish diversity in Danish harbors, identifying 15 species. The funding included several economically significant fish and a rare migratory species [13]. Sigsgaard conducted a year-long investigation, combining water sampling and snorkeling observations along the Danish coastline [14]. As result, the seasonal dynamics of fish community were identified via eDNA metabarcoding. When compared with historical data, the eDNA metabarcoding results showed partial consistency with snorkeling observations. Notably, most fish species detected through snorkeling were identified by eDNA metabarcoding. This research highlights the utility of eDNA metabarcoding in capturing seasonal shifts in the diversity of marine fish communities. eDNA metabarcoding technology has recently emerged as a powerful tool for biodiversity monitoring [15].

Dongshan Bay is situated in a subtropical latitude zone and is influenced by the Min-Zhe coastal current and the Taiwan Strait thermocline in the autumn, and by the South China Sea in the summer, presenting the typical characteristics of a subtropical bay ecosystem [16]. The region serves as a vital habitat for many aquatic species, particularly fish species, providing essential areas for their habitat, growth, fattening, and reproduction [16,17]. However, excessive human development has led to the scattering of aquaculture areas, worsening the eutrophication of seawater. Reclamation activities have resulted in the destruction of wetland habitats, and tidal volumes have consequently diminished [18]. Dongshan Bay is also subject to various impacts from the Zhangzhou Nuclear Power Plant located within the bay. Therefore, there is an urgent need to establish a rapid, effective, and environmentally friendly monitoring method for the protection and ecological restoration of fish diversity in Dongshan Bay, providing scientific support for fishery management and the development and implementation of ecological protection policies.

In this study, we employed eDNA metabarcoding technology to analyze the relationship between fish diversity and environmental factors in Dongshan Bay during December 2023 and June 2024. The objective was to conduct an initial assessment of the fish community status in Dongshan Bay, with the goal of providing foundational data to support the conservation and management of fish diversity in this region.

2. Methods

2.1. Field Site and Sample Collection

Environmental DNA samples were collected in Dongshan Bay on 9 December 2023 and 21 June 2024. A total of 12 sampling sites that comprehensively cover the marine area and the distribution of the sampling sites are shown in Figure 1. At each site, a 1 L water sample was collected using a water sampler and then placed in a disposable sterile sample bag. If a site had a water depth greater than 5 m, two samples were taken from the surface and bottom, while for sites with a depth of less than 5 m, only one sample was taken from the surface. To avoid cross-contamination, each collection bag (Labshark, Changde, China) was rinsed twice with 150 mL of local seawater before sampling. The rinse water was discarded, and disposable gloves were replaced. The collected samples were stored in a cooling box and filtered within 24 h using a circulating water vacuum pump (Greatwall, Zhengzhou, China) and a six-way filter (PALL, Port Washington, NY, USA). Filtration was carried out using a diameter of 47 mm and a polycarbonate membrane with a pore size of 0.2 μm (Millipore, Darmstadt, Germany), and the filter was disinfected with a 75% ethanol spray. We waited for a few seconds and then wiped the filter dry with a cotton ball. We performed this both before and after filtering to eliminate residual DNA and prevent contamination [7]. The filters were then placed in 4.5 mL cryovials (Axygen, Union City, CA, USA) and stored at −20 °C until genomic DNA extraction was performed in the laboratory. DNA was extracted using the DNeasy PowerWater Kit (Qiagen, Hilden, Germany) to enrich and recover the eDNA [13,19]. The quality of the extracted eDNA was assessed using 1% agarose gel electrophoresis, and the DNA samples were stored at −20 °C for subsequent PCR amplification. Each sample was independently processed, with all equipment disinfected using 75% ethanol spray before and after each step to prevent exogenous contamination [20].

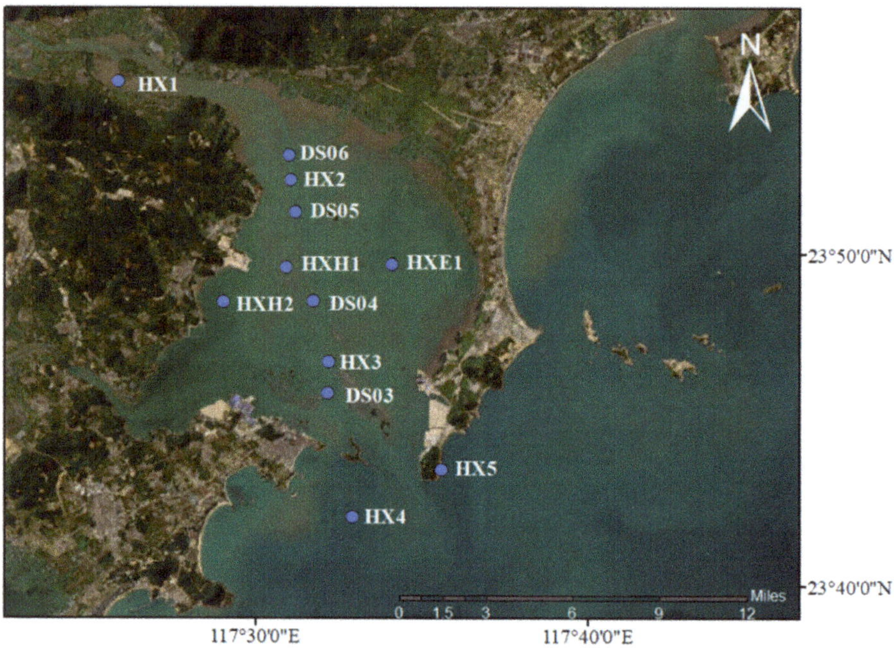

Figure 1. A map of the sampling sites in the Dongshan Bay.

2.2. Measurement of Environmental Factors

The collection of environmental parameters was carried out simultaneously with sample collection. During the sampling process, a multi-parameter water quality meter (YSI, Yellow Springs, OH, USA) was employed to measure the following environmental parameters, including chlorophyll (chlorophyll), dissolved oxygen (DO), salinity (Sal), Blue Green Algae Phycoerythrin (BGA PE), pH (pH), and water temperature (WT).

2.3. PCR, Sequencing and Annotation

eDNA metabarcoding, conducted using universal MiFish primer pairs, has been shown to amplify short fragments of fish DNA in various taxa from environmental samples [21]. Our samples were analyzed using the universal fish primer pairs (Mifish-U-F: 5′-GTCGGTAAAACTCGTGCCAGC-3′; Mifish-U-R: 5′-GTTTGACCCTAATCTATGGGGTGATAC-3′) in order to amplify the mitochondrial 12S rRNA gene [22]. The multiplex PCR volume was 25 μL. This included 12.5 μL of Taq 2× Master Mix (Vazyme, Nanjing, China), 1 μL of each primer (10 μmol/L), 1 μL of the DNA solution, and 9.5 μL of sterile distilled H_2O. The thermocycler used was an ABI2720 model (ThermoFisher, Waltham, MA, USA). The thermal cycle PCR process included an initial 2 min denaturation at 94 °C, followed by 35 cycles of denaturation at 98 °C for 5 s each, and then annealing at 50 °C for 10 s, extension at 72 °C for 10 s, and completion with a final extension at 72 °C for 5 min [23]. The PCR products were analyzed using 1% agarose gels. The electrophoresis procedure was as follows: 2 μL of the DNA sample and 2 μL of DL2000 DNA Ladder Marker (Takara, Osaka, Japan) were loaded into the sample wells of an agarose gel. Electrophoresis was conducted in a 1% TAE (Biosharp, Beijing, China) agarose gel at 100 V with a constant voltage and an approximately 80 mA current for 20 min. DNA fragment size and concentration were assessed by comparing the band positions and intensities with those of the Marker. Samples with a bright main strip of 297 ± 25 bp were selected and all negative controls showed no target bands, confirming the purity of the samples and the absence of contamination.

QIIME2 (v.2022.11) was employed, and the analysis workflow was modified and optimized following the official tutorial (https://docs.qiime2.org/2022.11/tutorials/) (accessed on 2 September 2024). The raw sequencing data were processed, using the demux plugin for decoding, the Cutadapt (v2.3) for primer removal, and the DADA2 (v1.26) for quality filtering, denoising, and merging. Sequences were clustered into Amplicon Sequence Variants (ASVs) using the uparse algorithm in VSEARCH (v2.7.1), and the resulting sequences were clustered at 100% similarity to generate ASVs. The ASVs feature sequences were, using the BLAST algorithm, compared against the reference sequences in the NCBI (https://www.ncbi.nlm.nih.gov/) (accessed on 10 September 2024) and MitoFish (http://mitofish.aori.u-tokyo.ac.jp) (accessed on 10 September 2024) database to obtain taxonomic information for each ASV. Rare ASVs were excluded from the abundance matrix for further analyses. Non-fish ASVs were excluded, and ASVs identified as belonging to the same species were merged. The relative abundance of each fish species' valid sequences was calculated in Excel, and species identification was cross-verified and refined using the Fishbase database (https://www.fishbase.de/search.php) (accessed on 22 September 2024) and historical catch data from Dongshan Bay [24], along with additional information on fish classification and life history.

2.4. Data Analysis

Data preprocessing was performed using Excel, and the differences in fish species between the two seasons were visualized using a Venn diagram created through the online platform Venny 2.1.0 (https://bioinfogp.cnb.csic.es/tools/venny/) (accessed on 27 September 2024). Initially, QIIME2 (v.2022.11) was used to randomly subsample the total sequence counts of each sample in the ASV abundance matrix at varying depths. Rarefaction curves were generated based on the number of sequences sampled at each depth and the corresponding number of ASVs. This process was used to assess whether the current sequencing depth for each sample was adequate to capture the microbial diversity of the community. Subsequently, to facilitate the comparison of diversity across samples, the ASV abundance matrix was rarefied to 95% of the sequence count from the sample with the lowest sequence count, thereby correcting for the sequencing of depth-related diversity discrepancies between samples. Then, Alpha diversity was assessed using several indices—the Shannon index (H') and Simpson index (D′) [25,26]—to represent inter-group diversity. We used the Chao1 index [27] to estimate inter-group richness, the Pielou_e index (J') [28] to measure species evenness, and the goods_coverage index to evaluate sample coverage [29]. The calculation formulas are as follows:

$$\text{Shannon index}: H' = -\sum_{i=1}^{S} p_i * ln(p_i) \quad (1)$$

$$\text{Simpson index}: \text{D}' = 1 - \sum p_i^2 \quad (2)$$

$$\text{Chao1 index}: \text{Chao1 index} = \text{S} + \frac{n_1(n_1-1)}{2(n_2+1)} \quad (3)$$

$$\text{Pielou\_e index}: J' = H'/\text{lnS} \quad (4)$$

$$\text{goods\_coverage index}: \text{Coverage index} = 1 - n_1/\text{N} \quad (5)$$

In the formulas, S is the number of ASVs; p_i denotes the relative abundance of the ASV for the i-th fish species as a proportion of the total fish abundance; n_1 represents the count of ASVs with only one sequence; n_2 represents the count of ASVs with exactly two sequences; and N indicates the total sequence count in the sample. All statistical analyses and visualizations were conducted in R (v4.3.3) [30]. The vegan package (v2.6.6.1) was used to calculate Alpha diversity indices, including the Shannon–Wiener index, Simpson index, Pielou_e index, and goods_coverage index. The Shapiro–Wilk normality test was performed on each index, revealing that none of the indices conformed to a normal distribution. Thus, the non-parametric Wilcoxon test was applied. Differences in Alpha diversity indices between groups were evaluated using either a t-test or the Wilcoxon test. In order to assess the similarity between samples, we employed pheatmap package (v1.0.12) and used hierarchical clustering. For both rows and columns, Euclidean Distance was used as the distance metric, and clustering was performed with the Complete Linkage method. Complete Linkage merges clusters based on the maximum distance between them, generating a dendrogram to illustrate the similarity relationships between samples and species. The clustering results were visualized as a heatmap, where the color of each cell indicated the standardized abundance of the samples and species. Detrended Correspondence Analysis (DCA) was conducted by using the decorana function from the vegan package to decide whether to apply Redundancy Analysis (RDA), which is based on linear models, or Canonical Correlation Analysis (CCA), which is based on unimodal models, in order to investigate the primary environmental factors influencing the distribution of fish communities. If the gradient length of the ordination axes in DCA exceeded 4.0, CCA was chosen; otherwise, RDA was used [31]. The plots were created using the ggplot2 package (v3.5.1).

3. Results

3.1. Species Composition

A total of 1,399,397 raw MiFish sequences were obtained. After initial quality filtering, low-quality sequences and chimeras were removed, leaving a total of 1,309,228 sequences and 30,881 chimeric sequences. High-quality sequences accounted for 93.56% of the total raw sequences.

A total of 76 fish species were detected across the winter and summer seasons, with 43 species identified in winter. These belonged to 23 orders, 30 families, and 41 genera. Among these, 13 species were commonly detected in both winter and summer, representing 17.1% of the total species detected, as shown in Figure 2. These species included *Acanthopagrus schlegelii, Pseudobalistes fuscus, Plectorhinchus cinctus, Acanthopagrus latus, Halichoeres notospilus, Decapterus maruadsi*, and so on. In winter, based on the ASVs detected, the five families with the highest relative abundance were Gobiidae, Dasyatidae, Cynoglossidae, Clupeidae, and Sparidae. In the summer, 45 fish species were identified, belonging to 19 orders, 24 families, and 40 genera. The top five families with the highest relative abundance in the summer were Gobiidae, Labridae, Clupeidae, Balistidae, and Carangidae (Table 1).

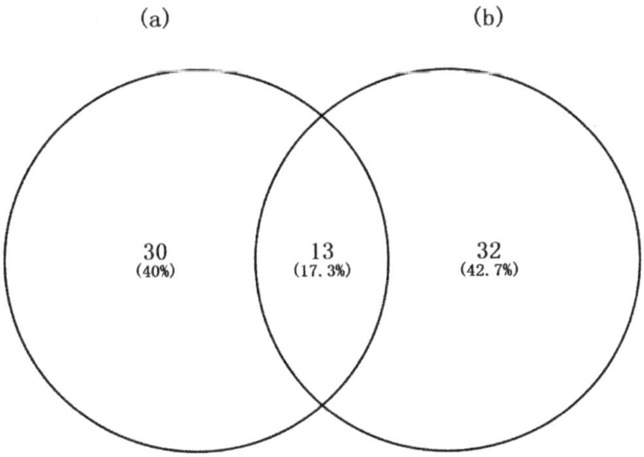

Figure 2. Venn diagram of fish species in winter (**a**) and summer (**b**).

Table 1. Detection of fish species in Dongshan Bay based on eDNA metabarcoding technology.

Order	Family	Species	Season	
			Winter	Summer
Mugiliformes	Mugilidae	*Planiliza affinis* (Günther, 1861)	*	*
		Valamugil speigleri (Bleeker, 1858)	*	
		Moolgarda engeli (Bleeker, 1858)	*	
		Mugil cephalus (Linnaeus, 1758)	*	*
Clupeiformes	Clupeidae	*Sardinops melanostictus* (Jenyns, 1842)	*	*
		Clupanodon thrissa (Linnaeus, 1758)	*	*
		Sardinella albella (Valenciennes, 1847)		*
		Sardinella hualiensis (Chu & Tsai, 1958)		*
	Engraulidae	*Encrasicholina punctifer* (Fowler, 1938)	*	
		Engraulis japonicus (Temminck & Schlegel, 1846)	*	
		Thryssa hamiltonii (Gray, 1835)		*
Centrarchiformes	Girellidae	*Girella punctata* (Gray, 1835)	*	
	Terapontidae	*Pelates quadrilineatus* (Bloch, 1790)		*
Chaetodontiformes	Leiognathidae	*Leiognathus brevirostris* (Valenciennes, 1835)	*	
		Leiognathus ruconius (Hamilton, 1822)		
Acropomatiformes	Banjosidae	*Banjos banjos* (Richardson, 1846)	*	

Table 1. *Cont.*

Order	Family	Species	Season	
			Winter	Summer
Anguilliformes	Congridae	*Uroconger lepturus* (Richardson, 1845)	*	
	Moringuidae	*Moringua javanica* (Kaup, 1856)	*	
	Muraenidae	*Uropterygius makatei* (Bleeker, 1864)	*	
	Nemichthyidae	*Nemichthys curvirostris* (Strömman, 1896)	*	
Atheriniformes	Atherinidae	*Hypoatherina valenciennei* (Bleeker, 1854)	*	
Beloniformes	Belonidae	*Strongylura strongylura* (van Hasselt, 1823)	*	
Blenniiformes	Blenniidae	*Parablennius yatabei* (Jordan & Snyder, 1900)		*
Caproiformes	Caproidae	*Antigonia rubicunda* (Ogilby, 1910)		*
Carangiformes	Carangidae	*Decapterus maruadsi* (Temminck & Schlegel, 1843)	*	*
		Selaroides leptolepis (Cuvier, 1833)	*	
		Seriola dumerili (Risso, 1810)	*	*
Gadiformes	Macrouridae	*Coelorinchus hubbsi* (Matsubara, 1936)	*	
Gobiiformes	Gobiidae	*Amblychaeturichthys hexanema* (Bleeker, 1853)	*	
		Drombus triangularis (Weber, 1909)	*	*
		Parachaeturichthys polynema (Bleeker, 1853)		*
		Rhinogobius giurinus (Rutter, 1897)	*	
		Trypauchen vagina (Bloch & Schneider, 1801)		*
Kurtiformes	Apogonidae	*Ostorhinchus pleuron* (Fraser, 2005)		*
Labriformes	Labridae	*Cheilinus oxycephalus* (Bleeker, 1853)	*	
		Halichoeres notospilus (Günther, 1864)	*	
		Halichoeres hartzfeldii (Bleeker, 1852)		*
		Hologymnosus doliatus (Lacepède, 1801)	*	
		Labroides bicolor (Fowler & Bean, 1928)	*	
		Labroides alleni (Randall, 1981)		*
		Leptojulis cyanopleura (Bleeker, 1853)	*	*
		Parajulis poecilepterus (Temminck & Schlegel, 1845)		*
		Coris pictoides (Randall & Kuiter, 1982)		*
Lutjaniformes	Haemulidae	*Plectorhinchus cinctus* (Temminck & Schlegel, 1843)	*	*
Perciformes	Platycephalidae	*Platycephalus indicus* (Linnaeus, 1758)	*	
	Cottidae	*Trachidermus fasciatus* (Heckel, 1837)		*
	Sebastidae	*Sebastiscus marmoratus* (Cuvier, 1829)		*
	Serranidae	*Epinephelus akaara* (Temminck & Schlegel, 1842)		*
		Epinephelus bruneus (Bloch, 1793)		*
		Epinephelus coioides (Hamilton, 1822)		*
	Stichaeidae	*Alectrias benjamini* (Jordan & Snyder, 1902)		*
Pleuronectiformes	Cynoglossidae	*Cynoglossidae quadrilineatus* (Bleeker, 1851)	*	
		Paraplagusia japonica (Temminck & Schlegel, 1846)		*
Scombriformes	Scombridae	*Scomber japonicus* (Houttuyn, 1782)	*	*
		Auxis thazard (Lacepède, 1800)		*
	Trichiuridae	*Trichiurus japonicus* (Temminck & Schlegel, 1844)	*	
Spariformes	Sparidae	*Acanthopagrus latus* (Houttuyn, 1782)	*	*
		Acanthopagrus schlegelii (Bleeker, 1854)	*	*
		Rhabdosargus sarba (Forsskål, 1775)		*
Mulliformes	Mullidae	*Upeneus japonicus* (Houttuyn, 1782)	*	
Myctophiformes	Myctophidae	*Benthosema fibulatum* (Gilbert & Cramer, 1897)		*
		Benthosema pterotum (Alcock, 1890)		*
		Myctophum aurolaternatum (Garman, 1899)		*
Syngnathiformes	Syngnathidae	*Hippichthys penicillus* (Cantor, 1849)	*	
		Hippichthys spicifer (Rüppell, 1838)	*	

Table 1. *Cont.*

Order	Family	Species	Season	
			Winter	Summer
Tetraodontiformes	Balistidae	*Pseudobalistes fuscus* (Bloch & Schneider, 1801)	*	*
	Monacanthidae	*Monacanthus chinensis* (Osbeck, 1765)		*
		Paramonacanthus otisensis (Whitley, 1931)		*
		Stephanolepis setifer (Bennett, 1831)		*
Trachichthyiformes	Monocentridae	*Monocentris japonicus* (Houttuyn, 1782)	*	
Zeiformes	Parazenidae	*Cyttopsis cypho* (Fowler, 1934)	*	
	Zeidae	*Zeus faber* (Linnaeus, 1758)	*	
Myliobatiformes	Dasyatidae	*Brevitrygon walga* (Müller & Henle, 1841)	*	
		Hemitrygon akajei (Müller & Henle, 1841)		*
Osmeriformes	Salangidae	*Neosalangichthys ishikawae* (Wakiya & Takahashi, 1913)		*
		Salangichthys microdon (Bleeker, 1860)		*

Note: * indicates that the fish was detected in the season.

Regarding species richness, the HX3 station had the highest number of species in winter, with 18 species detected, while the HXH1 station had the lowest, with only 5 species detected. In summer, the DS04 station had the highest number of species, with 22 species detected, and the HXH2 station had the lowest, with only 3 species detected, as shown in Figure 3. The top 10 dominant species by relative sequence abundance across the sampling sites are shown in Figure 3. *Gobius sinensis* and *Sphaeramia nematoptera* were detected at all sites during both winter and summer.

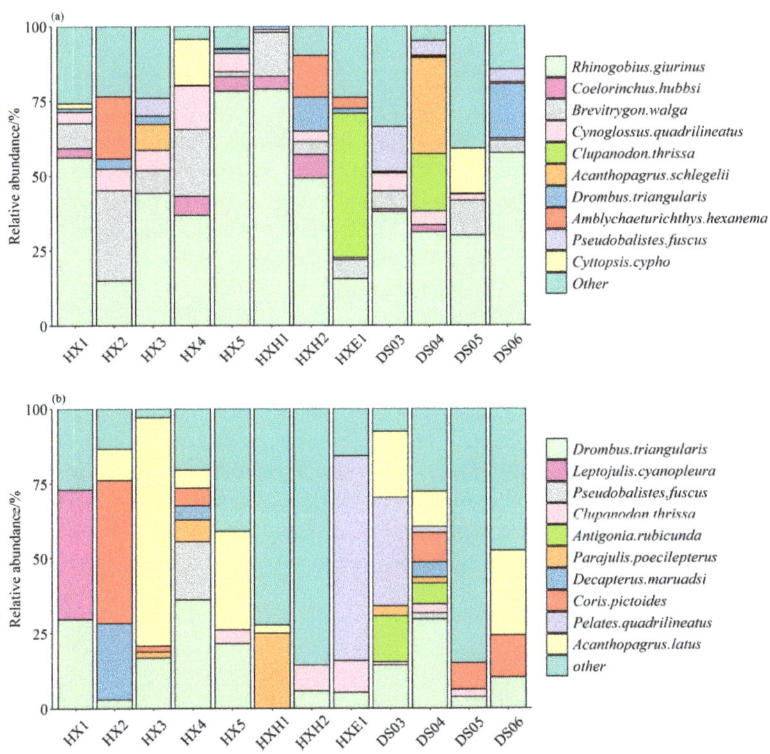

Figure 3. Composition of dominant fish species at each sampling station during (**a**) winter and (**b**) summer.

3.2. Alpha Diversity Analysis

To comprehensively assess fish species richness and diversity across sampling sites, 5 Alpha diversity indices were calculated. As shown in Table 2, the goods_coverage index for winter exceeded 0.99 across all sites, reflecting high sample coverage [32]. The average Shannon index for winter was 1.563, peaking at 2.093 at DS03 and dipping to 0.701 at HXH1. The average Simpson index was 0.678, ranging from 0.828 at DS05 to 0.354 at HXH1. Although the Shannon and Simpson indices showed general consistency, site-specific differences were evident. The Chao1 index was averaged at 11.028, with the highest value of 18.333 recorded at HX3 and the lowest value of 5.000 recorded at HXH1. The Pielou_e index averaged 0.459, with the highest value of 0.553 seen at HX2 and the lowest value of 0.287 seen at HX5.

Table 2. Alpha diversity index.

Site	Shannon Index		Simpson Index		Chao1 Index		Pielou_e Index		goods_coverage Index	
	W	S	W	S	W	S	W	S	W	S
HX3	1.979	1.151	0.768	0.414	18.333	19.200	0.475	0.295	0.998	0.931
DS03	2.093	1.823	0.807	0.783	17.000	12.500	0.512	0.527	1.000	0.967
DS04	1.645	2.545	0.756	0.878	11.000	25.500	0.476	0.571	0.999	0.931
DS05	1.953	0.591	0.828	0.268	12.000	5.000	0.545	0.254	0.998	0.988
HXH2	1.550	0.506	0.706	0.255	7.000	3.000	0.552	0.319	1.000	1.000
HXE1	1.716	0.977	0.719	0.494	13.000	5.000	0.464	0.421	1.000	0.982
DS06	1.408	1.396	0.623	0.682	11.000	7.000	0.407	0.497	1.000	1.000
HX1	1.232	1.218	0.608	0.678	7.000	4.000	0.439	0.609	1.000	1.000
HX2	1.978	1.311	0.822	0.678	12.000	5.000	0.553	0.565	1.000	1.000
HX5	0.861	1.643	0.377	0.715	9.000	13.000	0.287	0.458	0.994	0.972
HXH1	0.701	0.681	0.354	0.415	5.000	3.000	0.302	0.430	1.000	0.972
HX4	1.646	2.234	0.765	0.876	10.000	11.000	0.495	0.646	1.000	1.000

Note: W: winter; S: summer.

In summer, the average Shannon index was 1.340, with the highest value of 2.545 observed at DS04 and the lowest value of 0.506 seen at HXH2. The Simpson index averaged 0.595, ranging from 0.878 at DS04 to 0.255 at HXH2, following a similar trend. The Chao1 index averaged 9.433, with a maximum value of 25.5 seen at DS04 and a minimum value of 3.000 seen at HXH2. The Pielou_e index was 0.466 on average, with the highest value of 0.646 recorded at HX4 and the lowest of 0.254 at DS05.

The fish species composition heatmap at the species level reveals differences in species composition across various sampling sites, as shown in Figure 4. Clustering analysis of the sites indicates that winter sites HXE1 and DS03 share the greatest similarity. Regarding species clustering, the distributions of *Strongylura strongylura* and *Leiognathus brevirostris* are most similar across different sites. In terms of species abundance, DS03 demonstrates higher species richness and diversity, while HXH1 shows lower diversity, which is consistent with the Alpha diversity index results. In the summer, sites HXH2 and DS05 exhibit the greatest similarity. Species clustering shows that the distribution of *Pelates quadrilineatus* and *Scomber japonicus* across different sites is most similar. In terms of species diversity and richness, the DS04 and HX5 sites show higher levels of both, while HXH2 exhibits lower diversity, which corresponds closely to the Alpha diversity index findings.

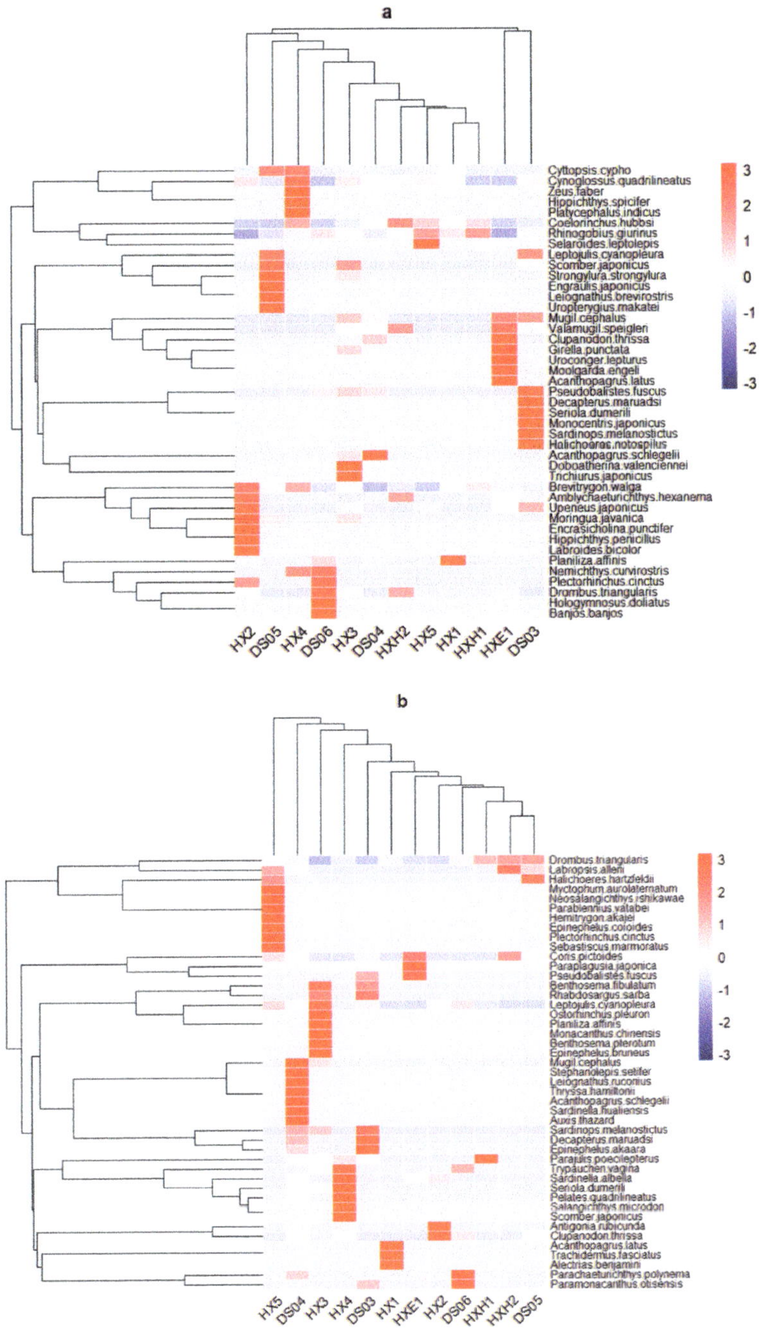

Figure 4. Heat maps of the fish species composition in the Dongshan Bay for the winter (**a**) and summer (**b**).

Figure 5 presents a comparison of Alpha diversity between the two seasons. Both the Shannon and Simpson indices in winter are higher than those in summer, suggesting that the fish diversity in winter is generally higher than that in summer. The Chao1 index for winter is also greater than that in summer, although summer shows two notable larger values at the HX3 and DS04 sites. The species richness at these two summer sites is significantly higher than that at the other sites, but overall, fish richness remains lower in summer compared to winter. There is no significant difference in the Pielou_e index between the two seasons ($p > 0.05$). The uniformity between sites in summer shows more variation, while the winter uniformity is relatively consistent, with the Pielou_e index predominantly around 0.5.

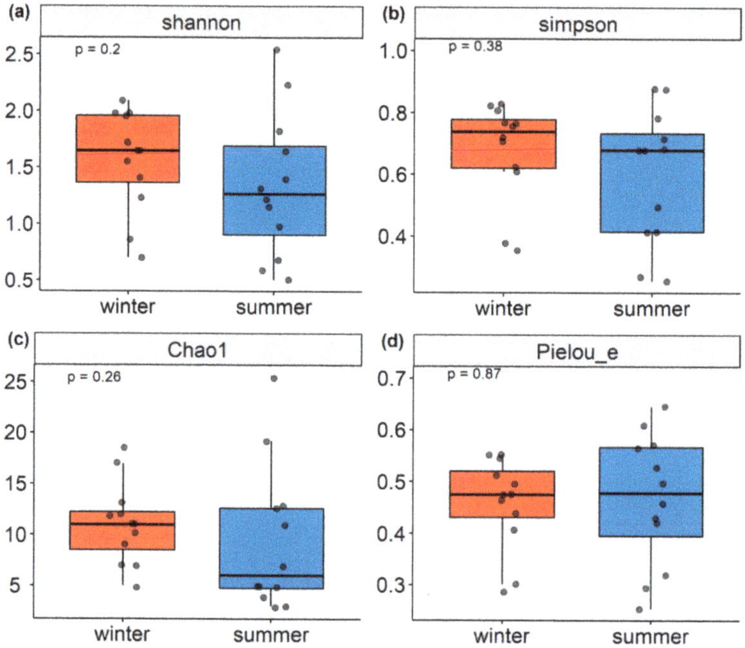

Figure 5. A comparison of winter and summer fish Alpha diversity index using the (**a**) Shannon index, (**b**) Simpson index, (**c**) Chao1 index, and (**d**) Pielou_e index.

3.3. Correlation Analysis of Environmental Factors

Figure 6 presents the environmental factors determining the fish distribution. In winter, the arrows for WT and Sal are relatively short, suggesting that these factors have a limited capacity to distinguish between samples during winter. In contrast, chlorophyll and BGA PE exhibit longer arrows, indicating that they have a more significant influence on sample distribution in winter. HX1 and HX2 are located closer to chlorophyll, suggesting that these sites are more strongly influenced by this environmental factor. DS05 appears as an isolated point, showing a weak correlation with most environmental variables. In summer, the arrows for WT and DO are notably extended, indicating that these factors have a significantly stronger influence on sample distribution in summer. Sal continues to have a significant effect in summer, consistent with the winter findings. HX3 and HX4 are positioned closer to WT and DO, suggesting that these sampling sites are located in areas with high temperature and high oxygen levels. In winter, the primary factors are chlorophyll and BGA PE, indicating that photosynthesis-related variables play a dominant role during this season. In contrast, WT and DO are the key factors in summer, highlighting that WT and DO are the decisive influences. The sample distribution in winter is more dispersed, suggesting the environmental factors have a lesser impact on the fish community structure across the sites. In summer, the sample distribution becomes more concentrated, indicating that environmental factors may have a stronger gradient effect.

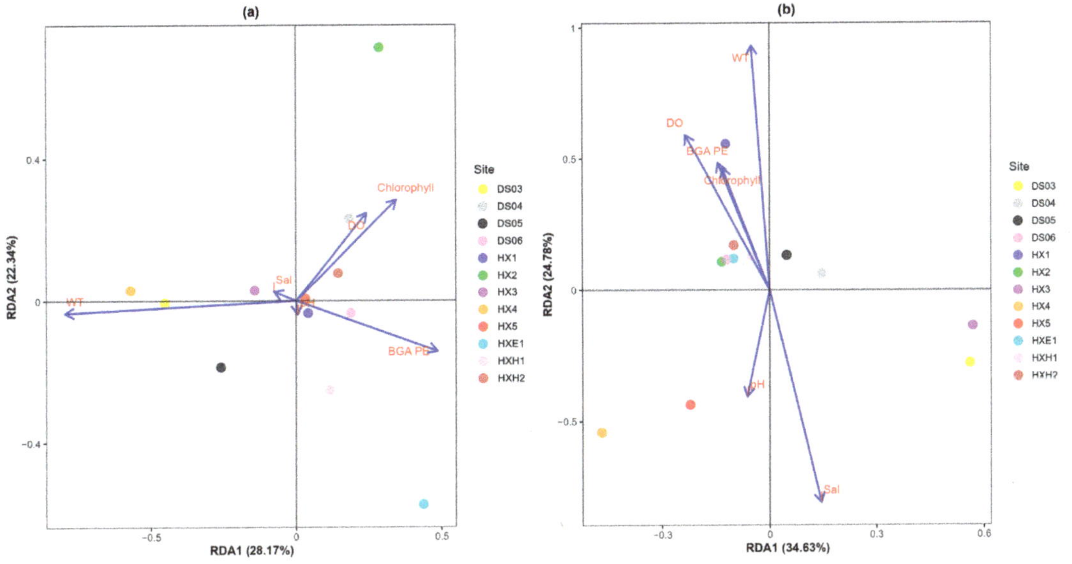

Figure 6. Redundancy analysis (RDA) of fish communities in each site and environmental factors in the Dongshan Bay during (**a**) winter and (**b**) summer.

4. Discussion

4.1. Fish Species Composition in Dongshan Bay During Winter and Summer

Due to the extensive area of cage aquaculture in Dongshan Bay, traditional sampling methods, such as bottom trawling, are difficult to implement in these areas [24]. As a result, we utilized eDNA metabarcoding technology to analyze the fish diversity in Dongshan Bay during winter and summer. The eDNA metabarcoding analysis of samples from 12 sites identified a total of 76 fish species. At the class level, the majority of species belonged to Actinopteri, while only one species from Chondrichthyes, *Brevitrygon walga*—was detected in winter. The Mifish-U primers used in the analysis are optimized for amplifying Actinopteri and show limited effectiveness for Chondrichthyes, which contribute to these phenomena [33]. A total of 43 species were identified in winter and 45 were found in in summer: the total number of species detected showed little variation. Overall, 13 species were detected in both winter and summer, representing 27.9% of the winter species and 26.7% of the summer species, highlighting distinct seasonal differences in species composition. In a gillnet survey conducted by our laboratory in November 2014 (unpublished), 115 fish species were captured, exceeding the 76 species identified in this study. A comparison shows that the 2014 survey involved more sampling sites and a longer duration, with a broader spatial distribution of sites. The differences in temporal and spatial factors were likely the main reasons for the discrepancy in the number of fish species identified between the two studies. Previous fishing surveys in Dongshan Bay and its adjacent waters identified 43 fish species in spring and detected 45 in autumn [34]. This was similar to the number of species identified through eDNA, but there was a notable difference in terms of dominant species composition. According to past fishing surveys, *Thryssa mystax* was the dominant species, with other key species including *Sardinella zunas*, *Leiognathus brevirostris*, and *Secutor ruconius*. In contrast, our eDNA survey revealed that the dominant species in Dongshan Bay were smaller species such as *Drombus triangularis*, *Rhinogobius giurinus*, and

Clupanodon thrissa. This results suggests a clear trend towards smaller dominant species in the fish community of Dongshan Bay, as well as a marked decrease in economically significant species. As a crucial fishery resource base in Fujian Province, Dongshan Bay has been experiencing increasing fishing pressure due to its growing development. An assessment in 2018 indicated that the overall carrying capacity of Dongshan Bay's fishery resources is reaching a critical threshold, and the impact of fishing may be a significant driver of the observed changes in fish species composition and diversity in this region [35].

4.2. Alpha Diversity Analysis and Correlation with Environmental Factors

In winter, the minimum values for the diversity and richness indices, including Shannon, Simpson, and Chao1, were observed at the HXH1 site, indicating that this site exhibited the lowest levels of fish diversity and richness. The HXH1 site is located at the intake of the Zhangzhou nuclear power plant, which generates significant heat during its operation and uses a cooling water system to dissipate this heat [36]. The intake can become obstructed due to the accumulation of aquatic organisms and sediment, and intercepting nets are installed to drive swimming species. Dispersal agents are also regularly used to drive away nearby aquatic life, which may be a significant factor contributing to the lower fish diversity and richness at HXH1 [37,38]. The DS03 site exhibited high diversity, while HX3 had the highest species richness. DS03 and HX3 are located in coral and aquaculture areas, respectively. These areas provide optimal ecological conditions and abundant food resources [39], attracting fish to inhabit and forage in these environments. Human activities in the aquaculture area also have negative effects, one of which is the onset of red tides caused by water eutrophication. The secretions from red tide algae can obstruct the gills of fish, impairing their ability to breathe and leading to suffocation and death. Moreover, the respiratory processes of the algae, along with the breakdown of their dead cells, significantly deplete the amount of DO in the seawater, resulting in severe hypoxia that further contributes to fish suffocation. Additionally, certain toxic algae produce substances that, once ingested by fish, accumulate in their bodies, ultimately leading to poisoning and death. Dongshan Bay, which is typically prone to red tide events, did not experience any during the two seasons observed in this study.

In summer, the highest and lowest diversity and richness values, as indicated by the Shannon, Simpson, and Chao1 indices, were found at the DS04 and HXH2 sites, respectively. The DS04 site recorded 22 fish species, while HXH2 recorded only 3 species, including *Labropsis alleni*, *Drombus triangularis*, and *Coris pictoides*. Despite the proximity of these two sites, there was a considerable difference in fish composition and diversity. This disparity could be explained by the fact that DS04 is located in an aquaculture area. Human activities in the aquaculture area, such as feeding and the discharge of domestic wastewater, significantly contribute to the increase in nutrients in the water. This nutrient enrichment fosters the growth of phytoplankton, which in turn boosts primary productivity at the site and attracts fish to congregate [40]. Another contributing factor may be the escape of cultured fish from nearby farms, with species such as *Acanthopagrus schlegelii*, *Decapterus maruadsi*, and *Mugil cephalus* commonly being cultured. In contrast, HXH2 is located near the discharge outlet of the nuclear power plant, where fish are impacted by thermal pollution from the warm discharge water [41], which may explain the lower diversity and richness at this site.

Fish community is influenced by both biotic and abiotic factors, and the relationship between environmental factors and fish communities varies significantly across different scales and ecosystem types [42,43]. In winter, environmental factors have a relatively balanced influence, with photosynthesis-related factors having a greater effect. Chlorophyll is strongly associated with primary productivity [44], and phytoplankton, which form the foundation of the aquatic food web, play a crucial role in providing food for fish. On the other hand, BGA PE is an important indicator in predicting red tide events. The massive proliferation of red tide algae can deplete nutrients in the water and even generate toxins, posing a threat to fish survival [45]. In the summer, WT and DO have a great impact on fish diversity. The influence of WT on fish communities is multifaceted, involving both physiological adaptation and ecological interactions [46]. An appropriate water temperature supports the stability and health of fish communities, while extreme temperatures, either too high or too low, can destabilize community structure, impact fish reproduction, food availability, and habitat conditions, ultimately leading to shifts in fish populations. DO supports the healthy circulation of substances in the water and reflects the suitability of the habitat for fish [47]. Dongshan Bay is a eutrophic bay, and during summer, rising temperatures lead to large cyanobacteria blooms, which can reduce DO levels in the water. However, according to the results of the environmental parameters, DO did not exhibit a negative correlation with temperature. This suggests that the summer survey may have occurred during the early stages of a red tide, when the oxygen produced by cyanobacterial photosynthesis outweighed the oxygen consumed through respiration. The level of DO in seawater affects fish respiration, swimming speed, and metabolic rate. Furthermore, chlorophyll and DO levels also influence zooplankton size and composition, which in turn affects fish species that predominantly feed on zooplankton, such as *Halichoeres notospilus* and *Decapterus maruadsi*. This could help explain the relatively higher sequence abundance of these species in our summer samples.

4.3. Impact of Thermal Discharge from Zhangzhou Nuclear Power Plant on Fish Communities

Water temperature plays a critical role in marine ecosystems and the biological activities of marine organisms. It significantly influences individual growth, metabolism, the maturation of reproductive cells, and the overall life cycle of species [48]. Compared to terrestrial and freshwater environments, ocean water temperature fluctuations are generally smaller, and marine organisms, including fish, have relatively low tolerance to temperature changes, making them more susceptible to the effects of thermal pollution [49,50]. Thermal discharge can alter the normal distribution of aquatic organisms, leading to shifts in community structure and abnormal developmental occurrences, and can also significantly impact migratory species. Thus, the impact of thermal discharge on fish is an issue that should not be underestimated [51].

DO levels in the aquatic environment significantly influence the life activities of aquatic organisms as DO is one of the essential factors required for metabolic processes. There is a strong negative correlation between water temperature and dissolved oxygen content. As the water temperature rises from 0 °C to 40 °C, the dissolved oxygen concentration decreases. However, in general, temperature increases in non-polluted water bodies do not lead to a reduction in dissolved oxygen levels below the minimum threshold required for fish survival [52]. However, at the discharge site during both seasons, no negative correlation was observed between DO and WT. The effect of thermal effluent diffusion on the environment is a complex issue. Given the elevated chlorophyll levels at the HXH2 site, we hypothesize that the higher water temperature at this site facilitated the growth of thermophilic algae [53], which in turn boosted the DO concentration in the water, counteracting the direct impact of WT on DO.

At the HXH2 site, located near the discharge outlet of the Zhangzhou Nuclear Power Plant, the water temperature is notably influenced by thermal discharge. The temperature at HXH2 (30.57 °C) is the highest among the 12 sampling sites, being 5 °C higher than the lowest temperature observed at HX4 and 3 °C above the average temperature at neighboring sites. Under high summer temperatures, thermal discharge further raises the water temperature at HXH2, surpassing the optimal living temperature for many fish species. This results in the avoidance of the area by fish during peak temperature periods. The three species detected at HXH2 are all warm-water species that prefer temperatures above 29 °C. Dongshan Bay, being a semi-enclosed bay, experiences seasonal variations in temperature, with winter temperatures primarily influenced by the Taiwan warm current [54]. During this period, freshwater input from the Zhangjiang River is minimal, and the warm water from the Taiwan current dominates, bringing additional warmth to the bay, which has limited water exchange. As a result, the environmental data collected during winter showed a temperature gradient, decreasing from south to north. The inflow of warm water during the winter months did not create a noticeable heating effect, and the water temperature at HXH2 did not differ significantly from surrounding areas. Moreover, there was no evidence of fish aggregating around thermal discharge areas during the colder months.

5. Conclusions

Traditional methods for fish resource surveys are time-consuming and labor-intensive. In contrast, eDNA metabarcoding technology outperforms traditional techniques in sensitivity, standardization, and species identification. It is also simple to use and has great potential for monitoring and conserving fish diversity. Despite the many advantages of eDNA metabarcoding technology, it cannot yet fully replace traditional fish survey methods. One limitation is that eDNA metabarcoding can only identify the presence of species based on genetic information from environmental samples, without providing key details such as population size, age structure, physiological condition, and the developmental stages of the target species. Additionally, the effectiveness of eDNA metabarcoding depends on the completeness of molecular databases; missing sequences for target species in the database can lead to false-negative results. Moreover, due to the complex mechanisms underlying eDNA presence in aquatic environments, further research is needed to improve the accuracy of estimating species relative biomass from eDNA sequence abundance.

In Dongshan Bay, where bottom trawl surveys are difficult to conduct, we applied eDNA metabarcoding technology to detect a total of 76 fish species over the winter and summer seasons, with 43 species detected in winter and found 45 in summer. Thirteen species were shared between the two seasons, revealing significant differences in species composition. We also analyzed the distribution of fish diversity across sampling sites, and the alpha diversity indices showed no significant seasonal differences, suggesting that the fish community in the region is relatively stable. Furthermore, we provide preliminary evidence of the impact of thermal discharge from the Zhangzhou Nuclear Power Plant on local fish communities. Fish exhibited a clear avoidance of high-temperature areas during the summer, while no significant changes were observed in the winter. Our findings can provide valuable scientific support for the conservation of fish diversity and the sustainable development of fishery resources in Dongshan Bay.

Author Contributions: Conceptualization, Y.Z., L.W., W.L. and H.H.; Methodology, W.H., D.O., J.Q. and W.L.; Software, Y.Z.; Validation, Y.Z.; Formal analysis, Y.Z., W.H., L.W., W.L. and H.H.; Investigation, W.H. and L.W.; Resources, D.O.; Data curation, W.H., D.O. and J.Q.; Writing—original draft, Y.Z.; Writing—review & editing, W.L. and H.H. All authors have read and agreed to the published version of the manuscript.

Funding: This work is supported by the Key research and development project of Ministry of Science and Technology (No. 2022YFF0802204), Natural Science Foundation of Fujian Province (No. 2022J05084) and Scientific Research Foundation of Third Institute of Oceanography, MNR (No. 201901).

Data Availability Statement: The data presented in this study are available on request from the corresponding author.

Conflicts of Interest: The authors declare no conflict of interest.

References

1. Levêque, C.; Oberdorff, T.; PAUGy, D.; Stiassny, M.; Tedesco, P.A. Global diversity of fish (Pisces) in freshwater. In *Freshwater Animal Diversity Assessment*; Springer: Dordrecht, The Netherland, 2008; pp. 545–567.
2. Liermann, C.R.; Nilsson, C.; Robertson, J.; Ng, R.Y. Implications of dam obstruction for global freshwater fish diversity. *BioScience* **2012**, *62*, 539–548. [CrossRef]
3. Holmlund, C.M.; Hammer, M. Ecosystem services generated by fish populations. *Ecol. Econ.* **1999**, *29*, 253–268. [CrossRef]
4. Yamamoto, S.; Minami, K.; Fukaya, K.; Takahashi, K.; Sawada, H.; Murakami, H.; Tsuji, S.; Hashizume, H.; Kubonaga, S.; Horiuchi, T. Environmental DNA as a 'snapshot'of fish distribution: A case study of Japanese jack mackerel in Maizuru Bay, Sea of Japan. *PLoS ONE* **2016**, *11*, e0149786.
5. Evans, N.T.; Lamberti, G.A. Freshwater fisheries assessment using environmental DNA: A primer on the method, its potential, and shortcomings as a conservation tool. *Fish. Res.* **2018**, *197*, 60–66. [CrossRef]
6. Taberlet, P.; Coissac, E.; Hajibabaei, M.; Rieseberg, L.H. Environmental DNA. *Mol. Ecol.* **2012**, *21*, 1789. [CrossRef] [PubMed]
7. Thomsen, P.F.; Willerslev, E. Environmental DNA–An emerging tool in conservation for monitoring past and present biodiversity. *Biol. Conserv.* **2015**, *183*, 4–18. [CrossRef]
8. Andruszkiewicz, E.A.; Yamahara, K.M.; Closek, C.J.; Boehm, A.B. Quantitative PCR assays to detect whales, rockfish, and common murre environmental DNA in marine water samples of the Northeastern Pacific. *PLoS ONE* **2020**, *15*, e0242689. [CrossRef]
9. West, K.M.; Stat, M.; Harvey, E.S.; Skepper, C.L.; DiBattista, J.D.; Richards, Z.T.; Travers, M.J.; Newman, S.J.; Bunce, M. eDNA metabarcoding survey reveals fine-scale coral reef community variation across a remote, tropical island ecosystem. *Mol. Ecol.* **2020**, *29*, 1069–1086. [CrossRef]
10. Shaw, J.L.; Clarke, L.J.; Wedderburn, S.D.; Barnes, T.C.; Weyrich, L.S.; Cooper, A. Comparison of environmental DNA metabarcoding and conventional fish survey methods in a river system. *Biol. Conserv.* **2016**, *197*, 131–138. [CrossRef]
11. Ficetola, G.F.; Miaud, C.; Pompanon, F.; Taberlet, P. Species detection using environmental DNA from water samples. *Biol. Lett.* **2008**, *4*, 423–425. [CrossRef] [PubMed]
12. Cilleros, K.; Valentini, A.; Allard, L.; Dejean, T.; Etienne, R.; Grenouillet, G.; Iribar, A.; Taberlet, P.; Vigouroux, R.; Brosse, S. Unlocking biodiversity and conservation studies in high-diversity environments using environmental DNA (eDNA): A test with Guianese freshwater fishes. *Mol. Ecol. Resour.* **2019**, *19*, 27–46. [CrossRef]
13. Thomsen, P.F.; Kielgast, J.; Iversen, L.L.; Møller, P.R.; Rasmussen, M.; Willerslev, E. Detection of a diverse marine fish fauna using environmental DNA from seawater samples. *PLoS ONE* **2012**, *7*, e41732. [CrossRef] [PubMed]
14. Sigsgaard, E.E.; Nielsen, I.B.; Carl, H.; Krag, M.A.; Knudsen, S.W.; Xing, Y.; Holm-Hansen, T.H.; Møller, P.R.; Thomsen, P.F. Seawater environmental DNA reflects seasonality of a coastal fish community. *Mar. Biol.* **2017**, *164*, 128. [CrossRef]
15. Stat, M.; Huggett, M.J.; Bernasconi, R.; DiBattista, J.D.; Berry, T.E.; Newman, S.J.; Harvey, E.S.; Bunce, M. Ecosystem biomonitoring with eDNA: Metabarcoding across the tree of life in a tropical marine environment. *Sci. Rep.* **2017**, *7*, 12240. [CrossRef]
16. Chen, B.; Ji, W.; Zhou, K.; He, Q.; Fu, T. Nutrient and eutrophication characteristics of the Dongshan Bay, South China. *Chin. J. Oceanol. Limnol.* **2014**, *32*, 886–898. [CrossRef]
17. Xu, W.; Gong, L.; Yang, S.; Gao, Y.; Ma, X.; Xu, L.; Chen, H.; Luo, Z. Spatiotemporal dynamics of Vibrio communities and abundance in Dongshan Bay, South of China. *Front. Microbiol.* **2020**, *11*, 575287. [CrossRef] [PubMed]
18. Pan, Z.; Liu, Q.; Jiang, R.; Li, W.; Sun, X.; Lin, H.; Jiang, S.; Huang, H. Microplastic pollution and ecological risk assessment in an estuarine environment: The Dongshan Bay of China. *Chemosphere* **2021**, *262*, 127876. [CrossRef]
19. Deiner, K.; Walser, J.-C.; Mächler, E.; Altermatt, F. Choice of capture and extraction methods affect detection of freshwater biodiversity from environmental DNA. *Biol. Conserv.* **2015**, *183*, 53–63. [CrossRef]
20. He, W.; Wang, L.; Ou, D.; Li, W.; Huang, H.; Ou, R.; Qiu, J.; Cai, L.; Lin, L.; Zhang, Y. Fish Diversity Monitoring Using Environmental DNA Techniques in the Clarion–Clipperton Zone of the Pacific Ocean. *Water* **2023**, *15*, 2123. [CrossRef]
21. Miya, M.; Gotoh, R.O.; Sado, T. MiFish metabarcoding: A high-throughput approach for simultaneous detection of multiple fish species from environmental DNA and other samples. *Fish. Sci.* **2020**, *86*, 939–970. [CrossRef]
22. Jia, H.; Wang, Y.; Yoshizawa, S.; Iwasaki, W.; Li, Y.; Xian, W.; Zhang, H. Seasonal variation and assessment of fish resources in the Yangtze Estuary based on environmental DNA. *Water* **2020**, *12*, 2874. [CrossRef]
23. Yamamoto, S.; Masuda, R.; Sato, Y.; Sado, T.; Araki, H.; Kondoh, M.; Minamoto, T.; Miya, M. Environmental DNA metabarcoding reveals local fish communities in a species-rich coastal sea. *Sci. Rep.* **2017**, *7*, 40368. [CrossRef] [PubMed]
24. Huang-Rong, W. Survey on status of fishery resources of Dongshan Bay in Fujian. *J. Fish. Res.* **2016**, *38*, 112.
25. Shannon, C.E. A mathematical theory of communication. *Bell Syst. Tech. J.* **1948**, *27*, 379–423. [CrossRef]
26. Simpson, E. Measurement of Diversity. *Nature* **1949**, *163*, 688. [CrossRef]
27. Chao, A. Nonparametric estimation of the number of classes in a population. *Scand. J. Stat.* **1984**, *11*, 265–270.

28. Pielou, E.C. The measurement of diversity in different types of biological collections. *J. Theor. Biol.* **1966**, *13*, 131–144. [CrossRef]
29. Jost, L. Partitioning diversity into independent alpha and beta components. *Ecology* **2007**, *88*, 2427–2439. [CrossRef]
30. Bolker, B. *Ecological Models and Data in R*; Princeton University Press: Princeton, NJ, USA, 2008; Volume 396.
31. Oksanen, J.; Minchin, P.R. Instability of ordination results under changes in input data order: Explanations and remedies. *J. Veg. Sci.* **1997**, *8*, 447–454. [CrossRef]
32. Beamish, R.; Leask, K.; Ivanov, O.; Balanov, A.; Orlov, A.; Sinclair, B. The ecology, distribution, and abundance of midwater fishes of the Subarctic Pacific gyres. *Prog. Oceanogr.* **1999**, *43*, 399–442. [CrossRef]
33. Nakagawa, H.; Yamamoto, S.; Sato, Y.; Sado, T.; Minamoto, T.; Miya, M. Comparing local-and regional-scale estimations of the diversity of stream fish using eDNA metabarcoding and conventional observation methods. *Freshw. Biol.* **2018**, *63*, 569–580. [CrossRef]
34. Jing, Z.; Pu-Qing, S.; Zhi-Hui, Z.; Long-Shan, L. Species composition and abundance temporal-spatial distribution of fish egg, larvae and juveniles in Dongshan Bay of Fujian. *J. Fish. Res.* **2013**, *35*, 1.
35. Sun-Zhong, Y.; Dong-Lian, L.; Jian-Di, C.; Yong, L.; Chao, M.; Fang, Y.; Shu-Yue, G.; Chang-Chun, S. Characteristics of community structure and biomass distribution of fishery resources in Dongshan Bay. *J. Fish. Res.* **2018**, *40*, 358.
36. Chen, X.; Ji, P.; Wu, Y.; Zhao, Y.; Zeng, L. Coupling simulation of overland flooding and underground network drainage in a coastal nuclear power plant. *Nucl. Eng. Des.* **2017**, *325*, 129–134. [CrossRef]
37. Ding, X.; Tian, W.; Zhai, A.; Kang, B.; Zhu, Q. Research on Water Intake, Usage and Drainage Impact Demonstration for Coastal Nuclear Power Plants. *IOP Conf. Ser. Earth Environ. Sci.* **2019**, *223*, 012004. [CrossRef]
38. Liu, N.; Huang, Z.; Fang, Y.; Dong, Z. Impacts of Thermal Drainage on Bacterial Diversity and Community Construction in Tianwan Nuclear Power Plant. *Microb. Ecol.* **2023**, *86*, 2981–2992. [CrossRef]
39. Friedlander, A.M.; Ballesteros, E.; Caselle, J.E.; Gaymer, C.F.; Palma, A.T.; Petit, I.; Varas, E.; Muñoz Wilson, A.; Sala, E. Marine biodiversity in Juan Fernández and Desventuradas Islands, Chile: Global endemism hotspots. *PLoS ONE* **2016**, *11*, e0145059. [CrossRef] [PubMed]
40. Downing, J.A.; Plante, C.; Lalonde, S. Fish production correlated with primary productivity, not the morphoedaphic index. *Can. J. Fish. Aquat. Sci.* **1990**, *47*, 1929–1936. [CrossRef]
41. Yoshimura, M.; Yokoduka, T. Radioactive contamination of fishes in lake and streams impacted by the Fukushima nuclear power plant accident. *Sci. Total Environ.* **2014**, *482*, 184–192. [CrossRef]
42. Rotherham, D.; Chapman, M.; Underwood, A.; Gray, C.; Johnson, D. Untangling spatial and temporal variation in abundances of estuarine fish sampled with multi-mesh gillnets. *Mar. Ecol. Prog. Ser.* **2011**, *435*, 183–195. [CrossRef]
43. Larson, J.H.; Trebitz, A.S.; Steinman, A.D.; Wiley, M.J.; Mazur, M.C.; Pebbles, V.; Braun, H.A.; Seelbach, P.W. Great Lakes rivermouth ecosystems: Scientific synthesis and management implications. *J. Great Lakes Res.* **2013**, *39*, 513–524. [CrossRef]
44. Wang, S.; Yan, Z.; Hänfling, B.; Zheng, X.; Wang, P.; Fan, J.; Li, J. Methodology of fish eDNA and its applications in ecology and environment. *Sci. Total Environ.* **2021**, *755*, 142622. [CrossRef]
45. Kirkpatrick, B.; Fleming, L.E.; Squicciarini, D.; Backer, L.C.; Clark, R.; Abraham, W.; Benson, J.; Cheng, Y.S.; Johnson, D.; Pierce, R. Literature review of Florida red tide: Implications for human health effects. *Harmful Algae* **2004**, *3*, 99–115. [CrossRef]
46. Agersborg, H. The influence of temperature on fish. *Ecology* **1930**, *11*, 136–144. [CrossRef]
47. Mallya, Y.J. The Effects of Dissolved Oxygen on Fish Growth in Aquaculture. The United Nations University Fisheries Training Programme, Final Project. 2007. Available online: https://www.grocentre.is/static/gro/publication/58/document/yovita07prf.pdf (accessed on 11 November 2024).
48. Zhang, H.; Yoshizawa, S.; Iwasaki, W.; Xian, W. Seasonal fish assemblage structure using environmental DNA in the Yangtze Estuary and its adjacent waters. *Front. Mar. Sci.* **2019**, *6*, 515. [CrossRef]
49. Osborne, L.L.; Wiley, M.J. Influence of tributary spatial position on the structure of warmwater fish communities. *Can. J. Fish. Aquat. Sci.* **1992**, *49*, 671–681. [CrossRef]
50. Eaton, J.G.; Scheller, R.M. Effects of climate warming on fish thermal habitat in streams of the United States. *Limnol. Oceanogr.* **1996**, *41*, 1109–1115. [CrossRef]
51. Neuman, E. Thermal discharge and fish fauna in Sweden. *Water Sci. Technol.* **1983**, *15*, 67–87. [CrossRef]
52. Krishnakumar, V.; Sastry, J.; Swamy, G.N. Implication of Thermal Discharges into the Sea—A Review. 1991. Available online: http://drs.nio.org/drs/handle/2264/3339 (accessed on 22 November 2024).
53. Tang, Y. Politics of the risk management for marine invasive alien species in China: Constructing a cooperation system. *Appl. Mech. Mater.* **2013**, *295*, 520–527. [CrossRef]
54. Du, Y.; Meng, F.; Fu, W.; Wang, Z. Distribution, speciation and bioaccumulation of Hg and As in mariculture sediments from Dongshan Bay, China. *Soil Sediment Contam. Int. J.* **2016**, *25*, 489–504. [CrossRef]

water

MDPI

Article

Water Quality and Its Influence on Waterbird Habitat Distribution: A Study Along the Lieve River, Belgium

Xingzhen Liu, Long Ho, Andrée De Cock *, Nancy De Saeyer, Kim Pham, Diego Panique-Casso, Marie Anne Eurie Forio and Peter L. M. Goethals

Department of Animal Sciences and Aquatic Ecology, Faculty of Bioscience Engineering, Ghent University, Coupure Links 653, 9000 Ghent, Belgium; xingzhen.liu@ugent.be (X.L.); long.tuanho@ugent.be (L.H.); nancy.desaeyer@ugent.be (N.D.S.); thikimdung.pham@ugent.be (K.P.); gustavo.paniquecasso@ugent.be (D.P.-C.); marie.forio@ugent.be (M.A.E.F.); peter.goethals@ugent.be (P.L.M.G.)

* Correspondence: andree.decock@ugent.be

Abstract: Freshwater ecosystems face increasing pressures from human activities, leading to degraded water quality and altered habitats for aquatic species. This study investigates the relationship between water quality and waterbird distribution along the Lieve River, Belgium, based on manually conducted waterbird counts and water quality data collected from 48 transects in March 2024. Localized eutrophication was evident, with TN (2.7–5.6 mg L^{-1}), TP (up to 0.46 mg L^{-1}), and chlorophyll-*a* (median 70 ppb) exceeding environmental thresholds. Prati index analysis revealed that 58.3% of the sampling points along the Lieve River were categorized as "polluted", reflecting extensive water quality degradation. Eurasian coots (71.4%) and wild ducks (72.4%) were predominantly found in polluted areas, thriving in nutrient-enriched habitats linked to high TP levels. In contrast, common moorhens (80.3%) preferred acceptable quality areas, indicating higher water quality requirements. These findings indicate that phosphate is a key driver of waterbody eutrophication, as evidenced by the TP concentrations measured on-site, which far exceed the thresholds set by environmental standards. Future research should explore advanced monitoring approaches to improve waterbird and water quality assessments, ensuring the conservation of the Lieve River as one of Europe's oldest artificial canals, and the protection of its waterbird habitats.

Keywords: water quality; waterbird; nutrient dynamics; Prati index; organic matter; habitat suitability; Lieve River; eutrophication; freshwater ecosystem

1. Introduction

Human activities are placing unprecedented pressures on global freshwater ecosystems, leading to significant declines in water quality and biodiversity [1]. Freshwater habitats are vital for supporting ecological functions, providing critical ecosystem services, and sustaining species dependent on specific environmental conditions [2]. Among these species, waterbirds can serve as bioindicators whose habitat preferences and distribution patterns are closely linked to water quality [3,4]. Environmental factors such as turbidity, nutrient concentrations, dissolved oxygen (DO), and pollutants create diverse habitat conditions within river systems, influencing where waterbirds forage, nest, and gather [5–7]. Conversely, waterbirds themselves can influence water quality through behaviors like foraging and excretion [8,9]. Understanding these interdependencies is critical for advancing ecological knowledge, informing effective conservation strategies, and identifying ecological drivers of species distribution, particularly in riparian zones where human pressures intersect with natural ecosystems [4,10,11].

The Lieve River in Belgium provides a valuable case study for examining the relationship between water quality and waterbird habitat distribution. Originally constructed in the 13th century as one of Europe's earliest trade canals, it is now a small yet ecologically and historically significant freshwater system that has evolved into a biologically rich habitat and a recreational corridor on the outskirts of Ghent, Belgium's third-largest city in the Flanders region [12]. This dual role as both a natural habitat for waterbirds and a public recreational area brings complex challenges, particularly in balancing habitat health with pressures from human use [13–15]. Urban runoff, agricultural drainage, and residential activities such as commuting, cycling, and horseback riding may influence waterbird distribution and water quality along the Lieve River [9,16,17]. Studying the local waterbird responses to water quality variations not only reveals insights specific to the Lieve River but also enhances our understanding of how biodiversity in urban freshwater systems is shaped by varying environmental conditions [18,19].

Water quality factors such as nutrient levels, organic matter, and chlorophyll-*a* concentrations play critical roles in shaping waterbird habitats. Previous studies have highlighted the importance of these parameters in influencing bird distribution patterns [17,20,21]. For instance, elevated chlorophyll-*a* levels, a proxy for trophic state, are often associated with eutrophication, creating productive habitats for species like the Eurasian coot and wild duck [22]. Similarly, DO is a key determinant of water health, with low levels often correlating with degraded environments that limit food availability for waterbirds [8]. Organic matter indicators, such as total organic carbon (TOC) and chemical oxygen demand (COD), also significantly influence habitat suitability by supporting food web productivity and vegetation growth [21,23]. Spatial variations in these water quality parameters along the Lieve River reflect the combined effects of agriculture, urban runoff, and direct discharge of domestic wastewater. These anthropogenic pressures not only create localized impacts but also cumulatively influence habitat conditions, making these parameters central to understanding bird distributions. Similar localized patterns have been documented in studies of ecological dynamics in freshwater systems [4,11,24].

In this study, we focused on the Lieve River, which traverses a landscape characterized by intensive agricultural use, small-town communities, and facilities such as animal feed factories that contribute to nutrient loading and organic pollution [12,25]. In overdeveloped and polluted river basins, habitat degradation has led to significant declines in waterbird populations, as evidenced by studies conducted in Europe and other regions with similar environmental conditions. Related studies include the impacts of organic micropollutants (OMPs) in Spain (which have driven waterbird population declines), river regulation in Poland (which reduced breeding-bird richness by 23% and abundance by 33%), and wetland restoration in Southern California (which improved water quality and boosted endangered bird populations), underscore the need for waterbird habitat protection and pollution control, as mandated by the EU Birds Directive 2009/147/EC, which designates IBAs as Special Protection Areas (SPAs) within the Natura 2000 network [26–31]. These findings underscore the critical need to investigate how water quality parameters affect waterbird habitat distribution along the Lieve River. Over an 11 km stretch, 48 monitoring locations were established to analyze key water quality parameters, such as dissolved oxygen, nutrient concentrations, and organic matter. Waterbird surveys targeted three principal species: the Eurasian coot (*Fulica atra*), the wild duck (*Anas platyrhynchos*), and the common moorhen (*Gallinula chloropus*), providing a comprehensive assessment of species abundance and distribution along the river. This study also considers the challenge of balancing ecological conservation with the preservation of historically significant freshwater systems like the Lieve River. Addressing these interconnected objectives helps bridge the gap between biodiversity conservation and cultural heritage management, offering insights into managing ecologically and culturally valuable systems [15].

The study addresses the following research questions: (1) What is the water quality along the Lieve River and are environmental limits exceeded? (2) How are birds distributed along the Lieve River? (3) Can correlations be found between water quality and bird distribution along the Lieve River? (4) What policy recommendations and future research directions can be proposed to enhance water quality management and the conservation of waterbird habitats along the Lieve River? The findings and the methodology of this study provide valuable insights into investigating similar river systems on the planet.

2. Materials and Methods

2.1. Study Area

This study was conducted along an 11 km stretch of the Lieve River, located in the municipality of Lievegem, near Ghent in Belgium's Flanders region (51°03′ N, 3°43′ E) (Figure 1). It serves as a critical habitat for diverse waterbird species and remains a recreational corridor for the surrounding communities. Based on elevation data from Figure 1, the Lievegem area features low-lying terrain (30 m) and flat topography, which is particularly suitable for agricultural use [32]. The river flows through urban, agricultural, and peri-urban landscapes, creating a natural gradient of environmental pressures, such as nutrient runoff, urban inputs, and recreational use [25,33,34]. These anthropogenic factors contribute to spatial variations in water quality, influencing the habitat suitability of waterbirds.

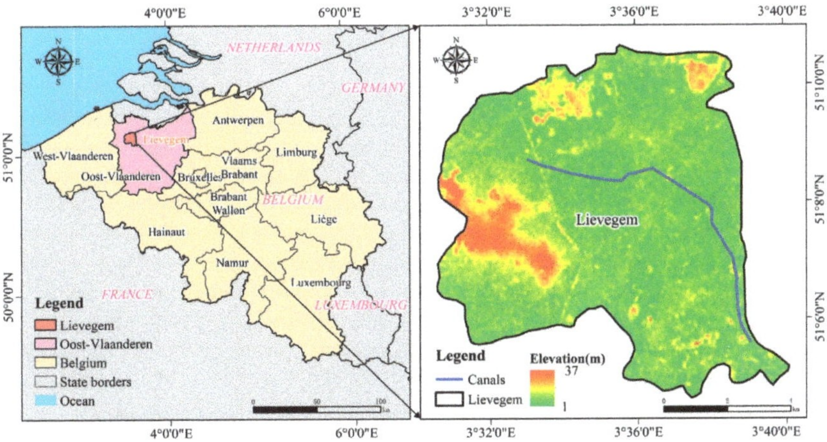

Figure 1. Study area. The blue line marks the Lieve River within Lievegem, where the sampling was conducted. Elevation data is sourced from NASA's 30 m Digital Elevation Model (DEM) [32].

To facilitate a detailed investigation of water quality and waterbird distributions, the study area was systematically divided into 24 monitoring regions, labeled F1, F2, F3, ... F24. Within each track, water samples were collected at 200 m intervals, resulting in 48 monitoring transects. Each transect is referred to as F1S1, F1S2, F2S1, F2S2, and so forth, where "S1" and "S2" indicate the first and second segments within each 200 m interval. A minor exception was made at F17, where F17S1 covers a 200 m segment, while F17S2 represents a slightly shorter distance of 150 m due to site-specific constraints. A visual representation of this division is provided in Supplementary Materials, Figure S1.

This adjustment ensured that the overall monitoring design remained consistent and comprehensive.

In addition to water quality sampling, waterbird abundance was recorded on-site at each of the 48 monitoring transects during field surveys. This systematic field recording was conducted to align with the spatial resolution of the water quality monitoring, enabling detailed correlations between water quality parameters and bird distribution patterns. This structured sampling design ensured consistent and high-resolution coverage of both environmental and biological data across the entire study area [35]. The segmentation framework facilitated the identification of localized patterns in water quality and their associations with waterbird abundance and habitat preferences [13,36].

2.2. Water Quality Measurement

The water quality assessment for this study on the Lieve River was conducted during a comprehensive sampling campaign in March 2024, encompassing measurements across 48 designated monitoring 200 m transects along an 11 km stretch of the river. Water quality parameters were analyzed through both in situ field measurements and laboratory analyses to provide a detailed evaluation of spatial variability in key water quality indicators.

Field measurements: DO, water temperature, pH and electrical conductivity (EC), were measured using a handheld multiprobe (WTW Multi, WTW Xylem Inc., Weilheim, Bavaria, Germany). Turbidity and chlorophyll-*a* were measured using an AquaFluor handheld fluorometer (San Jose, CA, USA) from Turner Designs. Each parameter was measured 3 times per site to ensure consistency and accuracy, with average values recorded. To avoid cross-contamination, all probes were thoroughly rinsed with deionized water between measurements. In addition, weather conditions, water flow characteristics, and any notable human activities (e.g., recreational use or agricultural drainage) were documented onsite to supplement field observations. Equipment was checked before and after sampling to ensure reliable performance and data integrity.

In addition to in situ measurements, water samples were collected for further laboratory analysis. For water sample collection, a bucket was rinsed 3 times with stream water, with the sides and bottom hand-washed, and the rinsing water discarded onshore or downstream. From the bucket, unfiltered samples were transferred into leakproof, freeze-resistant recipients for COD, TOC, TN, and TP analysis, while filtered samples (0.45 μm pore PES filter) were transferred into separate recipients for the analysis of NO_2^-, NO_3^-, NH_4^+, and PO_4^{3-}.

Laboratory Analysis: Water samples from all 48 transects were transported at 4 °C in the dark to preserve their integrity and then stored in a refrigerator at −20 °C until analysis. Nutrient concentrations, including NH_4^+-N, NO_3^--N, NO_2^--N, orthophosphate (PO_4^{3-}-P), total nitrogen (TN), and total phosphorus (TP), were measured spectrophotometrically using low-range Merck test kits with a spectrophotometer (Hach® DR6000™, Hach Company, Loveland, Colorado, USA). The organic content, including total inorganic carbon (TIC) and TOC, was analyzed at the Laboratory of AECO at Ghent University using a TOC analyzer (Shimadzu TOC-L, Shimadzu Corporation, Kyoto, Japan). COD was also measured to evaluate organic matter content. All laboratory analyses were conducted with instruments regularly calibrated using certified standard solutions to ensure high analytical precision and accuracy. We adhered to the quality assured by the American Public Health Association (APHA) and the United States Environmental Protection Agency (US EPA) [37,38]. The detailed measurement accessories and kits are provided in Table S5 of Supplementary Materials.

2.3. Waterbird Survey

Waterbird surveys were conducted simultaneously with water quality measurements in March 2024 to provide integrated insights into the relationship between water quality and bird distribution. The survey focused on quantifying populations of three principal waterbird species: the Eurasian coot (*Fulica atra*), the wild duck (*Anas platyrhynchos*), and the common moorhen (*Gallinula chloropus*), all of which are ecologically significant indicators of freshwater habitat health [39–41]. The survey was conducted at every 50 m transect within these transects to ensure detailed and fine-scale data collection. Each transect was systematically marked to prevent overlap and ensure precise counts. Field observers followed standardized ornithological protocols, manually recording the number and species of waterbirds encountered at each interval. This systematic approach facilitated the accurate capture of bird abundance and species distribution across the study area. Ethical considerations were integral to the survey design. Observers maintained a sufficient distance from the birds to avoid disturbance and ensured minimal disruption to their natural behaviors. Field documentation also included notes on habitat conditions and potential human activities in the area, providing additional context for the recorded distributions [42,43].

2.4. Data Analysis

To evaluate the relationships between water quality parameters and waterbird abundance, both the Mantel test and Pearson correlation analysis were performed using R Studio (version 2022.12.0, Posit Software, Boston, MA, USA, 2022) [44,45]. The Mantel test was applied to identify spatial correlations between environmental factors (e.g., water quality parameters) and bird distributions by calculating the correlation between two distance matrices: one for environmental variables and the other for bird abundances across the 48 monitoring transects. The Pearson correlation analysis was conducted to examine pairwise relationships between water quality variables and bird abundance [46,47]. In this study, the terms NH_4-N, NO_3-N, NO_2-N, and PO_4-P are used to represent the weight of nitrogen (N) or phosphorus (P) atoms within their respective molecular forms (e.g., NH_4^+, NO_3^-, NO_2^-, and PO_4^{3-}), serving as proxies for the concentrations of these ions. Additionally, for the measurement of COD, a value of "<10" was recorded at site F1 S2; to facilitate calculations and maintain consistency in the analysis, the boundary value of 10 was used for this measurement.

To assess the degree of pollution in the Lieve River, the Basic Prati water quality index (Prati WQI; hereafter referred to as the Prati index), developed by Prati et al. (1971) [48], was utilized. This index aggregates multiple pollutants, more specifically dissolved oxygen saturation (DO_sat), COD, and NH_4-N, into a single score. Using transformation formulas, measured values are standardized and their average provides an index score that simplifies complex water quality data into a clear and logical format [49]. In this study, the Basic Prati index was calculated for each sampling point, with specific methods detailed in Supplementary Materials. Water quality was classified into five categories based on the Prati Index evaluation: Excellent (0–1), Acceptable (1–2), Polluted (2–4), Heavily Polluted (4–8), and Extremely Polluted (>8). Based on this, a standardized framework for identifying pollution gradients and hotspots in the Lieve River was used [50–52].

To evaluate the ecological health of the Lieve River, measured water quality parameters were compared against established environmental standards for freshwater systems. These standards (Supplementary Materials Table S6) were compiled from established international and regional sources, including the World Health Organization (WHO, 2011), the European Union Water Framework Directive (WFD, 2000/60/EC), the United States Environmental Protection Agency (US EPA), and VLAREM (the Flemish environmental legislation in Belgium), which classifies this section of the Lieve River as a "small river", and supporting literature such as Williams (2001) and Kim (2021) [53–58].

3. Results and Discussion

3.1. Spatial Variability of Water Quality Parameters

The spatial variability of water quality parameters along the Lieve River was analyzed, focusing on environmental factors represented as in situ variables (Chlorophyll-*a*, DO, pH, EC), nitrogen species (TN, NO_3-N, NO_2-N, NH_4-N), phosphorus species (PO_4-N, TP), and organic matter (TIC, TOC, COD). Figure 2. illustrates the spatial distribution of these parameters, highlighting significant variations across monitoring locations. Other detailed values for those parameters are provided in Supplementary Materials Tables S1–S3.

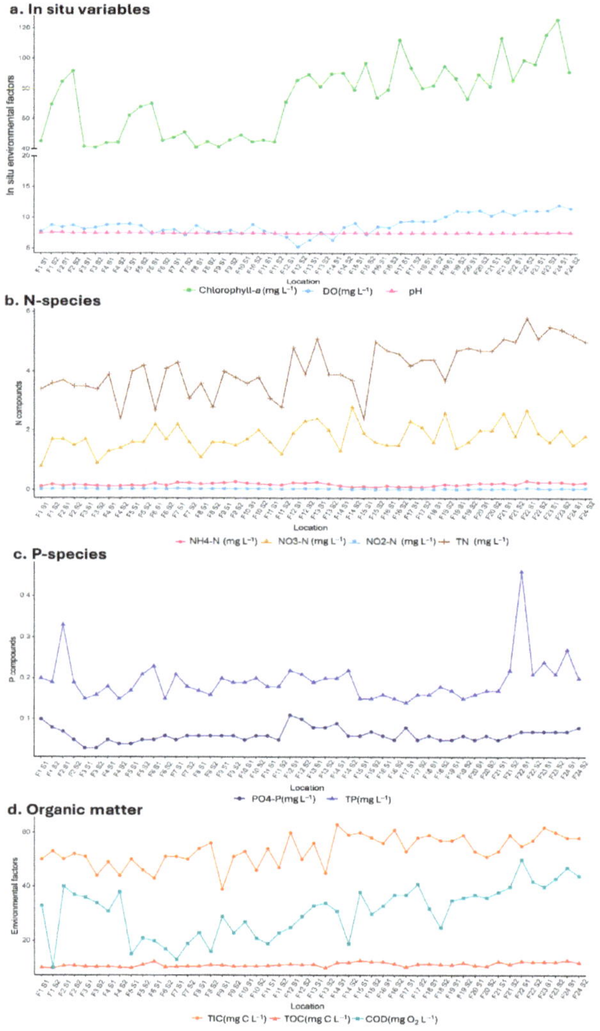

Figure 2. Spatial distribution of water quality parameters along the Lieve River. (**a**) In situ variables (Chlorophyll-*a*, ppb; pH; DO, mg L^{-1}); (**b**) Nitrogen species (TN, NO_3-N, NO_2-N, NH_4-N, mg L^{-1}); (**c**) Phosphorus species (PO_4-P, TP, mg L^{-1}); (**d**) Organic matter indicators (TIC, TOC, mg C L^{-1}; COD, mg O_2 L^{-1}).

Chlorophyll-*a* concentrations demonstrated a clear spatial gradient along the river. The values ranged from 41.1 ppb at F3 S2 to 126.1 ppb at F24 S1, the highest recorded value. Additional peaks were observed at F23 S2 (116.0 ppb) and F21 S2 (113.8 ppb). Areas closer to the F1 direction, such as F3 S2 (41.1 ppb) and F4 S1 (44.3 ppb), exhibited significantly lower chlorophyll-*a* levels. DO concentrations ranged from 5.30 mg L^{-1} at F12 S2 to 12.07 mg L^{-1} at F24 S1. The pH was very stable, with an average value hovering around 7.50. This spatial pattern indicates increased algal productivity towards the F24 direction (Figure 2a).

Nitrogen species displayed distinct spatial trends. TN concentrations ranged from 2.4 mg L^{-1} at F15 S1 to 5.8 mg L^{-1} at F22 S1. NO_3-N concentrations peaked at 2.7 mg L^{-1} at F22 S1 and 2.6 mg L^{-1} at F21 S1 and F18 S2. NH_4-N concentrations were highest at 0.31 mg L^{-1} at F22 S1, while NO_2-N remained consistently low, ranging between 0.02 and 0.08 mg L^{-1}. Higher nitrogen levels in the F24 direction suggest localized nutrient enrichment, possibly from agricultural or urban runoff (Figure 2b).

Phosphorus species exhibited spatial variability. PO_4-P concentrations ranged from 0.03 mg L^{-1} at F3 S1 and F3 S2 to 0.11 mg L^{-1} at F12 S1, the highest value recorded. TP concentrations varied between 0.14 mg L^{-1} at F17 S1 and 0.46 mg L^{-1} at F22 S1. Elevated phosphorus levels at F12 S1 and F22 S1 indicate localized nutrient input, likely driven by human activities. Lower phosphorus levels were observed near the F1 direction, such as at F4 S1 (Figure 2c).

Organic matter indicators showed substantial spatial variability. TOC concentrations ranged from 9.8 mg C L^{-1} at F1 S2 to 12.7 mg C L^{-1} at F24 S1, while COD varied between 10 mg O_2 L^{-1} at F1 S2 and mg O_2 L^{-1} at F22 S1. TIC values also showed variability, with higher TIC often observed in regions of elevated TOC and COD. Elevated TOC and COD levels at F22 S1 and F24 S1 suggest organic matter accumulation, potentially linked to decomposition processes (Figure 2d).

3.2. Prati Index-Based Water Quality Assessment

The water quality of the Lieve River, assessed using the Prati index, displayed spatial and categorical variability across the 48 sampling points, as shown in Figure 3. Results indicated that 41.7% (20 locations) of the sampling locations were categorized as "Acceptable quality" (Prati index $\leq$ 2.0), while 58.3% (28 locations) were classified as "Polluted" (Prati index > 2.0). The distribution of water quality classifications reveals a trend, with cleaner sites predominantly located near the F1 direction and more polluted sites concentrated closer to the F24 direction. This spatial variability highlights potential cumulative anthropogenic impacts, such as urban runoff, agricultural discharge, and point-source pollution, contributing to water quality degradation. A detailed summary of Prati index results is provided in Supplementary Materials Table S4.

Spatial analysis of the Prati index values revealed distinct differences between sites. Locations near the F1 direction, including F1 S2 (Prati index = 1.40) and F4 S1 (Prati index = 1.93), were associated with higher DO_sat ($\geq$80%), moderate COD ($\leq$31 mg O_2 L^{-1}), and low NH_4-N ($\leq$0.2 mg L^{-1}). These conditions reflect reduced levels of organic and nitrogenous pollution, which is indicative of favorable water quality. In contrast, sites near the F24 direction, such as F12 S2 (Prati index = 2.97) and F24 S1 (Prati index = 2.56), showed reduced DO_sat (e.g., F12 S2 = 49%) and elevated COD and NH_4-N levels. F24 S1, for instance, recorded high COD (47 mg O_2 L^{-1}) and NH_4-N (0.22 mg L^{-1}), suggesting significant organic and ammonia-related contamination at these locations. This spatial trend underscores the influence of cumulative pollution along the river.

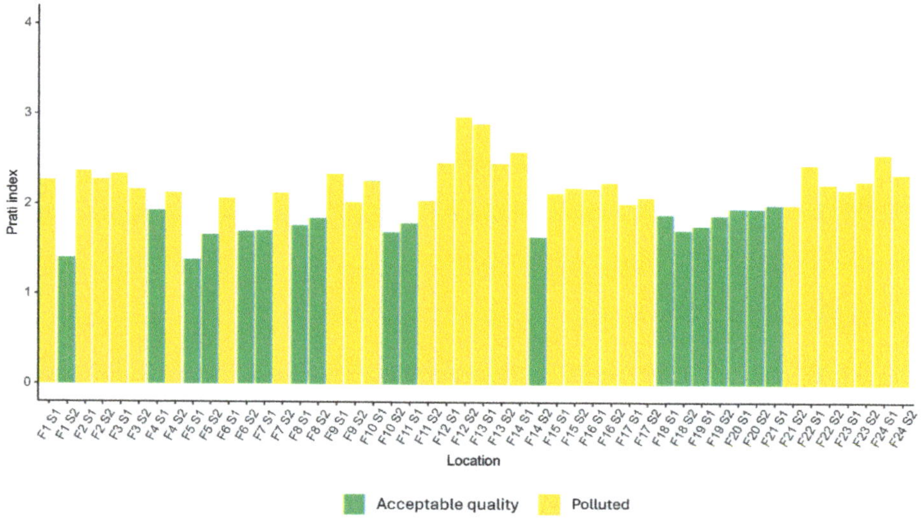

Figure 3. Spatial distribution of Prati index and water quality classification along the Lieve River. Green bars represent "Acceptable quality" (Prati index $\leq$ 2.0); Yellow bars represent "Polluted" (Prati index > 2.0).

Key water quality parameters strongly influenced the Prati index values. Dissolved oxygen saturation emerged as a critical factor, with higher DO_sat levels correlating with better water quality classifications. For example, F5 S1 (DO_sat = 85%, Prati index = 1.38) displayed favorable conditions, while reduced DO_sat in sites such as F13 S1 (DO_sat = 58.3%, Prati index = 2.90) was associated with poorer water quality. Similarly, elevated COD levels were a dominant contributor to higher Prati index values, with sites like F22 S1 (COD = 50 mg O_2 L^{-1}) and F24 S1 (COD = 47 mg O_2 L^{-1}) demonstrating substantial organic pollution. Ammonium concentrations (NH_4-N) also played a significant role, with locations like F23 S1 (NH_4-N = 0.27 mg L^{-1}) and F22 S1 (NH_4-N = 0.31 mg L^{-1}) corresponding to "Polluted" classifications.

The Prati index effectively captured the spatial variability and drivers of water quality along the Lieve River. Sites near the F1 direction exhibited better water quality conditions, likely due to natural purification processes and lower pollutant loads, while sites near the F24 direction reflected increasing anthropogenic pressures. These findings provide crucial insights into the spatial distribution of pollution hotspots and emphasize the need for targeted water quality management strategies in downstream (towards F24) sections of the river.

3.3. Waterbird Abundance and Distribution Patterns

The spatial distribution and abundance of waterbird species across the 48 monitoring transects along the Lieve River revealed distinct patterns in species composition and spatial variability (Figure 4). Three principal waterbird species were observed: Eurasian coot, common moorhen, and wild duck. Each species exhibited different abundance levels and spatial patterns, reflecting their habitat preferences and responses to localized environmental conditions. Strikingly, nearly in each section of 200 m, at least 2 individuals were found, in 16 sections there were more than 5 individuals recorded, and, on average, there were about 20 birds recorded per kilometer.

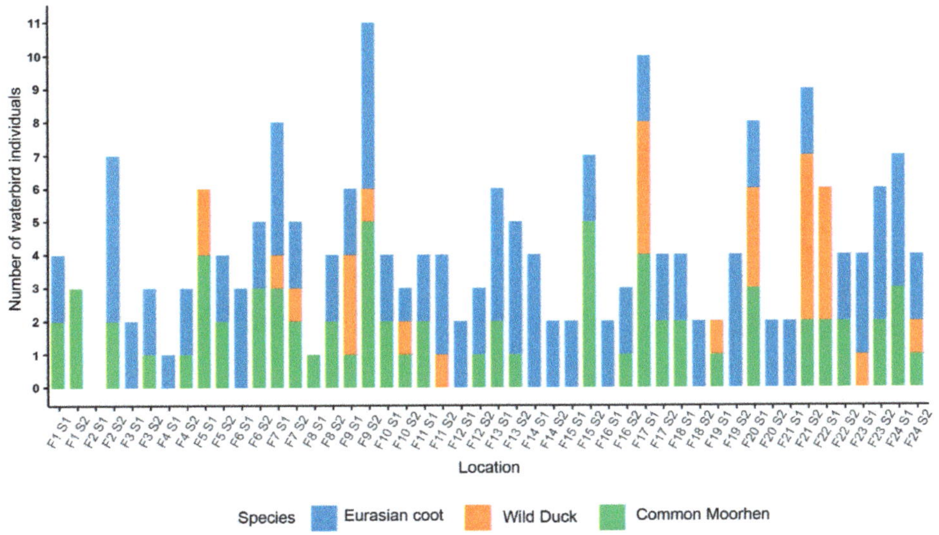

Figure 4. Spatial distribution of waterbird individuals (Eurasian coot, wild duck, and common moorhen) across the 48 monitoring locations along the Lieve River. The bar heights represent the total number of individuals recorded per species at each location.

The Eurasian coot was the most abundant species, with a total of 105 individuals recorded across all transects. Peak coot counts were observed at F2 S2 and F9 S2, where 5 individuals were recorded. These locations may reflect the availability of abundant aquatic vegetation and calmer water conditions, which provide optimal foraging opportunities and shelter [59,60]. Interestingly, towards the F24 sections such as F23 S2 and F24 S1, which exhibited elevated chlorophyll-*a* concentrations and higher organic matter inputs in the earlier water quality analysis, they also recorded moderate coot abundances. This may suggest that nutrient-enriched environments, while often associated with reduced water quality, can support increased productivity of aquatic vegetation and phytoplankton, indirectly benefiting coots [60,61].

The common moorhen displayed a variable distribution, with a total of 71 individuals recorded. Moorhens reached their peak abundances at F9 S2 and F15 S2, where 5 individuals were observed. These locations may offer shallow water habitats characterized by emergent vegetation, which are commonly associated with moorhen foraging and nesting preferences, as suggested by previous studies [59,62]. Regions like F9 S2 exhibited moderate DO levels and elevated TOC concentrations, conditions that may indicate moorhens' tolerance to localized organic matter accumulation. However, conclusions on their adaptability require further analysis across multiple sites.

The wild duck, the least abundant species, totaled 29 individuals and displayed a sporadic distribution. The highest number of 5 individuals was recorded at F21 S2, a towards-F24 section characterized by higher dissolved oxygen concentrations and moderate nutrient levels. The sporadic distribution of wild ducks, particularly their preference for regions with higher dissolved oxygen concentrations and moderate nutrient levels, may be attributed to the availability of submerged vegetation and aquatic invertebrates, critical food sources for this species [59,63]. However, variations in vegetation structure, predation pressure, and competition could explain their absence in other nutrient-rich, oxygenated areas [59,64,65].

The spatial distribution of waterbirds along the Lieve River reflects a combination of distinct patterns and relatively uniform trends influenced by species-specific habitat preferences and behavioral factors. While the dominance of Eurasian coots, localized peaks of common moorhens, and the sporadic presence of wild ducks indicate varying environmental tolerances and preferences, the overall distribution appears relatively homogeneous during the nesting season. This homogeneity can be partly attributed to territorial behavior, which influences how individuals spread across available habitats, and the absence of sites with extreme water quality conditions, either particularly favorable or unfavorable, which creates a more consistent environmental baseline. Regions with elevated chlorophyll-*a* concentrations and organic matter (TOC and COD) were associated with higher abundances of coots and ducks, demonstrating their affinity for nutrient-rich areas within this overall pattern.

3.4. Influence of Water Quality on Waterbird Distribution Patterns

The correlation analysis between water quality parameters and waterbird abundance revealed species-specific responses, providing insight into how water quality influences waterbird distribution along the Lieve River (Figure 5). Combined with the spatial patterns of water quality and waterbird abundance (3.3), these results highlight key drivers of habitat preference and species-specific ecological adaptations.

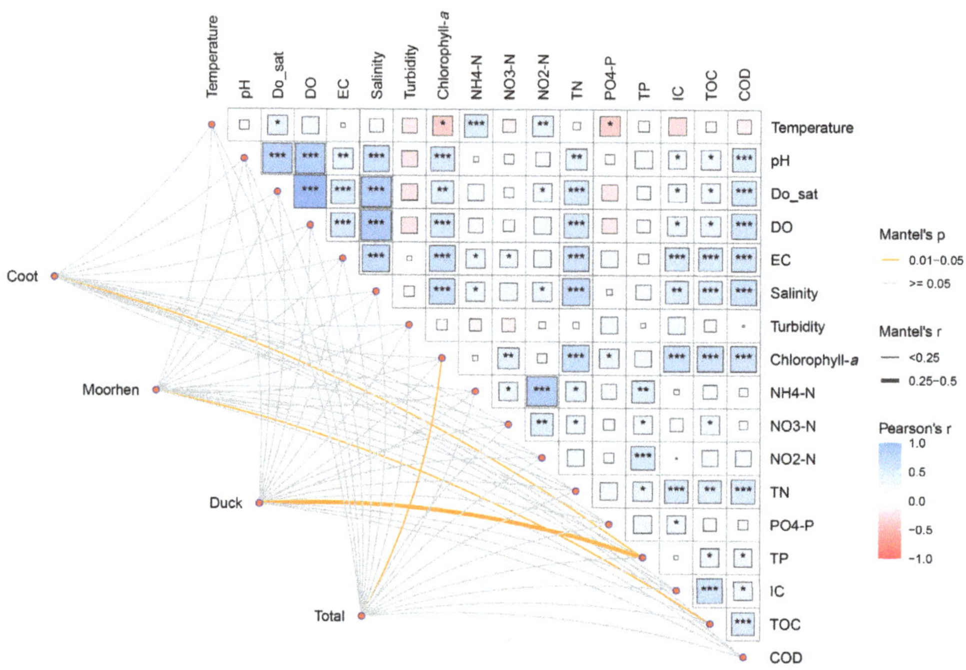

Figure 5. Relationships between water quality parameters and waterbird abundance along the Lieve River. The figure combines Pearson correlation coefficients and Mantel test results to highlight associations between water quality parameters and waterbird species abundance. Positive correlations are shown in blue, while negative correlations are in red. The line thickness represents Mantel's r values, with thicker lines indicating stronger correlations. Mantel's r: A measure of the strength of correlation between two distance matrices, reflecting how closely the spatial variation of water quality matches bird abundance. Mantel's *p*: The statistical significance of the Mantel test, with lines in orange representing significant relationships (p = 0.01–0.05) and lines in gray indicating non-significant relationships ($p \geq 0.05$). (*) indicate the significance level of Pearson's correlation coefficients: $p < 0.05$ (*), $p < 0.01$ (**), and $p < 0.001$ (***).

The Eurasian coot exhibited a weak but significant positive correlation with TP (r = 0.20, p = 0.048), indicating a clear association between changes in TP and the presence of coots. Among waterbirds, the Eurasian coot is known for producing a large TP mass, as highlighted by Scherer et al. (1995) [66]. This nutrient enrichment promotes the growth of aquatic vegetation, which in turn provides essential resources for coots, creating a feedback loop that can exacerbate localized eutrophication. This dynamic aligns with the findings of Scherer et al. (1995) and Clausen (2025), who emphasized that the excretion of omnivorous waterbirds like the Eurasian coot significantly alters phosphorus concentrations in aquatic systems [66,67]. Similarly, the study by Tóth et al. (2023) observed that, while the presence of waterbirds is not a decisive factor for overall water quality, it does have a substantial localized impact, particularly in nutrient-enriched habitats [9].

The common moorhen exhibits a weak but significant positive correlation with TOC (r = 0.10, p = 0.036), suggesting a preference for habitats with slightly elevated organic matter content. As an omnivorous bird, the common moorhen's diet includes aquatic plants, seeds, insects, and small invertebrates, many of which thrive in environments with higher levels of organic matter [68]. Elevated TOC levels typically indicate the accumulation of organic material, creating microhabitats rich in detritus and microbial activity, which serve as a food base for invertebrates [69]. This finding supports the moorhen's opportunistic foraging behavior, as it often feeds in shallow waters and along the edges of aquatic habitats rich in organic matter, aligning with observations from previous studies [68,70–73].

The wild duck exhibits weak but significant positive correlations with TP (r = 0.300, p = 0.046), suggesting a preference for habitats with slightly elevated concentrations of total phosphorus. Wild ducks feed on aquatic plants, seeds, insects, and small invertebrates, many of which thrive in environments rich in nutrients such as phosphorus and minerals [74]. Increased TP levels typically reflect nutrient enrichment [75,76]. This contributes to the growth of submerged vegetation and invertebrates, which are critical food resources for ducks. Additionally, studies have shown that wild ducks adapt to decreases in TP and increases in water clarity by altering their mating behaviors, which can influence breeding density and reproductive success [77,78]. These findings align with previous research, supporting the notion that wild ducks exhibit adaptive foraging behavior and reproductive strategies, as they frequently inhabit nutrient-rich aquatic environments that enhance food availability and facilitate their survival and reproduction [74,77–79].

By combining the observations from Figure 3 (Prati index distribution) in Section 3.2 and Figure 4 (spatial distribution of waterbirds) in Section 3.3, it was found that 71.4% of Eurasian coots were located in the yellow "polluted" areas. Wild ducks exhibited a similar trend, with 72.4% of their distribution also in polluted areas. In stark contrast, however, only 19.7% of common moorhens were found in polluted areas, with the majority of common moorhens choosing the green "acceptable quality" regions, indicating a preference for higher water quality. This observation aligns with the results of the correlation analysis, where Eurasian coots and wild ducks both showed correlations with TP, while common moorhens did not. Instead, common moorhens exhibited a weak relationship with TOC. Interestingly, the observed pollution is likely a result of water eutrophication, which is closely associated with phosphate concentrations.

These findings collectively point to the influence of eutrophication and organic matter enrichment and deposition. Such dynamics may be linked to land use practices in the monitored region, which include agricultural areas, animal feed factories, and densely populated residential zones along the Lieve River. The discharge of industrial and domestic wastewater, as well as agricultural activities, significantly contributes to water eutrophication and organic matter accumulation. Addressing these challenges requires targeted efforts to control wastewater discharge and manage the use of chemical fertilizers and organic manures in agriculture to mitigate their effects on water quality and aquatic ecosystems.

3.5. Water Quality Deviations from Environmental Standards and Ecological Implications

As shown in Figure 6, nutrient enrichment emerges as a significant driver of eutrophication in the Lieve River. TP levels peaked at 0.46 mg L^{-1} (median: 0.18 mg L^{-1}), frequently exceeding the environmental threshold of 0.14 mg L^{-1}. Similarly, TN concentrations ranged from 2.7 to 5.6 mg L^{-1} (median: 4.0 mg L^{-1}), with many locations meeting or surpassing the upper limit of 4.0 mg L^{-1}. These elevated nutrient levels correspond to chlorophyll-*a* concentrations (median: 70 ppb) far exceeding the upper threshold of 15 ppb. This nutrient accumulation promotes algal blooms and eutrophic conditions, providing fertile habitats for species such as the Eurasian coot. As previously noted, Eurasian coots are recognized for their role in increasing TP levels through excretion, reinforcing nutrient enrichment and contributing to feedback loops that exacerbate localized eutrophication [9,66].

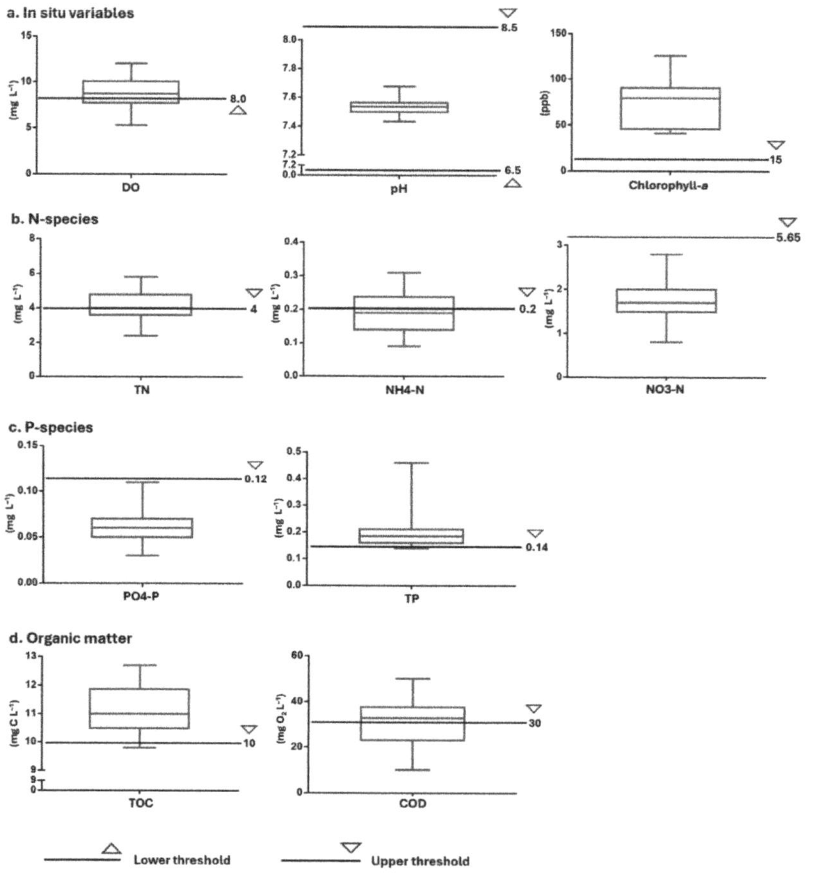

Figure 6. Comparison of water quality parameters in the Lieve River against environmental thresholds defined by WHO (2011), WFD (2000/60/EC), US EPA, and VLAREM, as outlined in Supplementary Materials Table S6.

Organic matter parameters, such as TOC and COD, also exhibit deviations from environmental standards. The median TOC value (11 mg C L^{-1}) exceeded the upper threshold of 10 mg C L^{-1}, while COD levels (median: 33 mg O_2 L^{-1}) consistently surpassed the limit of 30 mg O_2 L^{-1}. These results suggest significant organic matter accumulation, likely driven by agricultural runoff, wastewater discharges, and urban activities. Such conditions provide favorable environments for waterbirds, as the accumulation of organic material creates detritus-rich microhabitats that support microbial activity and invertebrate populations, which are critical resources for the opportunistic foraging behavior of waterbirds [74,75].

Overall, Figure 6 highlights severe deviations in TP, TN, chlorophyll-*a*, TOC, and COD from environmental standards, underscoring the combined impacts of nutrient enrichment and organic matter accumulation. These findings emphasize that agricultural runoff, untreated or insufficiently treated wastewater, and urban activities are the primary drivers of these deviations. Addressing these challenges requires targeted management strategies, including nutrient load reductions, improved wastewater treatment, and the implementation of riparian buffer zones, to restore water quality and support the ecological health of the Lieve River. These strategies are critical for balancing anthropogenic pressures with biodiversity conservation, ecosystem functionality, and sustainable development [80]. Future research should prioritize innovative monitoring technologies, such as drones and AI-assisted waterbird surveys, to strengthen conservation efforts and ensure long-term biodiversity sustainability [81].

4. Conclusions

This study provides an assessment of water quality and waterbird distribution along the Lieve River, highlighting the impacts of nutrient enrichment, organic matter accumulation, and anthropogenic activities on the river's ecological health and its role as a multifunctional resource. The main findings are as follows:

Nutrient enrichment: The Lieve River exhibits significant nutrient enrichment, with elevated levels of TN, TP, and chlorophyll-*a* serving as key drivers of eutrophication.

Organic matter accumulation and habitat dynamics: TOC and COD levels frequently exceed environmental thresholds, particularly in areas influenced by agricultural runoff, urban wastewater discharge, and other human activities.

The Prati index analysis revealed that 58.3% of the Lieve River sampling points were classified as "Polluted", with higher COD and NH_4-N levels and reduced DO_sat observed near the F24 direction. These results suggest that impacts from urban runoff, agricultural discharge, and point-source pollution influence water quality, emphasizing the need for targeted management in these areas.

The Eurasian coot was the most abundant species and showed a positive correlation with TP. The common moorhen correlated with TOC, while wild ducks showed weak correlations with TP. Total waterbird abundance was associated with chlorophyll-*a*, indicating nutrient dynamics as a key driver.

Management and monitoring recommendations: As one of Europe's earliest artificial canals, the historical and cultural significance of the Lieve River enhances its value as an ecological and recreational resource. Protecting this dual role requires sustainable management practices that address both conservation and human utilization. Key actions include reducing nutrient loads through improved wastewater treatment, promoting sustainable agricultural practices, and implementing riparian buffer zones to minimize runoff. Furthermore, a similar approach as presented in this study is valuable for investigating similar river systems in the world.

Supplementary Materials: The following supporting information can be downloaded at: https://www.mdpi.com/article/10.3390/w17040595/s1, Supplementary Materials provide a comprehensive overview of the water quality assessment conducted along the Lieve River, including detailed datasets, analysis methods, and a schematic representation of the monitoring design. The contents are as follows. Table S1: In situ parameters (Chlorophyll-*a*, DO, pH, EC) from the monitoring results of 48 transects along the Lieve River in March 2024. Table S2: Nutrient concentrations (NH_4-N, NO_3-N, NO_2-N, TN, PO_4-P, and TP) measured across 48 transects in March 2024. Table S3: Organic matter parameters, including TIC, TOC, and COD, derived from the same transects in March 2024. Table S4: Results of the Prati water quality index (Prati WQI) analysis, summarizing dissolved oxygen saturation (DO_sat), COD, NH_4-N levels, and corresponding water quality classifications for each transect in March 2024. Table S5: Summary of physicochemical water quality variables, measurement techniques, and associated standards. Table S6: Water quality standard thresholds for freshwater ecosystems. The Prati Index Methodology: A detailed explanation of the Prati index calculation, including its classification system and the transformation formulas used for key parameters (DO_sat, COD, NH_4-N). This section provides the basis for assessing pollution levels along the river. Figure S1: A schematic representation of the Lieve River monitoring design, illustrating the division of the river into 24 monitoring regions (F1–F24) and 48 transects (F1S1–F24S2) at 200 m intervals.

Author Contributions: X.L.: Conceptualization, Investigation, Writing—original draft preparation; L.H.: Conceptualization, Investigation, writing—review and editing; A.D.C.: Conceptualization, Investigation, writing—review and editing; N.D.S.: Investigation; K.P.: Investigation; D.P.-C.: Investigation; M.A.E.F.: Investigation, writing—review and editing; P.L.M.G.: Conceptualization, Investigation, writing—review and editing, resources, supervision. All authors have read and agreed to the published version of the manuscript.

Funding: This research was funded by the Special Research Fund (BOF) of Ghent University (grant number 01CD04724), BOF-BAF funding (BOF/BAF/4Y/24/1/522) and China Scholarship Council (CSC), No. 202006330023.

Data Availability Statement: Data are contained within the article.

Acknowledgments: Marie Anne Eurie Forio is supported by the EU Horizon2020 projects OPTAIN and MERLIN and Horizon Europe project OneAquaHealth. Long Ho and Kim Pham are supported by the Horizon Europe project OneAquaHealth.

Conflicts of Interest: The authors declare no conflicts of interest.

References

1. Haase, P.; Bowler, D.E.; Baker, N.J.; Bonada, N.; Domisch, S.; Marquez, J.R.G.; Heino, J.; Hering, D.; Jähnig, S.C.; Schmidt-Kloiber, A.; et al. The Recovery of European Freshwater Biodiversity Has Come to a Halt. *Nature* **2023**, *620*, 582–588. [CrossRef]
2. Sexton, A.N.; Beisel, J.N.; Staentzel, C.; Wolter, C.; Tales, E.; Belliard, J.; Buijse, A.D.; Fernández, V.M.; Wantzen, K.M.; Jähnig, S.C.; et al. Inland Navigation and Land Use Interact to Impact European Freshwater Biodiversity. *Nat. Ecol. Evol.* **2024**, *8*, 1098–1108. [CrossRef]
3. Alexander, N.; Dickinson, A.; Benson, T.J.; Ford, T.W.; Mateus-Pinilla, N.; Arneson, J.; Davis, M.A. Waterbird Disease in the United States Laurentian Great Lakes under Climate Change. *J. Great Lakes Res.* **2024**, *50*, 102450. [CrossRef]
4. Amano, T.; Székely, T.; Sandel, B.; Nagy, S.; Mundkur, T.; Langendoen, T.; Blanco, D.; Soykan, C.U.; Sutherland, W.J. Successful Conservation of Global Waterbird Populations Depends on Effective Governance. *Nature* **2017**, *553*, 199–202. [CrossRef] [PubMed]
5. Ho, L.; Barthel, M.; Panique-Casso, D.; Vermeulen, K.; Bruneel, S.; Liu, X.; Bodé, S.; Six, J.; Boeckx, P.; Goethals, P. Impact of Salinity Gradient, Water Pollution and Land Use Types on Greenhouse Gas Emissions from an Urbanized Estuary. *Environ. Pollut.* **2023**, *336*, 122500. [CrossRef]
6. Loiseau, N.; Mouquet, N.; Casajus, N.; Grenié, M.; Guéguen, M.; Maitner, B.; Mouillot, D.; Ostling, A.; Renaud, J.; Tucker, C.; et al. Global Distribution and Conservation Status of Ecologically Rare Mammal and Bird Species. *Nat. Commun.* **2020**, *11*, 5071. [CrossRef]
7. Sullivan, S.M.P.; Corra, J.W.; Hayes, J.T. Urbanization Mediates the Effects of Water Quality and Climate on a Model Aerial Insectivorous Bird. *Ecol. Monogr.* **2021**, *91*, e01442. [CrossRef]
8. Sibilia, C.D.; Aguirre-Gutiérrez, J.; Mowbray, L.; Malhi, Y. Effects of Submerged Aquatic Vegetation and Water Quality on Waterfowl Abundance by Foraging Guild. *Ecol. Solut. Evid.* **2022**, *3*, e12137. [CrossRef]
9. Tóth, P.; Tarcsay, B.L.; Kovács, Z.; Ionescu, D.T.; Németh, S.; Domokos, E. Assessment of the Correlation between the Nutrient Load from Migratory Bird Excrement and Water Quality by Principal Component Analysis in a Freshwater Habitat. *Environ. Sci. Pollut. Res.* **2023**, *30*, 66033–66049. [CrossRef] [PubMed]

10. Wurtsbaugh, W.A.; Miller, C.; Null, S.E.; Rose, R.J.D.; Wilcock, P.; Hahnenberger, M.; Howe, F.; Moore, J. Decline of the World's Saline Lakes. *Nat. Geosci.* **2017**, *10*, 816–821. [CrossRef]
11. Strubbe, D.; Jiménez, L.; Barbosa, A.M.; Davis, A.J.S.; Lens, L.; Rahbek, C. Mechanistic Models Project Bird Invasions with Accuracy. *Nat. Commun.* **2023**, *14*, 2520. [CrossRef] [PubMed]
12. Gelaude, F. De Lieve, de Realisatie van Een Opmerkelijk Kanaal in de 13e Eeuw. Een Illustratie van Stedelijke Machtsontplooiing Door Gent. *Handel. Maatsch. Voor Geschied. Oudheidkd. Gent* **2023**, *73*, 7–35. [CrossRef]
13. Laguna, C.; López-Perea, J.J.; Viñuela, J.; Florín, M.; Feliu, J.; Chicote, Á.; Cirujano, S.; Mateo, R. Effects of Invasive Fish and Quality of Water and Sediment on Macrophytes Biomass, and Their Consequences for the Waterbird Community of a Mediterranean Floodplain. *Sci. Total Environ.* **2016**, *551–552*, 513–521. [CrossRef] [PubMed]
14. Merckx, T.; Souffreau, C.; Kaiser, A.; Baardsen, L.F.; Backeljau, T.; Bonte, D.; Brans, K.I.; Cours, M.; Dahirel, M.; Debortoli, N.; et al. Body-Size Shifts in Aquatic and Terrestrial Urban Communities. *Nature* **2018**, *558*, 113–116. [CrossRef] [PubMed]
15. Sawakuchi, A.O.; Schultz, E.D.; Pupim, F.N.; Bertassoli, D.J.; Souza, D.F.; Cunha, D.F.; Mazoca, C.E.; Ferreira, M.P.; Grohmann, C.H.; Wahnfried, I.D.; et al. Rainfall and Sea Level Drove the Expansion of Seasonally Flooded Habitats and Associated Bird Populations across Amazonia. *Nat. Commun.* **2022**, *13*, 4945. [CrossRef]
16. Lens, L.; Dongen, S.V.; Norris, K.; Githiru, M.; Matthysen, E. Avian Persistence in Fragmented Rainforest. *Science* **2002**, *298*, 1236–1238. [CrossRef] [PubMed]
17. Mitchell, M.E.; Newcomer-Johnson, T.; Christensen, J.; Crumpton, W.; Richmond, S.; Dyson, B.; Canfield, T.J.; Helmers, M.; Lemke, D.; Lechtenberg, M.; et al. Potential of Water Quality Wetlands to Mitigate Habitat Losses from Agricultural Drainage Modernization. *Sci. Total Environ.* **2022**, *838*, 156358. [CrossRef]
18. Mikula, P.; Tomášek, O.; Romportl, D.; Aikins, T.K.; Avendaño, J.E.; Braimoh-Azaki, B.D.A.; Chaskda, A.; Cresswell, W.; Cunningham, S.J.; Dale, S.; et al. Bird Tolerance to Humans in Open Tropical Ecosystems. *Nat. Commun.* **2023**, *14*, 2146. [CrossRef] [PubMed]
19. Chyb, A.; Jedlikowski, J.; Włodarczyk, R.; Minias, P. Consistent Choice of Landscape Urbanization Level across the Annual Cycle in a Migratory Waterbird Species. *Sci. Rep.* **2021**, *11*, 836. [CrossRef]
20. Sorace, A.; Formichetti, P.; Boano, A.; Andreani, P.; Gramegna, C.; Mancini, L. The Presence of a River Bird, the Dipper, in Relation to Water Quality and Biotic Indices in Central Italy. *Environ. Pollut.* **2002**, *118*, 89–96. [CrossRef]
21. Xie, S.; Wang, X.; Ren, Y.; Su, Z.; Su, Y.; Wang, S.; Zhou, W.; Lu, F.; Qian, Y.; Gong, C.; et al. Factors Responsible for Forest and Water Bird Distributions in Rivers and Lakes along an Urban Gradient in Beijing. *Sci. Total Environ.* **2020**, *735*, 139308. [CrossRef]
22. Acuña, M.P.; Vukasovic, M.A.; Hernández, H.J.; Acuña, T.A.; Estades, C.F. Effects of the Surrounding Landscape on Waterbird Populations in Estuarine Ecosystems of Central Chile. *Wetl. Ecol. Manag.* **2019**, *27*, 295–310. [CrossRef]
23. Wu, H.; Dai, J.; Sun, S.; Du, C.; Long, Y.; Chen, H.; Yu, G.; Ye, S.; Chen, J. Responses of Habitat Suitability for Migratory Birds to Increased Water Level during Middle of Dry Season in the Two Largest Freshwater Lake Wetlands of China. *Ecol. Indic.* **2021**, *121*, 107065. [CrossRef]
24. Vallecillo, S.; Maes, J.; Polce, C.; Lavalle, C. A Habitat Quality Indicator for Common Birds in Europe Based on Species Distribution Models. *Ecol. Indic.* **2016**, *69*, 488–499. [CrossRef]
25. de Smet Willem, H.O.; Frans, M.J.C.E. A Hydrobiological Study of the Polluted River Lieve (Ghent, Belgium). *Hydrobiologia* **1972**, *39*, 91–154. [CrossRef]
26. Figarski, T.; Kajtoch, Ł. Alterations of Riverine Ecosystems Adversely Affect Bird Assemblages. *Hydrobiologia* **2015**, *744*, 287–296. [CrossRef]
27. Dulsat-Masvidal, M.; Ciudad, C.; Infante, O.; Mateo, R.; Lacorte, S. Water Pollution Threats in Important Bird and Biodiversity Areas from Spain. *J. Hazard. Mater.* **2023**, *448*, 130938. [CrossRef]
28. Gregory, R.D.; van Strien, A.; Vorisek, P.; Gmelig Meyling, A.W.; Noble, D.G.; Foppen, R.P.B.; Gibbons, D.W. Developing Indicators for European Birds. *Philos. Trans. R. Soc. B Biol. Sci.* **2005**, *360*, 269–288. [CrossRef] [PubMed]
29. Horne, A.; Beutel, M.; Woodside, G. The Prado Wetlands: 24 Years of River Diversion Treatment Wetlands Reducing Nitrate and Phosphate While Increasing an Endangered Bird. *Ecol. Eng.* **2024**, *200*, 107174. [CrossRef]
30. Rigal, S.; Dakos, V.; Alonso, H.; Auniņš, A.; Benkő, Z.; Brotons, L.; Chodkiewicz, T.; Chylarecki, P.; de Carli, E.; del Moral, J.C.; et al. Farmland Practices Are Driving Bird Population Decline across Europe. *Proc. Natl. Acad. Sci. USA* **2023**, *120*, e2216573120. [CrossRef]
31. Xu, Q.; Zhou, L.; Xia, S.; Zhou, J. Impact of Urbanisation Intensity on Bird Diversity in River Wetlands around Chaohu Lake, China. *Animals* **2022**, *12*, 473. [CrossRef] [PubMed]
32. Tucker, C.J.; Grant, D.M.; Dykstra, J.D. NASA's Global Orthorectified Landsat Data Set. *Photogramm. Eng. Remote Sens.* **2004**, *70*, 313–322. [CrossRef]
33. Boussauw, K. City Profile: Ghent, Belgium. *Cities* **2014**, *40*, 32–43. [CrossRef]
34. Boets, P.; Lock, K.; Goethals, P.L.M. Using Long-Term Monitoring to Investigate the Changes in Species Composition in the Harbour of Ghent (Belgium). *Hydrobiologia* **2011**, *663*, 155–166. [CrossRef]

35. Luo, K.; Wu, Z.; Bai, H.; Wang, Z. Bird Diversity and Waterbird Habitat Preferences in Relation to Wetland Restoration at Dianchi Lake, South-West China. *Avian Res.* **2019**, *10*, 21. [CrossRef]
36. Harmon, K.C.; Winter, K.B.; Kurashima, N.; Fletcher, C.H.; Kane, H.H.; Price, M.R. The Role of Indigenous Practices in Expanding Waterbird Habitat in the Face of Rising Seas. *Anthropocene* **2021**, *34*, 100293. [CrossRef]
37. Campisano, R.; Hall, K.; Griggs, J.; Willison, S.; Reimer, S.; Mash, H.; Magnuson, M.; Boczek, L.; Rhodes, E. *Selected Analytical Methods for Environmental Remediation and Recovery (SAM) 2017*; US Environmental Protection Agency: Washington, DC, USA, 2017.
38. Rice, E.W.; Bridgewater, L.; Association, A.P.H. *Standard Methods for the Examination of Water and Wastewater*; American Public Health Association: Washington, DC, USA, 2012; Volume 10.
39. Bassuoni, N.F.; Abumandour, M.M.A.; Morsy, K.; Hanafy, B.G. Ultrastructural Adaptation of the Oropharyngeal Cavity of the Eurasian Common Moorhen (Gallinula Chloropus Chloropus): Specific Adaptive Dietary Implications. *Microsc. Res. Tech.* **2022**, *85*, 1915–1925. [CrossRef]
40. Halupka, L.; Czyż, B.; Macias Dominguez, C.M. The Effect of Climate Change on Laying Dates, Clutch Size and Productivity of Eurasian Coots Fulica Atra. *Int. J. Biometeorol.* **2020**, *64*, 1857–1863. [CrossRef] [PubMed]
41. Pławińska-Czarnak, J.; Wódz, K.; Piechowicz, L.; Tokarska-Pietrzak, E.; Bełkot, Z.; Bogdan, J.; Wiśniewski, J.; Kwieciński, P.; Kwieciński, A.; Anusz, K. Wild Duck (Anas Platyrhynchos) as a Source of Antibiotic-Resistant Salmonella Enterica Subsp. Diarizonae O58—The First Report in Poland. *Antibiotics* **2022**, *11*, 530. [CrossRef]
42. Barr, J.R.; Green, M.C.; DeMaso, S.J.; Hardy, T.B. Drone Surveys Do Not Increase Colony-Wide Flight Behaviour at Waterbird Nesting Sites, But Sensitivity Varies Among Species. *Sci. Rep.* **2020**, *10*, 3781. [CrossRef] [PubMed]
43. Strang, V. Listening to the River: Representing Non-Human Needs and Interests in Debates about Water Governance and Management. *River Res. Appl.* **2024**, *40*, 1698–1709. [CrossRef]
44. He, Z.; Liu, D.; Shi, Y.; Wu, X.; Dai, Y.; Shang, Y.; Peng, J.; Cui, Z. Broader Environmental Adaptation of Rare Rather than Abundant Bacteria in Reforestation Succession Soil. *Sci. Total Environ.* **2022**, *828*, 154364. [CrossRef]
45. Somers, K.M.; Jackson, D.A. Putting the Mantel Test Back Together Again. *Ecology* **2022**, *103*, e3780. [CrossRef]
46. Saccenti, E.; Hendriks, M.H.W.B.; Smilde, A.K. Corruption of the Pearson Correlation Coefficient by Measurement Error and Its Estimation, Bias, and Correction under Different Error Models. *Sci. Rep.* **2020**, *10*, 438. [CrossRef] [PubMed]
47. Edelmann, D.; Móri, T.F.; Székely, G.J. On Relationships between the Pearson and the Distance Correlation Coefficients. *Stat. Probab. Lett.* **2021**, *169*, 108960. [CrossRef]
48. Prati, L.; Pavanello, R.; Pesarin, F. Assessment of Surface Water Quality by a Single Index of Pollution. *Water Res.* **1971**, *5*, 741–751. [CrossRef]
49. Şener, Ş.; Şener, E.; Davraz, A. Evaluation of Water Quality Using Water Quality Index (WQI) Method and GIS in Aksu River (SW-Turkey). *Sci. Total Environ.* **2017**, *584–585*, 131–144. [CrossRef] [PubMed]
50. Ho, L.; Jerves-Cobo, R.; Barthel, M.; Six, J.; Bode, S.; Boeckx, P.; Goethals, P. Greenhouse Gas Dynamics in an Urbanized River System: Influence of Water Quality and Land Use. *Environ. Sci. Pollut. Res.* **2022**, *29*, 37277–37290. [CrossRef]
51. Sutadian, A.D.; Muttil, N.; Yilmaz, A.G.; Perera, B.J.C. Development of River Water Quality Indices—A Review. *Environ. Monit. Assess.* **2015**, *188*, 58. [CrossRef] [PubMed]
52. Zotou, I.; Tsihrintzis, V.A.; Gikas, G.D. Performance of Seven Water Quality Indices (WQIs) in a Mediterranean River. *Environ. Monit. Assess.* **2019**, *191*, 505. [CrossRef]
53. Parry, R. Agricultural Phosphorus and Water Quality: A U.S. Environmental Protection Agency Perspective. *J. Environ. Qual.* **1998**, *27*, 258–261. [CrossRef]
54. Edition, F. Guidelines for Drinking-Water Quality. *WHO Chron.* **2011**, *38*, 104–108.
55. Lepper, P. *Manual on the Methodological Framework to Derive Environmental Quality Standards for Priority Substances in Accordance with Article 16 of the Water Framework Directive (2000/60/EC)*; Fraunhofer-Institute Molecular Biology and Applied Ecology: Schmallenberg, Germany, 2005.
56. Flemish Government. Title II of VLAREM-Appendix 2.3.1. Basic Environmental Quality Standards for Surface Water. Available online: https://navigator.emis.vito.be/pdfServlet?woId=10071&woLang=nl&version=2023-11-15&lang=nl (accessed on 4 January 2025).
57. Williams, W.D. Anthropogenic Salinisation of Inland Waters. *Hydrobiologia* **2001**, *466*, 329–337. [CrossRef]
58. Kim, H.G.; Hong, S.; Chon, T.-S.; Joo, G.-J. Spatial Patterning of Chlorophyll *a* and Water-Quality Measurements for Determining Environmental Thresholds for Local Eutrophication in the Nakdong River Basin. *Environ. Pollut.* **2021**, *268*, 115701. [CrossRef]
59. Ma, Z.; Cai, Y.; Li, B.; Chen, J. Managing Wetland Habitats for Waterbirds: An International Perspective. *Wetlands* **2010**, *30*, 15–27. [CrossRef]
60. Dittrich, J.; Dias, J.D.; de Paula, A.C.M.; Padial, A.A. Experimental Nutrient Enrichment Increases Plankton Taxonomic and Functional Richness and Promotes Species Dominance Overtime. *Hydrobiologia* **2023**, *850*, 4029–4048. [CrossRef]

61. Van Onsem, S.; Triest, L. Turbidity, Waterfowl Herbivory, and Propagule Banks Shape Submerged Aquatic Vegetation in Ponds. *Front. Plant Sci.* **2018**, *9*, 1514. [CrossRef] [PubMed]
62. Juliet, C. Common Moorhen and Common Gallinule Behavior and Habitat. In *The Worlds Rarest Birds*; Princeton University Press: Princeton, NJ, USA, 2023.
63. Chaichana, R. *Birds and the Eutrophication of a System of Small Lakes*; The University of Liverpool: Liverpool, UK, 2008; ISBN 979-8-6415-8947-3.
64. Holopainen, S.; Arzel, C.; Dessborn, L.; Elmberg, J.; Gunnarsson, G.; Nummi, P.; Pöysä, H.; Sjöberg, K. Habitat Use in Ducks Breeding in Boreal Freshwater Wetlands: A Review. *Eur. J. Wildl. Res.* **2015**, *61*, 339–363. [CrossRef]
65. Holopainen, S.; Miettinen, E.; Väänänen, V.-M.; Nummi, P.; Pöysä, H. Balancing between Predation Risk and Food by Boreal Breeding Ducks. *Ecol. Evol.* **2024**, *14*, e11011. [CrossRef] [PubMed]
66. Scherer, N.M.; Gibbons, H.L.; Stoops, K.B.; Muller, M. Phosphorus Loading of an Urban Lake by Bird Droppings. *Lake Reserv. Manag.* **1995**, *11*, 317–327. [CrossRef]
67. Clausen, P.; Lauridsen, T.L.; Pedersen, C.L.; Nielsen, H.H.; Jeppesen, E.; Søndergaard, M.; Schreven, K.H.T.; Nolet, B.A.; Madsen, J.; Fox, A.D. Are Increasing Roosting Waterbird Numbers Responsible for Eutrophication of Shallow Lakes? Examples from a Danish Ramsar Site. *Hydrobiologia* **2025**, *852*, 389–410. [CrossRef]
68. Wallau, G.L.; Della-Flora, F.; Bueno, A.S.; Corso, J.; Ortiz, M.F.; Cáceres, N.C. Behaviour of the Common Moorhen in Rio Grande Do Sul, Brazil. *Acta Ethologica* **2010**, *13*, 127–135. [CrossRef]
69. Lange, G.; Haynert, K.; Dinter, T.; Scheu, S.; Kröncke, I. Adaptation of Benthic Invertebrates to Food Sources along Marine-Terrestrial Boundaries as Indicated by Carbon and Nitrogen Stable Isotopes. *J. Sea Res.* **2018**, *131*, 12–21. [CrossRef]
70. Foxi, C.; Delrio, G. Larval Habitats and Seasonal Abundance of Culicoides Biting Midges Found in Association with Sheep in Northern Sardinia, Italy. *Med. Vet. Entomol.* **2010**, *24*, 199–209. [CrossRef] [PubMed]
71. Hamza, F.; Selmi, S. Habitat Features and Human Presence as Predictors of the Abundance of Shorebirds and Wading Birds Wintering in the Gulf of Gabès, Tunisia. *Mar. Ecol. Prog. Ser.* **2015**, *540*, 251–258. [CrossRef]
72. Lai, C.-H.; Lin, S.-H.; Tsai, C.-Y.; Chen, S.-H. Identifying Farm Pond Habitat Suitability for the Common Moorhen (Gallinula Chloropus): A Conservation-Perspective Approach. *Sustainability* **2018**, *10*, 1352. [CrossRef]
73. Soorae, P.; Sakkir, S.; Saji, A.; Khan, S.B.; Zaabi, R.A.; Shah, J.N.; Ali, A.; Al Omari, K.; Al Dhaheri, A.; Javed, S.; et al. A Review of the Flora and Fauna in the Al Wathba Wetland Reserve in Abu Dhabi, United Arab Emirates. *Wetlands* **2020**, *40*, 1505–1512. [CrossRef]
74. Dessborn, L.; Brochet, A.L.; Elmberg, J.; Legagneux, P.; Gauthier-Clerc, M.; Guillemain, M. Geographical and Temporal Patterns in the Diet of Pintail Anas Acuta, Wigeon Anas Penelope, Mallard Anas Platyrhynchos and Teal Anas Crecca in the Western Palearctic. *Eur. J. Wildl. Res.* **2011**, *57*, 1119–1129. [CrossRef]
75. Adjovu, G.E.; Stephen, H.; James, D.; Ahmad, S. Measurement of Total Dissolved Solids and Total Suspended Solids in Water Systems: A Review of the Issues, Conventional, and Remote Sensing Techniques. *Remote Sens.* **2023**, *15*, 3534. [CrossRef]
76. Sarnelle, O. Nutrient Enrichment and Grazer Effects on Phytoplankton in Lakes. *Ecology* **1992**, *73*, 551–560. [CrossRef]
77. Fox, A.D.; Jørgensen, H.E.; Jeppesen, E.; Lauridsen, T.L.; Søndergaard, M.; Fugl, K.; Myssen, P.P.; Balsby, T.J.S.; Clausen, P.; Musil, P.; et al. Breeding Waterbird Species as Ecological Indicators of Shifts from Turbid to Clear Water Conditions in Northwest European Shallow Eutrophic Lakes. *Hydrobiologia* **2025**, *852*, 411–423. [CrossRef]
78. Keller, V.; Korner-Nievergelt, P. Effect of Trophic Status of a Deep-Water Lake on Breeding Great Crested Grebes Podiceps Cristatus during a Phase of Recovery from Eutrophication: A Long-Term Study. *Bird. Study* **2019**, *66*, 1–10. [CrossRef]
79. Mishra, A.P.; Kumar, S.; Patra, R.; Kumar, A.; Sahu, H.; Chandra, N.; Pande, C.B.; Alshehri, F. Physicochemical Parameters of Water and Its Implications on Avifauna and Habitat Quality. *Sustainability* **2023**, *15*, 9494. [CrossRef]
80. Forio, M.A.E.; Goethals, P.L.M. An Integrated Approach of Multi-Community Monitoring and Assessment of Aquatic Ecosystems to Support Sustainable Development. *Sustainability* **2020**, *12*, 5603. [CrossRef]
81. Liu, X.; Ho, L.; Bruneel, S.; Goethals, P. Applications of Unmanned Vehicle Systems for Multi-Spatial Scale Monitoring and Management of Aquatic Ecosystems: A Review. *Ecol. Inform.* **2025**, *85*, 102926. [CrossRef]

water

MDPI

Article

Untangling the Characteristics and Ecological Processes of Microbial Community Assembly in the Source Area of the East Route of the South-to-North Water Diversion Project in China Under Different Water Periods

Wei Cai [1,2,*], Xin Wen [1], Yueru Zhao [1], Xiusen Wu [1], Haoran Zheng [1], Jiangtao Chen [1], Zhengyang Hu [1], Qin Zhong [3] and Jun Wu [1,*]

[1] College of Environmental Science and Engineering, Yangzhou University, Huayang West Road #196, Yangzhou 225009, China

[2] National Key Laboratory of Water Disaster Prevention, Nanjing Hyraulic Research Institute, Guangzhou Road #223, Nanjing 210029, China

[3] Dongzhu Ecological Environmental Protection Co., Ltd., Xihu Middle Road #90, Wuxi 214000, China

* Correspondence: caiwei@yzu.edu.cn (W.C.); j.wu@yzu.edu.cn (J.W.)

Abstract: This study presented a comprehensive analysis of the microbial ecology in water diversion rivers (WDRs) in the source area of the East Route of the South-to-North Water Diversion Project (ER-SNWDP) in China across various water periods. *Proteobacteria*, *Chloroflexi*, *Acidobacteriota*, and *Bacteroidota* were identified as the dominant microbial phyla in river sediment. During the wet period, microbial communities exhibited the highest richness, biodiversity, and the most intense antagonistic relationships compared to those in the dry and normal water periods. Generally, the microbial network predominantly existed in symbiotic models characterized by mutual benefit and symbiosis throughout all periods. During the dry period, the microbial co-occurrence network was found to be the most complex, with microbial OTUs showing the closest interconnections. The dominant mechanisms governing community diversity, succession, and biogeography were spatial turnover of species and stochastic processes. A more pronounced impact of stochastic processes on microbial community assemblages was observed during normal or wet periods than the dry period. Functional prediction of metabolic pathways indicated that the main ecological functions of microbial communities encompassed carbohydrate metabolism, amino acid metabolism, energy metabolism, etc. This study could provide essential scientific data for ecological regulation, ecological protection, and water resources management in WDRs.

Keywords: biodiversity; microbial network; *Proteobacteria*; stochastic processes; metabolic pathways

1. Introduction

Adequate and high-quality water resources are essential for the health and sustainable development of human societies and ecosystems [1]. However, factors such as rapid population growth, climate change, water contamination, urbanization, and industrial development are exacerbating water shortages [2]. To meet the growing demands of domestic, agricultural, and industrial water use, both developed and developing countries have implemented numerous water diversion projects (WDPs) [3]. WDPs refer to engineering works that involve the construction or modification of hydraulic facilities to divert water from one source to another area or region. Currently, there are over 160 large-scale, long-distance WDPs worldwide [4]. These WDPs have played a crucial role in alleviating the uneven distribution of water resources [5]. The North-to-South WDP in California in the United States has promoted the economic development of California; Egypt's West-East WDP has provided valuable water resources to the Sinai Peninsula; the South-to-North WDP in China has promoted the coordinated development of the local economy, society, population, resources, and environment [3]. Water diversion rivers (WDRs) refer to rivers that are subject to water resource allocation and management through WDPs. These rivers, serving as the primary arteries for water transportation, often undergo frequent environmental changes due to the operation of WDPs. There is increasing concern about their impact on water ecology in WDRs with the construction of WDPs. Therefore, it is essential to understand the effects of water diversion on the ecosystems of water diversion rivers for the long-term implementation of WDPs.

Microorganisms play a vital role in WDR ecosystems by contributing to various functions and exhibiting diverse characteristics [6]. River sediments, which serve as a major source of pollutants in water bodies and as sites for microbial aggregation, are essential components of water ecosystems [7]. Microorganisms in sediments make important contributions to river ecosystems, particularly in biogeochemical cycles, nutrient cycles, and energy flows [8]. Furthermore, sediment environments harbor a rich diversity of microbial species, intense metabolic activity, and show high responsiveness to environmental changes [9]. The operation of WDPs can result in complex ecological and hydrological changes in WDRs [10]. The consequent changes in environmental parameters such as pH, temperature (T), dissolved oxygen (DO), and nutrient concentration could have a profound effect on the microbial communities in the river ecosystem [11]. Luo et al. (2019) pointed out that there was spatial variability in the planktonic bacterial community in the main channel of the South-to-North WDP, driven by DO, pH, and T [12]. Liu et al. (2023) found that pH, Total Phosphorus (TP), and Nitrate Nitrogen (NO_3-N) were the main factors influencing the bacterial community in river sediments of the Li River Basin [13]. Environmental changes have the potential to increase microbial community heterogeneity in sediments, facilitate the dispersal of pathogenic bacteria, and disrupt the original interspecific relationships within microbial communities [14]. Therefore, understanding the ecological characteristics of microbial communities in river sediments of WDRs under changing environments is essential for effective water ecological management in these regions.

However, most studies on the effects of WDPs on water ecology have focused on water quality, fish, and plants; until recent years, more attention has been paid to the effects on microorganisms [6]. Qu et al. (2018) investigated the dominant microbes in the water body of Miyun Reservoir in China and observed an increase in the relative abundance of *Proteobacteria* and *Verrucomicrobiota* under the influence of the South-to-North WDP, leading to intensified interspecies competition for ecological niches [15]. Yao et al. (2019) found differences in bioavailability and microbial diversity in the area affected by the South-to-North WDP in Hongze Lake compared with other sites [16]. Liu et al. (2022) determined that the spatial heterogeneity of bacterial communities in Dongping Lake decreased during the south-to-north water diversion period, potentially increasing the risk of biological homogenization between rivers [17]. However, research on microbial communities in watersheds affected by WDPs has mainly focused on receiving lakes in the area and regulating lakes along the route, with limited attention given to microbial communities in WDRs. Lv et al. (2021) found that under the influence of the Yellow River diversion in Shanxi Province in China, the dominant position of the core bacterial community in the sediment of the Fen River decreased, and the abundance of *Comamonadaceae* and *Hydrogenophaga* decreased [14]. WDRs in the water source area play a critical role in water quality regulation during water transportation, directly impacting the safety of drinking water for millions of people along the diversion route. The ecological and physicochemical status of WDRs change in the normal, dry, and wet periods affected by WDPs. Therefore, it is necessary to study the microbial communities in WDRs in the source area of WDPs under different water conditions. However, research on microbial communities in WDRs in the water source area of WDPs is still lacking.

This study investigated the microbial communities in river sediments within the source area of the East Route of the South-to-North Water Diversion Project (ER-SNWDP) in China. The objectives of this study were (1) to analyze the composition, distribution, and diversity patterns of microbial communities in WDRs in the source area of ER-SNWDP under the normal, dry, and wet periods with varying hydrologic conditions; (2) to investigate the complexity and interactions of microbial networks across different water periods; (3) to figure out the metabolic potential and ecological functions of microbial communities in the WDR ecosystem; (4) to reveal the dominant mechanisms governing community diversity, succession, and biogeography in WDRs. This research could provide essential scientific data for ecological regulation, environmental protection, and water resources management in WDRs.

2. Materials and Methods

2.1. Study Area

The study area located in the water source area of ER-SNWDP is shown in Figure 1. It is also a transition zone between a subtropical humid monsoon climate and a temperate monsoon climate, characterized by four distinct periods, ample sunlight, and abundant rainfall. Since 2013, ER-SNWDP has officially transferred water from the Yangtze River along hundreds of water diversion rivers to the water deficit zone in northern China. Rivers in this area are subject to significant hydrological fluctuations due to the impacts of frequent water diversion activities and watershed precipitation changes. They are undertaking complex tasks. In addition to serving as water diversion rivers for ER-SNWDP, rivers in this area also serve as the water conveyance rivers from the Huai River to the Yangtze River in normal water periods and as the flood discharge channels in flood periods. Therefore, the hydrological conditions of these rivers vary in complexity throughout the year, with frequent changes in water volume, flow rate, even flow directions, etc.

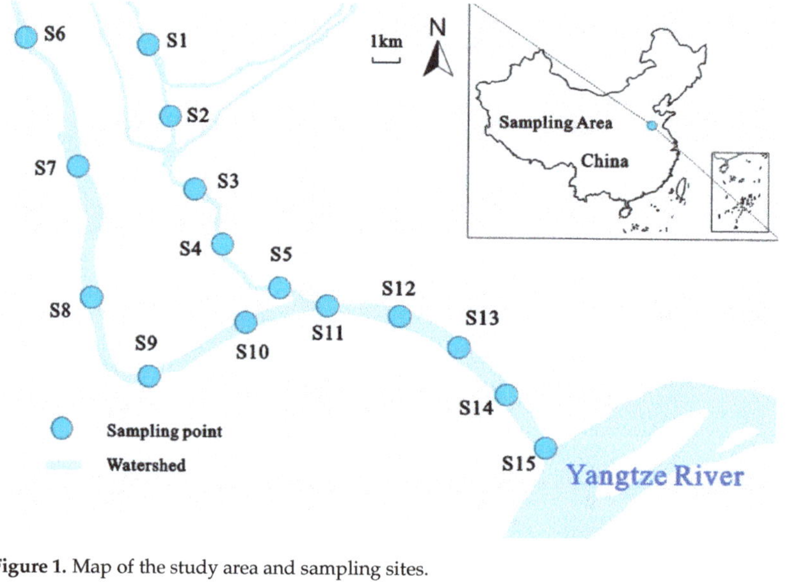

Figure 1. Map of the study area and sampling sites.

2.2. Sample Collection

According to the field investigation and survey, river sediment samples were taken from 15 sampling sites in three water diversion river channels in the source area of ER-SNWDP in China. Detailed location data are shown in Table S1 (Supporting Information). In the field of hydrology, the division of wet season, normal season, and dry season is determined based on the changes in water levels of rivers, lakes, and other water bodies. Generally speaking, the wet season refers to the period when the water level is high and the water volume is large, usually during the rainy season or the melting season of ice and snow (June to August). The normal water period refers to a time when the water level is moderate and the water volume is moderate, usually during seasons with relatively even rainfall (March to May; September to November). The dry season refers to the period when the water level is low and the water volume is small, usually during the dry or low rainfall season (December to March of the following year). The specific classification criteria may vary depending on factors such as region, climate, and hydrology. Therefore, sampling activities were conducted in three distinct water periods: January 2021 (the dry period), April 2022 (the normal period), and August 2022 (the wet period). Samples collected during these periods were marked as A1–A15, B1–B15, and C1–C15, respectively. Grab-type mud collectors were employed to collect river sediment samples, with each sample point being sampled in triplicate. The collected sediments in the same site were extracted from a depth of 0–20 cm, packed, and mixed in pre-sterilized plastic sealed bags. All sampling equipment used was pre-sterilized to prevent any external contamination. Furthermore, the samples were processed under controlled, contamination-free conditions. Samples were then kept in ice, transported to the laboratory immediately, and stored at −20 °C. Subsequently, the sediment samples were processed in the laboratory for total DNA extraction and subsequent analysis.

2.3. DNA Extraction, PCR Amplification, and Illumina MiSeq Sequencing

To reduce sampling errors, each sediment sample was thoroughly homogenized before subsampling and weighing. Biomass was extracted using a vacuum filtration method, followed by genomic DNA extraction with the FastDNA® Spin Kit (MP Biomedicals, Santa Ana, CA, USA) for Soil according to the manufacturer's instructions, with three replicates per sample. Microbial community composition was analyzed using Illumina MiSeq Sequencing (San Diego, CA, USA) with 16S rRNA. PCR amplification was performed using the bacterial 16S rRNA gene universal primers 341F (5′-CCTACGGGNGGCGWGCAG-3′) and 806R (5′-GGACTACNVGGGTWTCTAAT-3′). The amplification region of the 16S rRNA gene was the V3-V4 hypervariable region. Polymerase chain reaction (PCR) amplifications were carried out in triplicate for each sediment sample using a 20 μL volume. The reaction mixture included 4 μL of 5×FastPfu buffer, 2 μL of 2.5 mM dNTPs, 0.4 μL of each primer (5 μM), 0.8 μL of DNA template, and 0.4 μL of FastPfu polymerase. The amplification protocol was initiated at 94 °C for 5 min, followed by a series of cycles at 94 °C for 30 s (denaturation), 53 °C for 30 s (annealing), and 72 °C for 30 s (extension), culminating in a final extension at 72 °C for 8 min over 30 cycles. The purification of PCR amplicons was performed using AMPure beads to remove unused primers, which were then used to construct sequencing libraries quantitated using the Qubit 2.0 Fluorometer from Thermo Scientific (Waltham, MA, USA). The amplified products were placed on ice and immediately sent to Shanghai Biozeron Biotechnology Co., Ltd. (Shanghai, China) for sequencing. The bioinformatics analysis followed the previously described procedure [18]. DNA sequencing data are available at NCBI GenBank database under accession PRJNA1057511, PRJNA1057626, and PRJNA1058100.

2.4. Statistical Analysis

Alpha diversity, which encompasses richness and evenness, was evaluated using several indicators, including ACE, Chao1, Simpson, Shannon, and Pielou evenness indices. All data analysis was conducted using R software (version 4.5.0, http://www.r-project.org) accessed on September 2022. Differences in the structure and composition of microbial communities among samples were visualized through principal coordinate analysis (PCoA) and non-metric multidimensional scaling (NMDS) based on the distance matrix of normalized operational taxonomic units (OTUs). Periodical variation in microbial community composition was compared using histograms. For functional prediction analysis, the PICRUSt software was employed to predict sediment bacterial functions, which were subsequently compared with the Kyoto Encyclopedia of Genes and Genomes (KEGG) database to obtain functional prediction information. Detailed analysis steps were based on an online analysis platform (http://picrust.github.io/picrust/) accessed on September 2022.

3. Results

3.1. Species Richness, Evenness, and Diversity

A total of 1,997,209 sequence reads were obtained from 45 samples through 16S rRNA gene sequencing. Samples collected in the wet period showed the largest OTU numbers, with 73,794, followed by 68,033 OTUs in the normal period and 64,877 OTUs in the dry period, respectively (Table S2). These bacterial OTUs captured in the current sequencing depth were deemed representative of the microbial communities in all samples. The microbial alpha diversity showed different characteristics in different water periods. During the dry period, the Chao1, ACE, Shannon, and Simpson indices ranged from 2868.864 to 7625.035, 3012.299 to 7965.199, 6.395 to 7.894, and 0.9918 to 0.9992, respectively. During the normal period, the Chao1 index ranged from 6215.653 to 8371.831, the ACE index ranged from 6495.969 to 8616.480, the Shannon index ranged from 7.069 to 7.688, and the Simpson index ranged from 0.9970 to 0.9986. When at the wet period, Chao1, ACE, Shannon, and Simpson indices were from 6395.259 to 9644.657, 6698.417 to 9974.901, 6.978 to 7.849, and 0.9929 to 0.9989, respectively. These alpha diversity indexes exhibited the largest fluctuations in the dry period. The average values of Chao1 (8097.221) and ACE indexes (8445.687) were the highest in the wet period, indicating the richest ecological community in the wet period. The wet period also exhibited the highest average values for the Shannon index (7.585), Simpson index (0.998), and Pielou evenness index (0.893), suggesting the highest species biodiversity and evenness during this period.

3.2. Microbial Community Composition

Among the 45 samples, a total of 74 phyla and 1717 genera were identified and classified. The composition and distribution of the microbial communities at the phylum and genus level were characterized in Figure 2. Taxa with an average relative abundance of less than 2% were classified into the Remaining Group. Generally, *Proteobacteria*, *Chloroflexi*, *Acidobacteriota*, *Bacteroidota*, and *Nitrospirota* were the most dominant phylum, accounting for 29.95%, 10.05%, 8.29%, 7.67%, and 7.09% of the total sequences, respectively. In the dry period, the dominant phyla were *Proteobacteria* (28.12%), *Chloroflexi* (11.75%), *Bacteroidota* (8.92%), *Acidobacteriota* (8.81%), and *Desulfobacteriota* (6.22%). In the normal period, the dominant phyla were *Proteobacteria* (38.5%), *Desulfobacteriota* (7.15%), *Nitrospirota* (7.07%), *Acidobacteriota* (6.94%), and *Bacteroidota* (6.21%). In the wet period, the dominant phyla were *Proteobacteria* (23.24%), *Chloroflexi* (14.64%), *Acidobacteriota* (9.12%), *Nitrospirota* (8.63%), and *Bacteroidota* (7.88%). *Proteobacteria* was the dominant phylum throughout the three periods, with its relative abundance the highest in the normal period. *Chloroflexi* was obviously lower in the normal period than in the other periods. From the dry period to the wet period, the abundance of *Cyanobacteria* decreased, whereas that of *Nitrospirota* increased.

At the genus level, *Thermodesulfovibrionia, Anaerolineaceae, Steroidobacteraceae, Sva0485,* and *MBNT15* were the most abundant genera, accounting for 4.72%, 2.64%, 2.26%, 2.23%, and 2.07% of the total sequences, respectively (Figure 3). During the dry period, the dominant genera were *Thermodesulfovibrionia* (3.53%), *Anaerolineaceae* (3.32%), and *Chloroplast* (2.30%). In the normal period, the most abundant bacterial genera were *Thermodesulfovibrionia* (4.33%), *Steroidobacteraceae* (2.86%), and *Sva0485* (2.38%). In the wet period, *Thermodesulfovibrionia* also retained its status as the most abundant genus (6.29%), followed by *Anaerolineaceae* (4.02%) and *Sva0485* (2.87%). As the most dominant group in all periods, *Thermodesulfovibrionia* exhibited the highest relative abundance during the wet period and the lowest during the dry period. *Anaerolineaceae* was the second abundant genus both in the dry and wet periods but showed a much lower abundance (0.6%) in the normal period. Conversely, the relative abundance of *Steroidobacteraceae* was the highest in the normal period compared to that in the dry and wet periods.

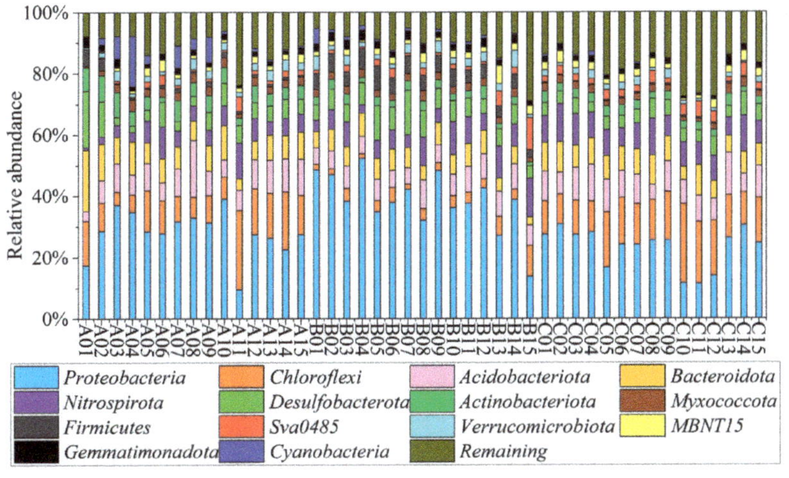

Figure 2. Microbial community composition at phylum level (A1–A15 represent samples collected in the dry period, B1–B15 represent samples collected in the normal period, and C1–C15 represent samples collected in the wet period).

3.3. PCoA Analysis and NMDS Analysis

The PCoA analysis was performed to analyze the distribution pattern of microbial communities across all sampling sites (Figure 4a). The first component (PCoA1) and the second component (PCoA2) accounted for 21% and 11% of the observed changes, respectively. The PCoA analysis revealed that the composition of microbial communities showed spatiotemporal variation among samples. Generally, the samples in the dry and wet periods were not well separated. However, those collected during the normal period were found to be distantly separated from those in the dry and wet periods, indicating relatively obvious differences between the microbial community structure in the normal period and other periods. Among different periods, samples in the dry period showed the most separation characteristics, while those in the normal and wet periods both exhibited a higher degree of similarity. These findings were further supported by NMDS analysis, which confirmed the distinctive distribution pattern of microbial communities across different sampling points (Figure 4b). These results provide evidence of discernible differences in the composition of microbial communities across different periods.

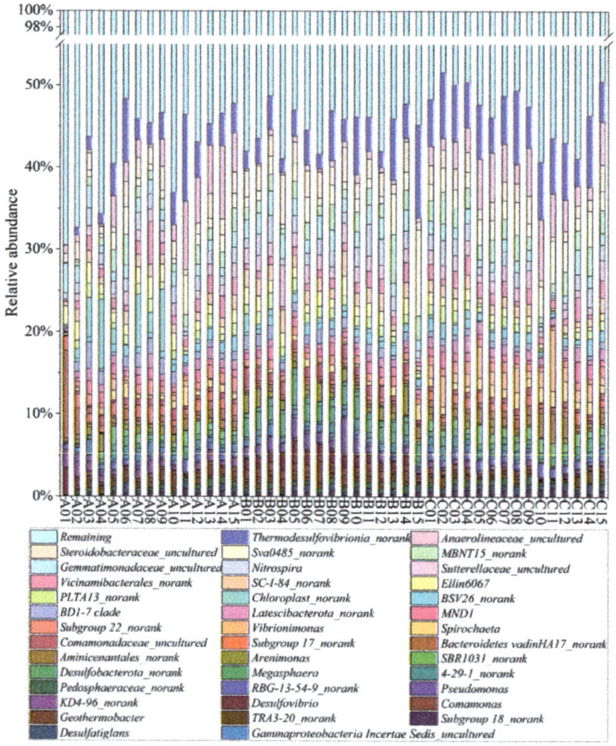

Figure 3. Microbial community composition at genus level (A1–A15 represent samples collected in the dry period, B1–B15 represent samples collected in the normal period, and C1–C15 represent samples collected in the wet period).

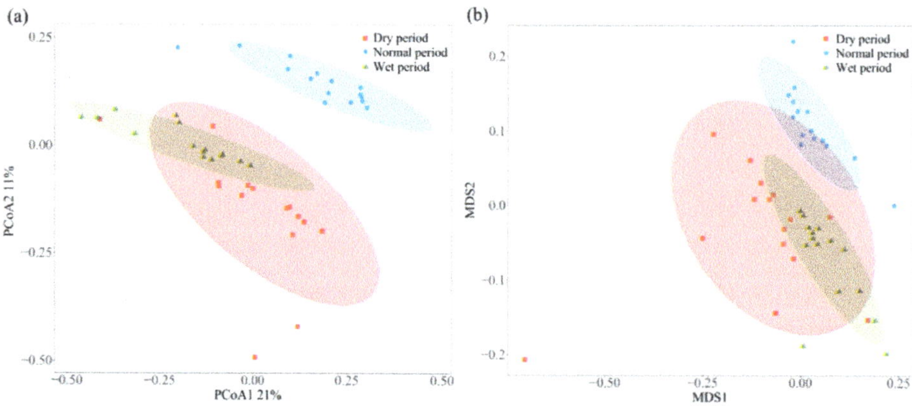

Figure 4. (**a**) PCoA and (**b**) NMDS analysis of the microbial communities (red squares represent samples collected in the dry period, blue circles represent samples collected in the normal period, and green triangles represent samples collected in the wet period).

3.4. *β*-Diversity Measurement

The divergence of causal mechanisms underlying biodiversity can be attributed to a series of environmental evolution, which is expected to give rise to two key phenomena in the beta diversity of microbial communities including nestedness and the spatial turnover of species assemblages [19]. The Simpson dissimilarity index (β_{SIM}), Sørensen dissimilarity index (β_{SOR}), and the nestedness-resultant dissimilarity index (β_{NES}) were used to differentiate the effects of species spatial turnover and nestedness components on the biotic similarity of microbial communities among multiple-site biotas. The phase dissimilarity indices across various sample groups are shown in Table S3 and Figure S1, according to Baselga's community assembly methodology [19]. β_{SOR} is dependent on the proportion of species shared among different communities, and β_{SIM} and β_{NES} were used to characterize species turnover and nestedness among the biotas of multiple sites. Overall, β_{SOR} was the highest, with an average value of 0.8282 among all samples, followed by β_{SIM} (0.8018) and β_{NES} (0.0202). Among the three water periods, the dry period showed the highest β_{SOR} value (0.8517), followed by the wet period (0.8210) and the normal period (0.8119). β_{SIM} values were much higher than β_{NES} values in all three periods, indicating the species spatial turnover in multiple-site biotas contributed much more to shaping the bacterial communities in different water periods.

3.5. Co-Occurrence Networks of Microbial Community

Network analysis has been adopted for elucidating microbial interactions in the WDRs in the source area of the ER-SNWDP under different water periods. Statistics for the topological parameters of each network are displayed in Table 1. In the dry period, *Proteobacteria, Acidobacteria,* and *Chloroflexi* were the dominant keystone species in the network diagram (Figure 5a). The positive correlation between species is 95.62%, which is much higher than the negative correlation (4.38%). The highest degree node is *Desulfobacteriota*, which has a high connection and is positively correlated with other phyla. The proportion of connections among different members of *Proteobacteria* was 16.1%, while that among *Chloroflexi* members was 6.79%. The connection between *Proteobacteria* and *Chloroflexi* was the closest among different phyla, with a connection proportion of 3.66%. In the normal period, the network consisted of the lowest abundance of nodes and edges, indicating the lowest degree of interconnectedness between OTUs in the normal period (Figure 5b). The main network nodes are *Proteobacteria, Bacteroidota, Nitrospirota*, etc. The positive correlation between species was 80.80%, which was higher than the negative correlation (19.20%). The highest degree node is *Proteobacteria*. The proportion of connections among *Proteobacteria* members and *Nitrospirota* members was 25.71% and 3.41%, respectively. The closest connection was found between *Proteobacteria* and *Bacteroidota,* with a connection proportion of 5.57%. In the wet period, *Proteobacteria* was also the most dominant keystone species as the network nodes, followed by *Chloroflexi* and *Acidobacteria* (Figure 5c). The positive correlation between species is 77.11%, also higher than the negative correlation (22.89%). *Chloroflexi* was the highest degree node and is mainly positively correlated with other phyla. The proportion of connections between members of *Proteobacteria* was 9.45%, the lowest among the three periods. *Proteobacteria* and *Chloroflexi* showed a relatively strong connection with a connection proportion of 4.51%.

Table 1. Topological properties of microbial networks in different water periods.

Network Features	Dry Period	Normal Period	Wet Period
Total nodes	485	343	650
Todal edges	845	323	1197
Positive correlations	808 (95.62%)	261 (80.80%)	923 (77.11%)
Negative correlations	37 (4.38%)	62 (19.20%)	274 (274)
Average clustering coefficient (avgCC)	0.298	0.284	0.303
Average degree (avgK)	3.48	1.883	3.683
Average path distance (GD)	6.691	7.927	5.547
Density	0.007	0.006	0.006

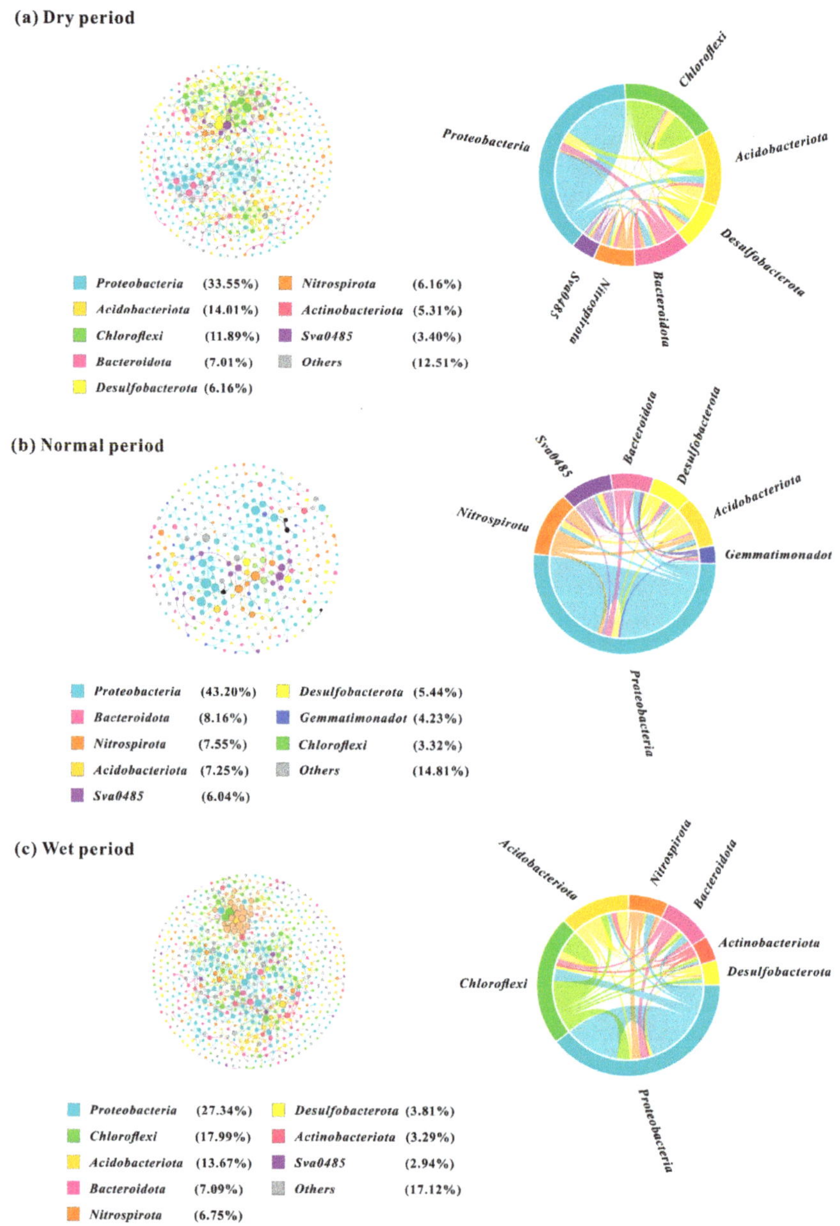

Figure 5. Co-occurrence network analysis of microbial communities in different water periods. The co-occurrence network of microbial communities in the dry period (**a**), normal period (**b**), and wet period (**c**).

3.6. Fit to the Neutral Model of Community Assembly

The neutral community model (NCM) was used to evaluate the effects of stochastic processes on shaping microbial community assembly in the WDRs in the source area of the ER-SNWDP under different water periods (Figure 6). In this study, the values of NCM parameter R^2 for communities in all periods, i.e., the dry period, normal period, and wet period, were 0.7973, 0.7048, 0.7686, and 0.7596, respectively, suggesting that the trait community model fitted for the microbial communities. The Nm-value and the dispersibility of microbial communities were the lowest in the dry period (Nm = 8946, m = 0.4601) and the highest in the normal period (Nm = 17,458, m = 0.8979).

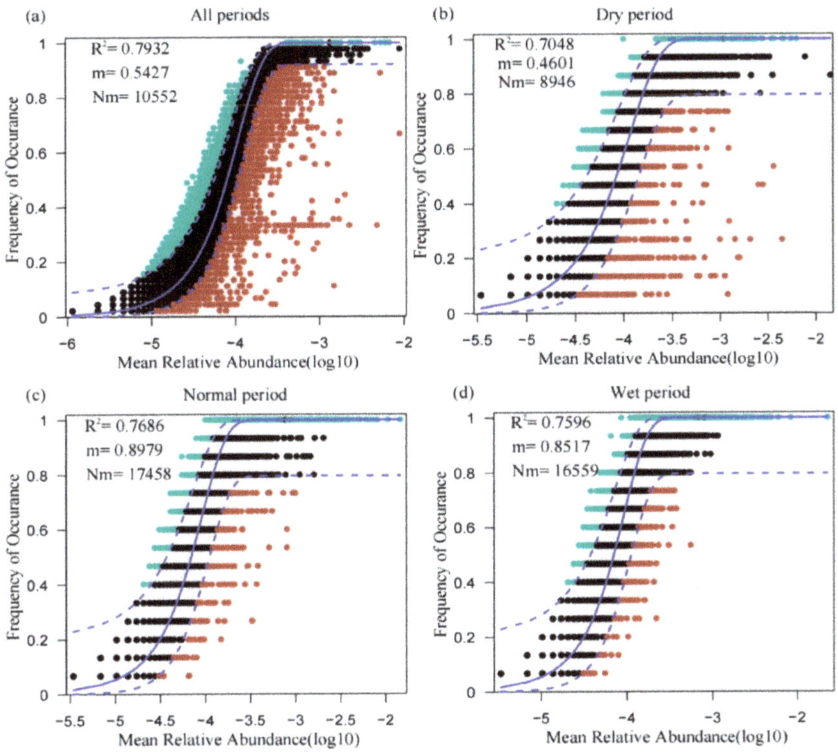

Figure 6. Fit of the neutral community model (NCM) of microbial community assembly. The predicted frequencies of occurrence for all periods, dry period, normal period, and wet period groups representing microbial communities from all periods (**a**), dry period (**b**), normal period (**c**), and wet period (**d**), respectively. Nm indicates the estimates of the metacommunity size times immigration rate, N demonstrates the metacommunity size, m is the immigration rate, and the coefficient of determination (R^2) is the goodness of fit of the neutral model.

3.7. Functional Analysis from PICRUSt

To determine the function of sediment microbial communities in the source area of the ER-SNWDP, PICRUSt was employed for analysis in this study. The results based on the KEGG database indicated that the metabolic pathways were mainly divided into six categories, including metabolism (68.32%), genetic information processing (15.90%), environmental information processing (9.15%), cellular processes (4.04%), organizational systems (1.16%), and human diseases (1.43%). The relative abundance of major metabolic pathways was depicted through a heat map (Figure 7). Among the 41 pathways detected, carbohydrate metabolism, amino acid metabolism, energy metabolism, cofactor and vitamin metabolism, and translation were the most predominant groups. These pathways were present in all samples, with an average abundance exceeding 1%.

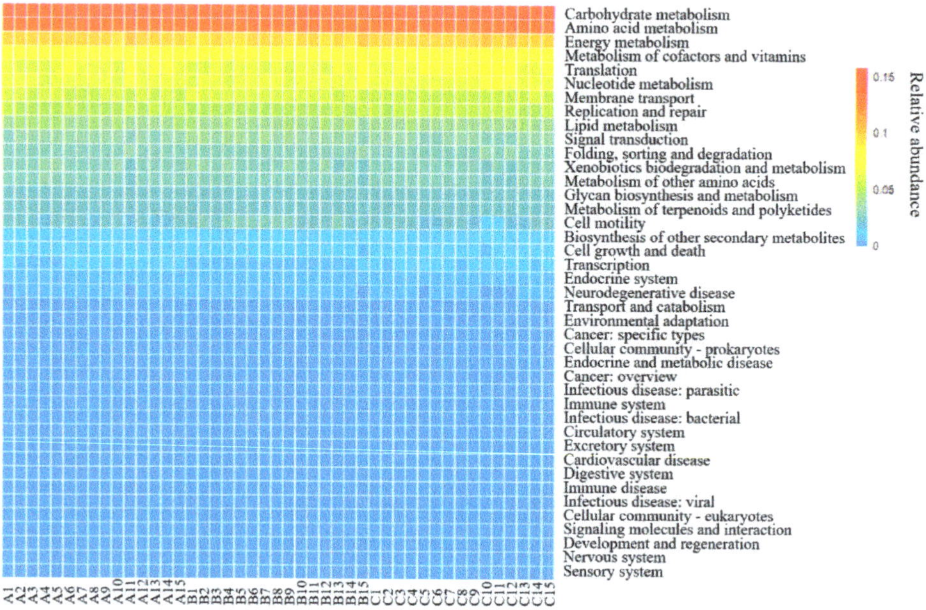

Figure 7. Relative abundance heatmap of predicted metagenomes using KEGG genes (A1–A15 represent samples collected in the dry period, B1–B15 represent samples collected in the normal period and C1–C15 represent samples collected in the wet period).

4. Discussion

Plentiful rainfall, floods in summer, water scarcity, and droughts in winter were the typical characteristics of the basic water regime in China, which led to the uneven spatial and temporal distribution of water resources. To mitigate water shortages, numerous WDPs including ER-SNWDP, have been implemented nationwide. The operation of these WDPs inevitably impacts the ecosystems of their source areas in various ways. Microbial communities play a crucial role in maintaining ecological balance within river ecosystems [11]. Therefore, this study showed a comprehensive investigation of the distribution and assembly patterns of microbial communities in the WDRs in the source area of ER-SNWDP.

The microbial communities in the sediment in the WDRs of the source area of ER-SNWDP are mainly composed of *Proteobacteria*, *Chloroflexi*, *Acidobacteriota*, *Bacteroidota*, *Nitrospirota*, and *Desulfobacterota* at the phylum level. These phyla are common in inland rivers and are usually the dominant bacterial species in natural river sediments [2]. *Proteobacteria* was the most prevalent phylum detected, showing the highest relative abundance in the normal period. This phylum is functionally diverse and widely involved in biogeochemical processes such as carbon, nitrogen, and sulfur cycling in sediments [20]. Its functional versatility enables *Proteobacteria* to adapt to different hydrological conditions, allowing it to maintain high abundance in various environments. The relatively stable hydrological conditions in the normal period may have provided a more favorable environment for *Proteobacteria*, allowing it to dominate the community structure. *Chloroflexi* was the second dominant phylum, with higher levels in the dry and wet periods than in the normal period. *Chloroflexi* is a photoautotrophic bacterium mainly involved in biological denitrification, organic pollutant degradation and transformation, and sediment nitrogen and sulfur cycling [21]. Their preference for nutrient-rich environments might suggest higher nutrient levels in the sediments in the dry and wet periods [22]. *Desulfobacterota* was abundant during the normal period but showed a relatively lower proportion in the other two periods. Members of *Desulfobacterota* can metabolize sulfides into sulfates, indicating higher sulfide concentrations during the normal period, which likely created a favorable environment for their growth. The stable hydrological conditions in this period may have facilitated sulfide accumulation in the sediments, supporting *Desulfobacterota* proliferation. This also suggests a more active sulfur biogeochemical cycle during the normal period. [23]. *Thermodesulfovibrionia*, *Anaerolineaceae*, and *Steroidobacteraceae* were the most abundant genera detected in the study at the genus level. *Thermodesulfovibrionia*, a genus within *Proteobacteria*, is active in the sulfur cycle and is known for its dissimilatory sulfate reduction capabilities. The abundant *Thermodesulfovibrionia* in the WDRs in the source area of the WDPs might also be related to the sulfide concentrations in this area. The stable presence of *Thermodesulfovibrionia* across all sampling periods indicates that sulfate reduction remains a continuous and essential process regardless of hydrological fluctuations. *Thermodesulfovibrionia* showed the highest relative abundance during the wet period and the lowest during the dry period, suggesting its strong adaptability to fluctuating hydrological conditions. During the wet period, enhanced water flow may promote the resuspension of sulfide-rich sediments, increasing sulfide availability and supporting *Thermodesulfovibrionia* proliferation. *Anaerolineaceae*, part of the *Chloroflexi* phylum, is a facultative anaerobic bacterium that degrades various organic compounds such as amino acids and lipids found in domestic wastewater and agricultural pollution [24]. Its presence in the sediment might indicate a high nutrient level in the region, which boosts bacterial cycling and oxygen consumption, aiding their proliferation and survival. *Steroidrobacteraceae* is usually the dominant genus of bacteria in activated sludge with a certain degree of denitrification ability [25]. Its abundance might further suggest a potential presence of specific pollutants in the study area.

The dynamic water conditions induced by water diversion across different periods may play a crucial role in shaping microbial ecology by altering hydraulic dynamics and water quality parameters in WDRs. Fluctuations in water volumes and flow rates brought by the operation of WDPs also contribute to alterations in the water body's physical and chemical properties, such as DO, nutrient concentrations, organic matter content, and temperature [13]. Shifts in the hydraulic and physicochemical conditions in different water periods could be pivotal in controlling the microbial communities in river ecosystems [26]. Microbial diversity is considered a key indicator of ecosystem stability in rivers [27]. Results showed that the diversity of microbial communities was the highest during the wet period of all sampling sites. Zhang et al. (2019) also found a higher microbial diversity in rivers flowing into Chaohu Lake during the wet period compared to that in the dry period [28]. Hydrological factors, such as temperature variability and water level fluctuations, are known to influence microbial diversity [11]. During the wet period, the rainfall and high river flow could probably cause disturbance of the river water and surface sediment, promoting the fusion of bacteria in the water and sediment [14]. The PCoA analysis also exhibited a temporal variation of the microbial communities in WDRs in different water periods. Samples collected during the normal period displayed a pronounced clustering, while those from the dry and wet periods exhibited varying degrees of intersection. This observation was further substantiated by the NMDS analysis. Furthermore, different microbial communities showed differential distribution characteristics in different water periods. *Chloroflexi*, *Firmicutes*, *Cyanobacteria*, *Anaerolineaceae*, and *Chloroplast* are more susceptible to changes in water conditions in different water periods. The relative abundance of *Chloroflexi* was much higher during the dry (11.75%) and wet periods (14.64%), contrasting sharply with a mere 3.75% in the normal period. *Cyanobacteria* seemed to be more adaptable to the environment of the dry period, showing the highest abundance in this period but the lowest in the wet period. It is suggested that a mild hydraulic condition with low water flow and velocity may favor the proliferation of *Cyanobacteria*. This could be due to enhanced hydrodynamics, sediment dilution, and changes in light conditions in the wet period, which may inhibit the growth of *Cyanobacteria*. In contrast, *Cyanobacteria* is highly adaptable to low nutrient conditions, and its relatively high abundance during the dry period reflects the positive effect of nutrient accumulation in the water during this time. It is also confirmed by the study in the Amazon River, which indicated that one of the main reasons for the decrease in blue-green algae abundance is the increased flow in the river [29,30]. The abundance variation patterns of *Chloroplasts* are similar to *Cyanobacteria*, with a higher relative abundance during the dry period (2.3%) compared to less than 1% in the normal and wet periods. This disparity may be attributed to the reduced water volume and ample sunlight in the dry period, which promotes aquatic plant growth and may favor *Chloroplast* distribution [31]. Conversely, *Anaerolineaceae* showed a pronounced presence in the wet period (4%) but was substantially lower (0.6%) during the normal period. This genus has demonstrated adaptability to sediment environments post-water diversion, thriving in both anaerobic and aerobic conditions, and could respond to environmental alterations brought about by water diversion [32]. Despite changes in microbial ecology in WDRs under different water periods with different water conditions observed in this study, the mechanisms of changing water conditions on the microbial communities remain unclear. More extensive studies on the impact mechanism of hydraulic factors (such as flow rate, flow direction, and flow volume.) coupling physicochemical factors on microbial communities in WDRs are needed for further investigation.

The complexity and correlation of microbial networks in different water periods were exhibited by the co-occurrence network analysis. *Proteobacteria*, *Acidobacteria*, *Chloroflexi*, *Bacteroidota*, and *Nitrospirota* were identified as the most dominant keystone species in the network diagram. The co-occurrence network patterns of microbial communities varied among different water periods. The microbial co-occurrence network in the dry period is the most complex, as indicated by the network density of microbes. During the dry period, the microbial network displayed clear modularity, with microbial OTUs showing the strongest interconnections. Most of the correlations were positive correlations, and it was greater than the negative correlation in all periods, indicating that the network probably mainly existed in symbiotic models characterized by mutual benefit and symbiosis. Species that tend to have symbiotic relationships in the community would gradually occupy the dominant position, causing changes in the structure of microbial communities [33]. This relationship could also contribute to the maintenance of the stability and functional diversity of ecosystems. The highest ratio of negative correlations in the wet period suggested the most intense antagonistic relationships, such as competition, existed compared to other periods.

The ecological processes governing microbial community assembly in WDRs in the source area of ER-SNWDP were determined by assessing the contributions of spatial turnover or nestedness to the β-diversity patterns and the evaluation of stochastic processes by NCM. The causes of the disparities in the processes promoting biodiversity may be explained by the changes in the redox state, nutritional composition, and electron acceptors, which could lead to nestedness and the spatial turnover of species assemblages of microbial communities. Nestedness leads to variances in species richness between communities, stemming from a nonrandom process of species loss, while spatial turnover involves the replacement of species across different communities [15]. In this study, both processes contributed to the formation of the microbial community assembly, especially the spatial turnover of species, which refers to the replacement of species and is caused by historical and geographical constraints as well as environmental factors. Results showed that the microbial community assemblies and different water periods in the source area of ER-SNWDP were all mainly shaped by the spatial turnover of species, suggesting that microbial communities benefited more from species replacement across regions. Microbial community assemblages in this typical area in different water periods were well described by a neutral-based model, indicating that stochastic processes explained a large fraction of the variation in shaping microbial community assembly. Varied Nm-value and the dispersibility in different water periods indicated a potential effect of water diversion on the dispersion of microbial communities. The better fit of the NCM in normal and wet periods than that in dry periods could probably suggest a stronger impact of stochastic processes on microbial community assemblages in normal and wet periods than that in dry periods. It is concluded that the impact of stochastic processes on microbial community assemblages could be changed with the water conditions in WDRs. To advance the understanding of the ecological processes governing microbial community assembly in WDRs, a series of river ecology theories should be further applied to the analysis. Furthermore, microbial parameters should be incorporated into the river dynamics models and water quality evaluation models to build a comprehensive ecological evaluation system in rivers.

5. Conclusions

WDRs function as specialized waterways and play a pivotal role in optimizing water resource allocation. The ecosystem status of WDRs in the source area of WDPs is basic insurance for the ecological health of rivers and residents' livelihood. Under the operation of water diversion projects, WDR ecology is constantly affected by external factors, resulting in frequent changes in water quality and hydraulic conditions in different water periods. Microorganisms play a crucial role in river ecosystems and are highly responsive to environmental changes. Frequent environmental changes in WDRs would probably have a considerable impact on the microbial ecology and ultimately affect the ecology of the river. This study investigated the microbial communities in sediments in the source areas of ER-SNWDP during three typical water periods for the first time. A total of 1,997,209 sequence reads were obtained by 16S rRNA gene sequencing. Generally, microbial communities in this area were dominated by *Proteobacteria*, *Chloroflexi*, *Acidobacteriota*, and *Bacteroidota*. Various microbial communities are more susceptible to changes in water conditions and exhibit distinct distribution characteristics across different water periods, notably *Chloroflexi*, *Firmicutes*, *Cyanobacteria*, *Anaerolineaceae*, and *Chloroplast*. During the wet period, microbial communities exhibited the highest richness, biodiversity, and the most intense antagonistic relationships compared to those in the dry and normal water periods. Throughout all periods, the microbial network predominantly existed in symbiotic models characterized by mutual benefit and symbiosis. During the dry period, the microbial co-occurrence network was found to be the most complex, with microbial OTUs showing the closest interconnections. The functional analysis determined carbohydrate metabolism, amino acid metabolism, energy metabolism, cofactor and vitamin metabolism, and translation as the most predominant functions of microbial communities. Microbial communities in this period also showed the highest nestedness-resultant dissimilarity index, followed by those in the normal period and the wet period. Spatial turnover of species and stochastic processes were determined as the dominant mechanisms controlling community diversity, succession, and biogeography in the WDRs in the source area of ER-SNWDP. A more pronounced impact of stochastic processes on microbial community assemblages was observed during normal or wet periods, as opposed to the dry period. This study could provide a comprehensive understanding of microbial ecology in WDRs under different water periods and is pivotal for the maintenance of river ecological health and the conservation of water resources in WDRs.

Supplementary Materials: The following supporting information can be downloaded at: https://www.mdpi.com/article/10.3390/w17050649/s1, Figure S1: The β-diversity pattern indexes of microbial communities in different water periods; Table S1: Data of longitude and latitude of each sampling point; Table S2: Statistics of the data for sequence reads, OTUs and gene alpha diversity indices among samples; Table S3: β-diversity pattern indexes of microbial communities. (Group A represents samples collected in the dry period, Group B represents samples collected in the normal period, and Group C represents samples collected in the wet period).

Author Contributions: Conceptualization, W.C. and J.W.; Methodology, W.C., X.W. (Xin Wen) and J.W.; Software, X.W. (Xin Wen) and Y.Z.; Validation, Y.Z., X.W. (Xiusen Wu), H.Z. and Z.H.; Investigation, W.C. and J.W.; Data curation, J.C. and Q.Z.; Writing—original draft, W.C., X.W. (Xin Wen) and J.W.; Supervision, W.C. and J.W.; Funding acquisition, W.C. and J.W. All authors have read and agreed to the published version of the manuscript.

Funding: This research was funded by the National Natural Science Foundation of China (Grant No. 51909229), the Natural Science Foundation of the Jiangsu Higher Education Institutions of China (Grant No. 24KJB610020), the Belt and Road Special Foundation of the National Key Laboratory of Water Disaster Prevention (Grant No. 2023nkms04), and the Student Science and Technology Innovation Foundation of Yangzhou University.

Data Availability Statement: The data are available from the corresponding author upon reasonable request.

Conflicts of Interest: We have no financial and personal relationships with other people or organizations that can inappropriately influence our work; there is no professional or other personal interest of any nature or kind in any product, service, and/or company that could be construed as influencing the position presented in the manuscript entitled "Untangling the characteristics and ecological processes of microbial community assembly in the source area of the East Route of the South-to-North Water Diverion Project in China under different water periods". We declare that the author Qin Zhong was employed by the company Dongzhu Ecological Environmental Protection Co., Ltd., Xihu Middle Road #90, Wuxi 214000, China. The remaining authors declare that the research was conducted in the absence of any commercial or financial relationships that could be construed as a potential conflict of interest.

References

1. Faraji, H.; Shahryari, A. Estimation of Water Quality Index and Factors Affecting Their Changes in Groundwater Resource and Nitrate and Fluoride Risk Assessment. *Water Air Soil Pollut.* **2023**, *234*, 608. [CrossRef]
2. Eliasson, J. The rising pressure of global water shortages. *Nature* **2015**, *517*, 6. [CrossRef] [PubMed]
3. Zhao, Z.-Y.; Zuo, J.; Zillante, G. Transformation of water resource management: A case study of the South-to-North Water Diversion project. *J. Clean. Prod.* **2017**, *163*, 136–145. [CrossRef]
4. Zhuang, W. Eco-environmental impact of inter-basin water transfer projects: A review. *Environ. Sci. Pollut. Res.* **2016**, *23*, 12867–12879. [CrossRef]
5. Davies, B.R.; Thoms, M.; Meador, M. An assessment of the ecological impacts of inter-basin water transfers, and their threats to river basin integrity and conservation. *Aquat. Conserv. Mar. Freshw. Ecosyst.* **1992**, *2*, 325–349. [CrossRef]
6. Zhang, S.; Pang, Y.; Xu, H.; Wei, J.; Jiang, S.; Pei, H. Shift of phytoplankton assemblages in a temperate lake located on the eastern route of the South-to-North Water Diversion Project. *Environ. Res.* **2023**, *227*, 115805. [CrossRef]
7. Funes, A.; de Vicente, J.; Cruz-Pizarro, L.; Álvarez-Manzaneda, I.; de Vicente, I. Magnetic microparticles as a new tool for lake restoration: A microcosm experiment for evaluating the impact on phosphorus fluxes and sedimentary phosphorus pools. *Water Res.* **2016**, *89*, 366–374. [CrossRef] [PubMed]
8. Pusch, M.; Fiebig, D.; Brettar, I.; Eisenmann, H.; Ellis, B.K.; Kaplan, L.A.; Lock, M.A.; Naegeli, M.W.; Traunspurger, W. The role of micro-organisms in the ecological connectivity of running waters. *Freshw. Biol.* **1998**, *40*, 453–495. [CrossRef]
9. Mohapatra, M.; Behera, P.; Kim, J.Y.; Rastogi, G. Seasonal and spatial dynamics of bacterioplankton communities in a brackish water coastal lagoon. *Sci. Total Environ.* **2020**, *705*, 134729. [CrossRef] [PubMed]
10. Wang, K.; Wang, Z.; Liu, K.; Cheng, L.; Wang, L.; Ye, A. Impacts of the eastern route of the South-to-North Water Diversion Project emergency operation on flooding and drainage in water-receiving areas: An empirical case in China. *Nat. Hazards Earth Syst. Sci.* **2019**, *19*, 555–570. [CrossRef]

11. Bagra, K.; Bellanger, X.; Merlin, C.; Singh, G.; Berendonk, T.U.; Klümper, U. Environmental stress increases the invasion success of antimicrobial resistant bacteria in river microbial communities. *Sci. Total Environ.* **2023**, *904*, 166661. [CrossRef] [PubMed]
12. Luo, Z.; Li, S.; Hou, K.; Ji, G. Spatial and seasonal bacterioplankton community dynamics in the main channel of the Middle Route of South-to-North Water Diversion Project. *Res. Microbiol.* **2019**, *170*, 24–34. [CrossRef] [PubMed]
13. Liu, B.; Guo, F.; Xie, P.; Zhang, W.; Hu, H.; Wu, J.; Yang, Z. Differences in sediment microbial community structure and co-occurrence network in different seasons. *J. Soils Sediments* **2023**, *23*, 3539–3549. [CrossRef]
14. Lv, J.; Niu, Y.; Yuan, R.; Wang, S. Different Responses of Bacterial and Archaeal Communities in River Sediments to Water Diversion and Seasonal Changes. *Microorganisms* **2021**, *9*, 782. [CrossRef] [PubMed]
15. Qu, J.; Jia, C.; Liu, Q.; Li, Z.; Liu, P.; Yang, M.; Zhao, M.; Li, W.; Zhu, H.; Zhang, Q. Dynamics of Bacterial Community Diversity and Structure in the Terminal Reservoir of the South-To-North Water Diversion Project in China. *Water* **2018**, *10*, 709. [CrossRef]
16. Yao, Y.; Wang, P.; Wang, C. The Influence on Contaminant Bioavailability and Microbial Abundance of Lake Hongze by the South-to-North Water Diversion Project. *Int. J. Environ. Res. Public Health* **2019**, *16*, 3068. [CrossRef] [PubMed]
17. Liu, Y.; Pan, B.; Zhu, X.; Zhao, X.; Sun, H.; He, H.; Jiang, W. Patterns of microbial communities and their relationships with water quality in a large-scale water transfer system. *J. Environ. Manage.* **2022**, *319*, 115678. [CrossRef] [PubMed]
18. Da Costa, C.; Colin, Y.; Debret, M.; Copard, Y.; Gardes, T.; Jacq, K.; Ayrault, S.; Berthe, T. Shifts in sediment bacterial communities reflect changes in depositional environments in a fluviatile context. *Sci. Total Environ.* **2023**, *885*, 163890. [CrossRef] [PubMed]
19. Baselga, A. Partitioning the turnover and nestedness components of beta diversity. *Glob. Ecol. Biogeogr.* **2010**, *19*, 134–143. [CrossRef]
20. Liu, J.; Sun, X.; Zuo, Y.; Hu, Q.; He, X. Plant species shape the bacterial communities on the phyllosphere in a hyper-arid desert. *Microbiol. Res.* **2023**, *269*, 127314. [CrossRef] [PubMed]
21. Newton Ryan, J.; Jones Stuart, E.; Eiler, A.; McMahon Katherine, D.; Bertilsson, S. A Guide to the Natural History of Freshwater Lake Bacteria. *Microbiol. Mol. Biol. Rev.* **2011**, *75*, 14–49. [CrossRef] [PubMed]
22. Colatriano, D.; Tran, P.Q.; Guéguen, C.; Williams, W.J.; Lovejoy, C.; Walsh, D.A. Genomic evidence for the degradation of terrestrial organic matter by pelagic Arctic Ocean Chloroflexi bacteria. *Commun. Biol.* **2018**, *1*, 90. [CrossRef] [PubMed]
23. Roberto, A.A.; Van Gray, J.B.; Leff, L.G. Sediment bacteria in an urban stream: Spatiotemporal patterns in community composition. *Water Res.* **2018**, *134*, 353–369. [CrossRef] [PubMed]
24. Kelly, J.J.; Tate Iii, R.L. Effects of Heavy Metal Contamination and Remediation on Soil Microbial Communities in the Vicinity of a Zinc Smelter. *J. Environ. Qual.* **1998**, *27*, 609–617. [CrossRef]
25. Abiriga, D.; Jenkins, A.; Klempe, H. Microbial assembly and co-occurrence network in an aquifer under press perturbation. *Ann. Microbiol.* **2022**, *72*, 41. [CrossRef]
26. Santos, A.; Rachid, C.; Pacheco, A.B.; Magalhães, V. Biotic and abiotic factors affect microcystin-LR concentrations in water/sediment interface. *Microbiol. Res.* **2020**, *236*, 126452. [CrossRef] [PubMed]
27. Meziti, A.; Tsementzi, D.; Rodriguez-R, L.M.; Hatt, J.K.; Karayanni, H.; Kormas, K.A.; Konstantinidis, K.T. Quantifying the changes in genetic diversity within sequence-discrete bacterial populations across a spatial and temporal riverine gradient. *ISME J.* **2018**, *13*, 767–779. [CrossRef] [PubMed]
28. Zhang, M.; Wu, Z.; Sun, Q.; Ding, Y.; Ding, Z.; Sun, L. The spatial and seasonal variations of bacterial community structure and influencing factors in river sediments. *J. Environ. Manag.* **2019**, *248*, 109293. [CrossRef]
29. Ma, X.; Li, Y.; Niu, L.; Shang, J.; Yang, N. Microbial community structure and denitrification responses to cascade low-head dams and their contribution to eutrophication in urban rivers. *Environ. Res.* **2023**, *221*, 115242. [CrossRef] [PubMed]
30. Doherty, M.; Yager, P.L.; Moran, M.A.; Coles, V.J.; Fortunato, C.S.; Krusche, A.V.; Medeiros, P.M.; Payet, J.P.; Richey, J.E.; Satinsky, B.M.; et al. Bacterial Biogeography across the Amazon River-Ocean Continuum. *Front. Microbiol.* **2017**, *8*, 882. [CrossRef]
31. Baldassi, A.C.; Balbuena, T.S. The Eucalyptus grandis chloroplast proteome: Seasonal variations in leaf development. *PLoS ONE* **2022**, *17*, e0265134. [CrossRef] [PubMed]
32. Lv, J.; Yuan, R.; Wang, S. Water diversion induces more changes in bacterial and archaeal communities of river sediments than seasonality. *J. Environ. Manag.* **2021**, *293*, 112876. [CrossRef] [PubMed]
33. Horton, D.J.; Theis, K.R.; Uzarski, D.G.; Learman, D.R. Microbial community structure and microbial networks correspond to nutrient gradients within coastal wetlands of the Laurentian Great Lakes. *FEMS Microbiol. Ecol.* **2019**, *95*, fiz033. [CrossRef] [PubMed]

MDPI AG
Grosspeteranlage 5
4052 Basel
Switzerland
Tel.: +41 61 683 77 34

Water Editorial Office
E-mail: water@mdpi.com
www.mdpi.com/journal/water

www.ingramcontent.com/pod-product-compliance
Lightning Source LLC
Chambersburg PA
CBHW061547210926
51850CB00009B/2031

* 9 7 8 3 7 2 5 8 6 2 6 2 7 *